Out of Nowhere

Out of Nowhere

The Emergence of Spacetime in Theories of Quantum Gravity

NICK HUGGETT
AND
CHRISTIAN WÜTHRICH

CLARENDON PRESS

OXFORD
UNIVERSITY PRESS

OXFORD
UNIVERSITY PRESS

Great Clarendon Street, Oxford, OX2 6DP,
United Kingdom

Oxford University Press is a department of the University of Oxford.
It furthers the University's objective of excellence in research, scholarship,
and education by publishing worldwide. Oxford is a registered trade mark of
Oxford University Press in the UK and in certain other countries

© Nick Huggett and Christian Wüthrich 2025

The moral rights of the authors have been asserted.

All rights reserved. No part of this publication may be reproduced, stored in a retrieval system,
transmitted, used for text and data mining, or used for training artificial intelligence, in any form or
by any means, without the prior permission in writing of Oxford University Press, or as expressly
permitted by law, by licence or under terms agreed with the appropriate reprographics rights
organization. Enquiries concerning reproduction outside the scope of the above should be sent
to the Rights Department, Oxford University Press, at the address above.

You must not circulate this work in any other form
and you must impose this same condition on any acquirer.

Published in the United States of America by Oxford University Press
198 Madison Avenue, New York, NY 10016, United States of America

British Library Cataloguing in Publication Data

Data available

Library of Congress Control Number: 2024941628

ISBN 9780198758501

DOI: 10.1093/oso/9780198758501.001.0001

Printed and bound by
CPI Group (UK) Ltd, Croydon, CR0 4YY

Cover Image: Getty Images

Links to third party websites are provided by Oxford in good faith and
for information only. Oxford disclaims any responsibility for the materials
contained in any third party website referenced in this work.

The manufacturer's authorised representative in the EU for product safety is
Oxford University Press España S.A. of El Parque Empresarial San Fernando de Henares, Avenida
de Castilla, 2 – 28830 Madrid (www.oup.es/en or
product.safety@oup.com). OUP España S.A. also acts as importer into Spain
of products made by the manufacturer.

To Fritz and Margeth. To Robert Weingard.

Preface

The story of this book closely follows the story of the development of the study of quantum gravity in philosophy. In the early 1990s, as the 'first string revolution' was passing from revolutionary to canonical, NH was introduced to the subject by his advisor, the great, and much missed Robert Weingard, who was the pioneer of the field. As a result of that exposure, NH and another of Weingard's advisees, Craig Callender, edited a collection of essays by philosophers and physicists, *Physics Meets Philosophy at the Planck Scale* (2001b), reflecting on the philosophical foundations of quantum gravity as they then stood. This work helped focus the attention of philosophers of physics; one of the essays, 'Spacetime and the philosophical challenge of quantum gravity' by Chris Isham and Jeremy Butterfield, has been particularly influential on the field.

In the early 2000s, CW was steered by his advisor, the legendary John Earman, towards loop quantum gravity (LQG) for his dissertation. Although Earman was working on the problem of time in Hamiltonian general relativity and canonical quantum gravity at the time (a topic he began to work on in the late 1990s, together with another of his advisees, Gordon Belot—they jointly contributed to the collection edited by NH and Callender), he sent CW to learn LQG from its founding fathers. Thus, funded by the Swiss National Science Foundation and armed with unbounded enthusiasm and many questions, CW spent the academic year 2003–2004 at the Perimeter Institute in Waterloo (Canada) visiting Lee Smolin, and at the Centre de Physique Théorique in Luminy (France) working with Carlo Rovelli. The time with Rovelli and the close interactions with him and the members of his group turned out to be a formative influence. In 2006, CW defended what was the first dissertation in philosophy of quantum gravity.

In 2000, NH presented a paper written with Callender (2001a) at the Biennial Philosophy of Science Association (PSA) Conference, in one of the very first sessions to have a focus on quantum gravity. They criticized the tradition of arguments purporting to show the *necessity* of quantizing the gravitational field; that it is inconsistent to have quantum matter with classical gravity as a fundamental theory. They found those arguments to be inconclusive, and argued that empirical evidence would be needed to support that conclusion (evidence that perhaps is now on the horizon; Huggett et al. 2023). This line of thought was developed further by CW in a paper given at the 2004 edition of the conference (Wüthrich 2005), which first brought us into contact, and led to NH inviting CW to present at the New PhDs Seminar in Physics and Philosophy at UIC in 2005—and to the conversation that we have had ever since.

In 2010, CW forged a connection with his colleague at UCSD, David Meyer, a founder of CST and advisee of Rafael Sorkin, the spiritus rector of CST, which quickly evolved into regular and meaningful meetings between the philosophers of physics, including Callender, and Meyer's group when David Rideout joined UCSD in 2011. After dabbling in it during his PhD, it was thus only natural for CW to take up the philosophy of CST in earnest at that time, particularly also given that David Malament—who wrote his PhD under Earman's supervision and, starting out from Earman's 1972 article on causal theories of time (Earman 1972), proved an important result instrumental in the development of CST (as we will elaborate in chapter 3)—was a faculty member at UC Irvine, just up the road from San Diego. The regular meetings, talks, seminars, and conversations with Callender, Malament, Meyer, Rideout, and our students John Dougherty, Nat Jacobs, and Sebastian Speitel, as well as with the philosophers of physics at UC Irvine, proved to be an important enabler for CW's work. After having initially met at Perimeter in 2003, CW's intensifying contact with Fay Dowker, another leader in CST, and exchanges with other leading theorists at the 2015 Edinburgh meeting and other venues proved similarly crucial in developing a full grasp of the theory and its history.

It was at the 2008 Geneva Summer School in the Philosophy of Physics when the idea to write a joint book on quantum gravity—and especially on the emergence of spacetime—was first hatched, in a hotel bar in Arolla in the Swiss Alps. We kicked around the idea for a couple more years, at the 2008 PSA conference and other meetings, before successfully obtaining a Collaborative Fellowship from the (incredibly patient) ACLS to support the writing of this monograph, which began in earnest at the start of 2011 (and was supposed to be completed in 2012). NH started working on string theory and non-commutative geometry, while CW took on causal set theory (CST) and LQG.

The fruits of this work started to appear at meetings that we organized at UIC (in 2013), at UCSD (2015), at a memorable summer school on Lake Geneva, WI (2016), and at the University of Geneva (2017), and that Daniele Oriti organized at the Albert Einstein Institute for Gravitational Physics in Potsdam, (2012, and 2014), which started to bring together philosophers and physicists interested in dialogue about the foundations of quantum gravity; and importantly started to garner the interest of the next generation of scholars. We also edited a special edition of *Studies in History and Philosophy of Modern Physics* (Huggett and Wüthrich 2013b) devoted to the question of spacetime emergence: our essay in that volume (Huggett and Wüthrich 2013a) was something of a manifesto for subsequent philosophical work on quantum gravity, and first posed the questions of this book.

viii PREFACE

Then the book—which we started calling *Out of Nowhere* because of its concern with the emergence of the spatiotemporal from the non-spatiotemporal—developed some kind of time dilation, seeming to be about a year (or less!) from completion for around a decade. On the one hand, the more we dug, and the more we talked with experts, the more it became clear that we had underestimated what needed to be said. On the other hand, inspired by the interest that people were starting to show in the field, we twice successfully applied to the Sir John Templeton Foundation for support of, not only the book, but also young researchers and interaction between physicists and philosophers working in quantum gravity. A stream of speakers and visitors joined us in Chicago and Geneva, and (before it became an everyday occurrence) we organized live-streaming between our two centers. (These talks are archived on the Beyond Spacetime YouTube channel, and they show a general trend of increasing success in using the technology.) We organized an annual essay contest (leading to a pair of edited volumes, Huggett et al. 2020; Wüthrich et al. 2021), and several meetings, most recently a Conference on the Foundations of Cosmology and Quantum Gravity at NYU Abu Dhabi (2020), and a Summer Institute in Morzine (2022). These activities had several effects: first, they helped create a community of scholars working in the field by connecting people and helping set an agenda. Second, they further broadened and deepened our appreciation of the foundational issues. And third, despite the help of many people, they took time to organize and participate in.

We also recruited pre- and postdoctoral fellows into the project, who brought vital new knowledge, talent, insight, and energy, which strengthened the book in untold ways, but which produced further time dilation!

Which is all to say 'thank you' (and 'apologies') to our publisher, OUP, and especially our editor for most that time, Peter Momtchiloff, for their forbearance.

Several years into the project it came to light that Karen Crowther had independently hit on the same title as us for a dissertation on similar topics at the University of Sydney. By mutual agreement we decided that there would be no problem if it was used by both her and us, and so that is what we did.

Our plan was to cover a range of approaches to quantizing gravity: NH to take the lead with chapters on string theory, non-commutative geometry, and group field theory, CW with chapters on LQG and CST. In the end, while CW completed his brief, string theory turned out to require rather more pages than initially conceived, and the other approaches were dropped (but see Huggett et al. 2021 and Huggett 2022).

Most of the work on CST (chapters 3 and 4) had been completed (though not fully written out) by 2015, when CW moved to Geneva and started to shift focus back to LQG, returning to his earlier dissertation work on which the chapters on LQG (5 and 6) are partly based. CW deeply benefitted from many exchanges over the years with several leading exponents of LQG such as Daniele Oriti,

Francesca Vidotto, and particularly Carlo Rovelli, with whom he has enjoyed regular discussions (not just over the foundations of LQG!) for over twenty years now.

NH's first job was to achieve sufficient mastery of the string theory formalism. Building on what he had learned from Weingard, he set about working through various textbooks and online lectures with students and colleagues: Kevin Davey, Josh Norton, and Tiziana Vistarini especially. He also benefitted greatly from the generosity of Jeff Harvey, who patiently straightened out a number of early confusions (those that remain are of course no fault of his!). The result is found in chapter 7, in which the basic technical framework is presented in terms suited for philosophers of physics (here special thanks go to Tushar Menon for meticulous feedback).

The most prominent philosophical issue initially concerned the nature and significance of string theoretic 'dualites', which have been a focus of philosophical interest, with scores of papers on the subject. NH started working on this topic in 2007, presenting in a joint session on the topic with Brian Greene at a string theory meeting at Rutgers Philosophy, at a 2008 quantum gravity workshop at Pittsburgh, a session that he and CW organized at the 2010 PSA conference, and later at a duality conference organized by Elena Castellani in Florence in 2014. This work led to a paper (Huggett 2017) that has been revised as chapter 8 of this work.

The next two chapters deal directly with the 'emergence' of classical spacetime from quantum string theory. At the heart of this discussion are the ideas that: classical gravitational fields can be understood as collective states of quanta of the gravitational field, 'gravitons'; and gravitons can in turn be understood as strings in appropriate states. On the former point NH have benefitted especially from discussions with James Read and Kian Salimkhani; on the latter Jeff Harvey helped appreciate the remarkable results in sigma models that underpin the claim. But most important, NH benefitted from a collaboration with Tiziana Vistarini on a co-authored paper on these central ideas (Huggett and Vistarini 2015).

The introduction(s) and conclusion we wrote together. The first chapter develops our 2013 paper to explain what we take the problem of spacetime emergence to be. The second lays out the functionalist approach that we adopt, in part drawing on work that CW undertook with Vincent Lam (Lam and Wüthrich 2018, 2021); this work benefitted from a workshop on spacetime functionalism in Geneva in 2018, a workshop in Ovronnaz in 2021, discussions at the 2022 QISS conference and the 2023 conference on spacetime functionalism in Pittsburgh.

So that, in outline, is how the book you are now looking at came into existence, and some of the key people and events that have led to it, in relation to the development of the field as a whole. Many others of course contributed to the field, and influenced our work; our apologies for leaving them from this very partial list, but we did not aim for a comprehensive history of the philosophy of quantum gravity.

X PREFACE

The following owes a great deal to generous assistance of many kinds from many sources. Without it the work would have been at best a shadow of what it is.

First, thanks to the pre- and postdoc members of the Beyond Spacetime Project: Eugene Chua, Radin Dardashti, Juliusz Doboszewski, John Dougherty, Saakshi Dulani, Baptiste Le Bihan, Lucy James, Keizo Matsubara, Tushar Menon, James Read, and Mike Schneider. And also to the students with whom we have worked on the issues of the book: Elisa Ballabio, Nicola Bamonti, Sam Bysh, Enrico Cinti, Lorenzo Cocco, Nat Jacobs, Niels Linnemann, Lorenzo Lorenzetti, Emilia Margoni, Tannaz Najafi, Josh Norton, Marta Pedroni, Diana Taschetto, Annica Vieser, Tiziana Vistarini, Niranjana Warrier, and Charlotte Zito.

Many others contributed, in ways big and small: Dave Baker, Sam Baron, Gordon Belot, Philipp Berghofer, Eve Boles, Robert Brandenberger, Jeremy Butterfield, Craig Callender, Claudio Calosi, Elena Castellani, Valeriya Chasova, Karen Crowther, Erik Curiel, Shamik Dasgupta, Kevin Davey, Richard Dawid, Sebastian de Haro, Neil Dewar, Fay Dowker, John Earman, Philip Ehrlich, Astrid Eichhorn, Rawad El Skaf, Michael Esfeld, Sam Fletcher, Doreen Fraser, Henrique Gomes, Brian Greene, Jeff Harvey, Rasmus Jaksland, Arno Keppens, Claus Kiefer, Eleanor Knox, Vincent Lam, Lew Licht, Fedele Lizzi, David Malament, Tim Maudlin, Vera Matarese, Kerry McKenzie, David Meyer, Mike Miller, Peter Momtchiloff, Alyssa Ney, John Norton, Oliver Pooley, Daniele Oriti, A W Peet, Dean Rickles, David Rideout, Joshua Rosaler, Carlo Rovelli, Kian Salimkhani, Tracy Sikorski, Chris Smeenk, Lee Smolin, Sebastian Speitel, Sumati Surya, Nic Teh, Karim Thébault, Francesca Vidotto, David Wallace, Al Wilson, Bob Weingard, David Yates, and Eric Zaslow. A very special thanks to the two anonymous referees at OUP for support and some very helpful feedback. To all those others who should have been on this list, our apologies as well as our thanks.

Various sources of funding and support made this work possible: American Council of Learned Societies, Benjamin Meaker Visiting Professor program at the University of Bristol, Center for Time at the University of Sydney, Swiss National Science Foundation (IZSEZO-214119), John Templeton Foundation (56314, 61387), Centre for Philosophy of Natural and Social Science at the London School of Economics, National Science Foundation (SES 1257354), and our home institutions University of California San Diego, University of Geneva, and University of Illinois Chicago. Some of the work was also carried out when we were fellows or visitors at a number of institutions: University of Barcelona, University of Bristol, Center for Philosophy of Science at the University of Pittsburgh, Institute for Philosophy in the School for Advanced Studies at the University of London, and Magdalen and Merton Colleges at the University of Oxford. Our sincere thanks to all these more material sources of support (and for the spiritual support of the people involved), and we are contractually obliged to point out that the following work may not reflect the views of any of them.

Finally, we could not have completed this work without the support of our families. NH thanks Kai and Phoebe, and especially Joanna, for not merely putting up with him, but actively nurturing, supporting, and encouraging him. CW thanks Ava, Lana, Barbara, Eveline, Hansueli, and particularly his parents Fritz and Margreth, for their steadfast support and their unwavering curiosity regarding the book's status.

Contents

1. Introduction: the emergence of spacetime — 1
  1.1 Quantum gravity and philosophy — 2
  1.2 Worlds without spacetime? — 4
  1.3 The role of philosophy in physics — 9
  1.4 The plan for the book — 11

2. Spacetime functionalism — 14
  2.1 Challenges to spacetime emergence — 15
  2.2 Physical salience — 17
  2.3 Non-commutative geometry — 22
  2.4 Spacetime functionalism — 25
  2.5 Out of Nowhere — 33

3. Spacetime from causality: causal set theory — 35
  3.1 Motivation and background — 36
    3.1.1 Historical prelude: causal theories of time and of spacetime — 36
    3.1.2 Earman's criticism of causal theories of spacetime — 38
    3.1.3 Malament's theorem — 43
  3.2 The basic plot: kinematic causal set theory — 47
    3.2.1 The kinematic axiom — 47
    3.2.2 Discussion of the basic assumptions — 49
    3.2.3 Causation? — 52
  3.3 The problem of space — 54
    3.3.1 The 'essence' of space — 56
    3.3.2 Space in a causal set — 57
    3.3.3 The dimensionality of a causal set — 60
    3.3.4 The spatial topology of causal sets — 63
    3.3.5 Finding distances — 67
    3.3.6 Wrapping up — 69

4. The emergence of spacetime from causal sets — 71
  4.1 Identifying necessary conditions — 72
    4.1.1 The basic setup — 72
    4.1.2 Classical sequential growth dynamics to the rescue — 75
    4.1.3 Manifoldlikeness — 80
    4.1.4 The Hauptvermutung — 83
  4.2 The case for spacetime functionalism — 87
    4.2.1 The road to spacetime functionalism — 87

xiv CONTENTS

	4.2.2 Spacetime functionalism in causal set theory	90
4.3	Causal set theory and relativistic becoming	91
	4.3.1 The basic dilemma of relativistic becoming	92
	4.3.2 Taking growth seriously	94
	4.3.3 Some bizarre consequences	97
4.4	Lorentz symmetry and non-locality	100
	4.4.1 Lorentz symmetry and discreteness entail non-locality	101
	4.4.2 Implications: hierarchy of scales and phenomenological signatures	105
	4.4.3 Swerving particles: putting matter on causets	107
	4.4.4 The ever-present Λ	108
4.5	Summing up	109

5.	The road to loop quantum gravity	111
5.1	Learning from general relativity for quantum gravity	111
5.2	Background independence and general covariance	113
5.3	Hamiltonian general relativity	122
	5.3.1 Reformulating general relativity	122
	5.3.2 Hamiltonian systems with constraints	124
	5.3.3 Gauge freedom and the problems of time and change	127
5.4	Loop quantum gravity	130
	5.4.1 The canonical quantization of Hamiltonian GR	130
	5.4.2 LQG: the basics	132
	5.4.3 Spin network states: the technical background	137
5.5	The covariant perspective	139
	5.5.1 Canonical-covariant methodology	140
	5.5.2 Covariant dynamics	142
5.6	Conclusion	145

6.	The disappearance and emergence of spacetime in loop quantum gravity	146
6.1	The non-spatiotemporality of canonical LQG	147
	6.1.1 Spin network states: interpretation	149
	6.1.2 Indeterminate space	152
	6.1.3 Disordered locality	155
	6.1.4 No change, no time?	158
	6.1.5 Objections	159
	6.1.6 The disappearance of space and time	162
6.2	Interpretation of covariant LQG	164
6.3	The emergence of spacetime in LQG	169
	6.3.1 Emergence, reduction, and the classical limit	169
	6.3.2 The Butterfield-Isham scheme of emergence	172
6.4	Spacetime functionalism	177
6.5	Eternalism, presentism, or neither?	184
6.6	Conclusion	187

7. A string theory primer	188
7.1 Classical string basics	190
7.1.1 Equation of motion	190
7.1.2 String actions and symmetries	193
7.2 Physics in the string	198
7.2.1 Conformal symmetry	199
7.2.2 String mass	201
7.3 Quantization	205
7.3.1 Canonical quantization	205
7.3.2 Lightcone coordinates	207
7.3.3 Lightcone gauge	208
7.3.4 Quantization and stringy particles	210
7.3.5 26 dimensions	216
7.4 Fermionic strings and supersymmetry	218
8. Duality	222
8.1 T-duality	222
8.2 Two interpretive forks	229
8.2.1 The first fork: physical equivalence?	229
8.2.2 Three senses of 'space'	231
8.2.3 The second fork: factual or indeterminate geometry?	233
8.3 Interpretation: physical equivalence	237
8.3.1 Analogy: the harmonic oscillator	238
8.3.2 On the concept of physical equivalence?	243
8.4 Interpretation: indeterminate geometry	247
8.5 Beyond closed string T-duality	251
8.5.1 T-Duality for open strings	252
8.5.2 Gauge-gravity duality	255
8.6 Conclusions	257
9. The string theoretic account of general relativity	259
9.1 Path integral string theory	260
9.2 Building a curved background	266
9.3 The Einstein field equation	271
9.4 Anomalies	276
9.5 Summary	282
9.A Appendix: coherent states of quantum fields	282
9.A.1 Massless scalar field	283
9.A.2 Coherent states of a simple harmonic oscillator	286
9.A.3 Coherent states of the scalar field	289
9.A.4 The classical limit	291
9.B Appendix: Noether's theorem	296
10. The emergence of spacetime in string theory	302
10.1 Deriving general relativity	303
10.2 Whence spacetime?	310

xvi CONTENTS

10.3	Whence *where*?	313
	10.3.1 The worldsheet interpretation	314
	10.3.2 T-duality and scattering	316
	10.3.3 Scattering and local topology	319
10.4	Whence the metric?	321
	10.4.1 'Background independence'	322
	10.4.2 Is there a Minkowski background?	325
	10.4.3 T-duality	329
10.5	Quantum field theoretic considerations	330
	10.5.1 The graviton concept	330
	10.5.2 Graviton coherent states	333
	10.5.3 GR from QFT	335
10.6	Conclusions	336
11.	Conclusion: whence spacetime?	338
11.1	Causal set theory	339
11.2	Loop quantum gravity	341
11.3	String theory	343
11.4	Out of nowhere	347
Bibliography		348
Index		363

Chapter 1
Introduction: the emergence of spacetime

"Big Bang Machine Could Destroy Earth"

... ran an attention grabbing headline in *The Sunday Times* (Leake 1999), regarding the new Relativistic Heavy Ion Collider (RHIC) at Brookhaven National Laboratory. To be fair, the main apocalyptic concern of the paper was that the RHIC would create a form of matter in which strange quarks eat the up and down quarks found in ordinary matter. But it also discussed the possibility that experiments involving high-energy collision of gold ions could create microscopic black holes, which would pull all the matter in the world into them. Such scenarios were taken seriously enough that they were evaluated by a panel of elders, who concluded that the chances of any such events were utterly minuscule (Busza et al. 2000)— they have not been contradicted by events at Brookhaven!

Let's look into the business of black hole formation more carefully to explain why physics needs an account of 'quantum gravity', a theory that combines quantum mechanics (QM) with general relativity (GR), and the subject of this book. First, the RHIC was built to probe how matter behaves under intense temperatures and pressures—in effect recreating in a tiny region the state of the universe within the first second of its existence, when quarks and gluons flowed in a plasma rather than binding to form particles. The predictions tested here are largely those of quantum chromodynamics, the quantum theory of the strong force binding nucleons and their constituents. That is, the collisions between heavy nuclei such as gold in the RHIC are governed by the laws of QM.

The concern over black holes, however, arises when one asks how GR—the classical, non-quantum theory of gravity, gets into the picture. Suppose that a mass M is located inside a sphere of radius $r = 2GM/c^2$ (where G is Newton's gravitational constant, c is the speed of light, and r the distance from the center). Then, according to GR, there will be an 'event horizon' at that radius, from which neither matter nor light can escape—the boundary of a black hole. However, if the mass is not located inside such a sphere, then no black hole forms, so it is a necessary condition for a mass M to form a black hole that it be contained within a radius less than $2GM/c^2$, a fact important for the RHIC analysis.[1]

---

[1] The spacetime metric outside a sphere of mass M takes the form:

$$ds^2 = c^2 dt^2 \left(1 - \frac{2GM}{rc^2}\right) - dr^2 / \left(1 - \frac{2GM}{rc^2}\right). \tag{1.1}$$

Out of Nowhere. Nick Huggett and Christian Wüthrich, Oxford University Press.
© Nick Huggett and Christian Wüthrich (2025). DOI: 10.1093/oso/9780198758501.003.0001

2 OUT OF NOWHERE

So finally, the questions posed by the panel at Brookhaven were 'how much energy would be created in collisions, and in how small a region would it be located?' (See page 7 of the report.) Acting cautiously, they assumed the best conditions for black hole formation, supposing that all the energy produced by the collision contributes to the mass: about 50 times that of a gold atom. For a black hole of this mass the event horizon has a radius of 10^{-39} m. On the other hand, a gold atom has a radius of around 10^{-12} m, so even supposing that all the energy is concentrated in a region the size of a suitably Lorentz-contracted nucleus, GR predicts that collisions will be many, many orders of magnitude from creating a black hole.[2]

Phew.

1.1 Quantum gravity and philosophy

What we have then is an argument that the physics of Brookhaven lies within the domain of relativistic QM, but that the gravitational effects of the collisions are utterly negligible. So perhaps the world is just fragmented: in some domains, such as the motions of the planets (GR explains the perihelion of Mercury, for instance), GR holds and QM is irrelevant; in others, as in the RHIC, it is QM that holds sway, with GR entering only to provide a background geometry determined by ambient bodies, not by the system under consideration. But the very argument here shows that it is sometimes necessary to bring considerations from both theories to bear on a single system, making clear that one can sensibly ask whether there are domains in which both theories apply. This point has been elaborated and argued forcefully by Wallace (2022b), who reviews many cases in which physicists successfully combine QM and GR: to understand stellar evolution, and to predict Hawking radiation, for instance (Burgess 2004). However, almost all of these examples lie in the same low energy, 'semi-classical' regime as Brookhaven. More extreme is the usual treatment of inflation: between the first 10^{-33} to 10^{-32} s quantum matter fields provide the energy which drives the expansion of classical spacetime, but fluctuations observed in the cosmic microwave background as a result of this process involve quantum fluctuations of the gravitational field itself.[3] Still, even this process only requires treating gravity as a quantum field, an

---

(ds^2 is the infinitesimal spacetime 'distance' squared—the 'interval'—between two radial points separated in time by dt and space by dr, in suitable co-ordinates; though an exact understanding is not crucial here.) What is important here is that this quantity blows up when $2GM = rc^2$, or $r = 2GM/c^2$. Understanding this occurrence, not as a true singularity but as an event horizon, was an important issue in the early development of GR (Earman 1999).

[2] Don't be confused if you have read of black holes being created at RHIC. In fact what has (perhaps) been observed is the Unruh effect. Assuming that the local equivalence of acceleration and gravity—the 'equivalence principle'—holds in the quantum domain, this provides an indirect test of the Hawking radiation predicted to occur around black holes. See Nastase (2005).

[3] See Huggett et al. (2023) for discussion of a proposed laboratory experiment to demonstrate the quantum nature of gravity.

INTRODUCTION: THE EMERGENCE OF SPACETIME 3

approach widely believed to fail in truly extreme situations (see Weinberg 2016 for some dissent).

The theories discussed in this book are intended to provide a more fundamental (perhaps *the* fundamental), unified account, unlike existing physics, applicable in the most extreme phenomena known to exist. First among them is the big bang, which is entailed by GR given the current state of the universe, in which matter and energy become so hot and dense that even a quantum field version of GR is not expected to suffice. Similarly, many also think that the singularities in black holes, and the Hawking radiation that they are predicted to produce (Hawking 1974), reveal them to have a fundamental nature that requires a deeper theory of the quantum and gravity than we now have. Thus, we say, it is to understand such phenomena, in which even the quantum field treatment of gravity is expected to fail, that we need a theory that goes beyond current known physics, yet unifies QM and GR by explaining their success in their appropriate domains—a theory of *quantum gravity* (QG).[4]

There are then good reasons for physicists to investigate QG. This book is predicated on the view that QG also has an important call on the efforts of philosophers. Indeed, we would like this work to encourage, by example, our colleagues to be more adventurous in their choice of topics of enquiry. Philosophy of physics, we suggest, has a tendency to look too much to the past, and to the metaphysics of well-established physics (and of course to internecine disputes): classical statistical mechanics, classical spacetime theory, and non-relativistic quantum mechanics are 'so twentieth (or even nineteenth) century', and yet have a virtual lock on the discipline. While quantum field theory (QFT) is becoming a significant topic, that is still at least half a century behind the physics!

We're overstating things somewhat for effect here: of course, even old theories do face important foundational problems, and their consequences for our broader understanding of the world take considerable elucidation. And of course it is unfair to suggest that no philosophers of physics show an interest in contemporary physics. Indeed, since we started writing this book, there has been an explosion of interest in QG, especially amongst a younger generation of scholars, which we find very exciting. Still, we do say that collectively the discipline pays insufficient attention to cutting-edge physics, and hope this book in some way serves as an impetus to greater engagement.

Our point is not that novelty is good for its own sake, nor that philosophy is a 'hand-maiden', who should dutifully follow the fashions of physics. Rather, we are inspired by recent work in the history and philosophy of science to believe that it is central to the business of philosophy to engage with developing physical

---

[4] There are other, more theoretical, arguments for such a theory and concerning the form it should take. For a critical evaluation of these, arguing that overlap is the best reason to seek quantum gravity, and that empirical considerations best dictate its form, see Callender and Huggett (2001a); for further discussion Wüthrich (2005).

4   OUT OF NOWHERE

theories—both because the search for philosophical knowledge must be responsive to empirical discoveries, and because philosophy has important contributions to make to the development of physics (and other sciences). We will discuss this point at greater length below (§1.3), to explain our aims and motivations.

First, in part to make that discussion more concrete, in §1.2 we will very briefly introduce some approaches to QG. We especially want to focus on a rather generic feature of them—that in various ways they do not contain familiar spacetime at a fundamental level, but rather it 'emerges' (in a sense to be discussed) in a higher, non-fundamental domain. In a final section (§1.4) we will give an overview of the different strategies one might take toward quantizing gravity, to relate the different proposals that we will consider in the book.

1.2 Worlds without spacetime?

All approaches to QG are, to a large extent, speculative and partial, so while we follow standard practice and call them 'theories' they are not articulated and established like true theories such as Newtonian mechanics, GR, or QM; referring to them as 'approaches' emphasizes this point. Moreover, some have had much more intellectual effort devoted to them than others, and are more fully articulated (in part) as a result. In particular, by a long way, string theory has attracted the attention of the most physicists in the QG community, especially during the 1990s and 2000s; of the remaining approaches, loop quantum gravity has had the most effort devoted to it. But despite being speculative and incomplete, there are good reasons to think that such theories teach important lessons in the search for QG, as we would now like to explain. In particular, they all point to the conclusion that *spacetime is in some way not fundamental, but only appears in higher level descriptions of a non-spatiotemporal reality*—the topic of this book.

The first three of the following examples are the focus of the following chapters. The remaining two are also illuminating for the present point and help to emphasize that the lessons drawn are indeed general, and not just restricted to the approaches investigated in this book (we discuss them elsewhere, the following summarizes those discussions). They are more technical, and might be skimmed if necessary.

- **Causal Set Theory (CST):** As we will see in chapters 3 and 4, CST makes liberal use of GR as a vantage point for its research programme. In fact, it takes its most important motivation from theorems stating that given the causal structure of a spacetime, its metric is determined up to a conformal factor: the structure that determines the paths along which light can travel fixes the full geometry of a spacetime, except its 'size'. Taking this cue, CST posits that the fundamental structure is a set of elementary events which is locally finite, and partially ordered by a basic causal relation. In other words, the fundamental structure is a *causal set*. The assumption of local finitarity is nothing but the formal demand that the fundamental structure—whatever else it is—is discrete. Together with the demand of Lorentz invariance at the derived level, the discreteness of causal sets forces a rather odd locality structure onto the elementary events of the causal set (§4.4). Furthermore, although the fundamental relation of causal precedence can double up as something akin to temporal precedence, space is altogether lost in a causal set (§3.3). Jointly, these facts entail that the structure we are facing in causal set theory is also rather different from the spacetime encountered in GR. In fact, the quantum nature of the causal sets yet to be incorporated into CST is bound to further complicate the picture and to remove the resulting structure from that of relativistic spacetimes.

- **Loop Quantum Gravity (LQG):** LQG starts out from a Hamiltonian formulation of GR and attempts to use a recipe for cooking up a quantum from a classical theory that has been utilized with great success in other areas of physics. This recipe is the so-called *canonical quantization*. The goal of applying the canonical quantization procedure is to find the physical Hilbert space, i.e., the space of admissible physical states, and the operators defined on it that correspond to genuinely physical quantities. As will be seen in chapters 5 and 6, following this recipe leads rather straightforwardly into a morass of deep conceptual, interpretative, and technical issues concerning the dynamics of the theory as well as on time quite generally. We find that the states in the 'kinematic' Hilbert space afford a natural geometric interpretation: its elements are states that give rise to physical space, yet are discrete structures with a disordered locality structure. At least in one decomposition, these states appear to represent a granular structure, welding together tiny 'atoms' of space(time). It is crucial to this picture that these atoms of space(time) are atoms in the original meaning of the word: they are the truly indivisible smallest pieces of space(time). The smooth space(time) of the classical can thus be seen to be supplanted by a discrete quantum structure. Moreover, since generically a state of this structure will be a superposition of basis states with a determinate geometry, generic states will not possess determinate geometric properties. If continuity, locality, or determinate geometry was an essential property of spacetime, then whatever the fundamental structure is, it is not spacetime. In this sense, spacetime is eliminated from the fundamental theory.
- **String Theory:** According to string theory (chapter 7) tiny one-dimensional objects move around space, wiggling as they go—the different kinds of vibration correspond to different masses (charges and spins) and hence to different subatomic particles. So it sounds as if spacetime is built into the theory in a pretty straightforward way; however, we will see that things are not so

simple. First, in chapter 8 we will see that various versions which are intuitively very different in fact correspond to the same physics—they are '*dual*'. For instance, suppose that at least one of the dimensions of space is 'compactified', or circular. Then it turns out that a theory in which the circumference of the dimension is C has the same collection of values for physical quantities as a theory in which the circumference is $1/C$: i.e., that theories in which compactified dimensions are small are physically indistinguishable from—or 'dual' to—those in which they are large. Other dualities relate spaces of different topologies. These facts raise important questions about whether the space in which the string lives is the one we observe, since that is objectively large, while the string space circumference could be small, or even conventional. Second, in chapters 9 and 10, we will see how the geometrical structure of spacetime—and indeed GR—arises from the behavior of large collections of strings in a 'graviton' state, the quantum particle that mediates gravitational forces.

- **Non-Commutative Geometry:** Picture a Euclidean rectangle whose sides lie along two coordinate axes, so that the lengths of its sides are x and y—its area is $x \cdot y = y \cdot x$. But what if such products fail to commute, $xy \neq yx$, so that area is no longer a sensible quantity? How can we understand such a thing—a *non-commutative geometry*? By abandoning ordinary images of geometry in terms of a literal space (such as the plane) and presenting it in an alternative, algebraic way. In fact, our example already starts to do so: even thinking about areas as products of co-ordinates uses Descartes' algebraic approach to Euclidean geometry. Once we have entered the realm of algebra, all kinds of possible modifications arise. Especially, an abstract algebra $\mathcal{A}$ requires an operation of 'multiplication', $\star$, but this can be a quite general map from pairs of elements, $\star : \mathcal{A} \times \mathcal{A} \to \mathcal{A}$, *which need not be commutative*! For instance, one could define 'multiplication' to satisfy $x \star y - y \star x = i\theta$ (a small number), and use it to generate polynomials in x and y; these carry geometric information. Such a thing is perfectly comprehensible from the abstract point of view of algebra, but it cannot be given a familiar Euclidean interpretation via Cartesian geometry. So, such a theory seems to describe a world that is fundamentally algebraic, not spatial (in the ordinary sense)—there is $x \star y$ and $y \star x$ but no literal rectangle. If there is thus fundamentally nothing 'in' space, is the ultimate ontology 'structural', based on algebraic relations only? And how could an appearance of familiar (commutative!) space arise; especially, what significance could point-valued quantities have? We will return to this example in §2.3.

- **Group Field Theory:** Consider rotations, of angle θ, in the plane: a different one for each value of $0 \leq \theta < 2\pi$. These form a group under composition: a rotation by α followed by an angle β is just a rotation by $\alpha + \beta$ (and for example, a rotation by $2\pi - \alpha$ undoes a rotation by α). Similarly for rotations in three dimensions, and indeed similarly for Lorentz transformations (which are in fact nothing but rotations in Minkowski spacetime), and so on. In each case, composition is some function from any two elements to a third. We can thus characterize the abstract group structure simply by the action of this function of an entirely arbitrary set of elements—forget that we started with rotations, and let the elements be anything, with a composition rule isomorphic to that of the rotations. One could take then a set of such blank elements, which compose like rotations in the plane, but which should not be thought of as literal rotations: no plane at all is postulated, the only manifold is that formed by the group elements themselves—the circle of points with labels $0 - 2\pi$, not a 2-dimensional plane.[5] And finally one can introduce a field on this group manifold, a real number for each group element, with a dynamical law for its evolution; and indeed quantize this field. As we have emphasized, while physical space was used to guide the construction of this theory, it is not an explicit component of it, and yet models of GR can be derived from it. This example is discussed in greater detail in Oriti 2014 and Huggett 2022.

All of these examples are speculative to some extent or other, and none can claim to be a complete quantum theory of gravity (and none has convincing, currently testable, novel predictions!), yet all have some claim to model relevant physical features of QG, worth exploring. In particular, we have emphasized in each case how spacetime features are missing in the theories: space is discrete, locality starts to dissolve, topology and distance become indeterminate, the very manifold is banished. From our survey, this situation appears to be a common condition of many approaches to QG.

That this is so should not be surprising, if one thinks about what it might take to have a quantum theory of spacetime. In the first place, observing anything smaller than a given distance requires wavelengths shorter than that distance, while in QM the shorter the wavelength the greater the energy; the smaller the region observed, the greater the energy we must concentrate in it. But we know from our earlier discussion that if enough energy (equivalently mass) is concentrated in a small enough region then a black hole forms, and *no* information can get out! In other words, there is a minimum size that can be observed if both QM and GR apply: this is the 'Planck length', around 10^{-35} m. Or again, relativistic quantum mechanics depends heavily on the causal structure of spacetime, but just as the position of a quantum particle can be indeterminate in many interpretations of QM, so the causal structure of quantum spacetime is indeterminate. In the quantum field approach this can be finessed, but at some scale—likely the same Planck scale—this solution will fail, and it appears that spacetime will have to give. Arguably, the reason that spacetime fades away in the examples that we have given is that

---

[5] More accurately, the space that is assumed is not the space of relativistic physics, or that of planar rotations, but four copies of the group of Minkowski rotations.

8   OUT OF NOWHERE

they are facing these challenges head on, and so reflecting the extreme difficulty of combining QM and GR in a fundamental way.

The ways in which spacetime disappears in the first three theories (and to a lesser extent in the fourth) will be explored in much greater depth in the remainder of the book. But to the extent that that conclusion is correct, we say that classical, relativistic spacetime is 'emergent'. We emphasize (as we have elsewhere) that we do not use this term in its strongest philosophical sense to indicate the *inexplicability* of X from Y, as some have claimed life or mind emerges from matter. On the contrary, we argue that classical spacetime structures *can* be explained in more fundamental terms: indeed, it was largely to explicate how physicists do so that we wrote the book.[6]

We speak of emergence because of the huge gulf between a fundamental theory that does not assume spacetime and a higher-level one that does. Having spacetime or not makes a huge formal and conceptual difference, in particular because in almost all theories prior to QG classical spacetime has apparently been one of the most basic posits. Indeed, this very gulf makes one wonder what it could mean to derive spacetime, and whether it is possible at all. Here we will appeal to a 'spacetime functionalism' based on that developed in the philosophy of mind. We will discuss this approach in detail in §2.4, so here we only sketch the idea.

The general situation is that we accept two theories, one describing less fundamental states and entities as behaving in a certain way, and another describing more fundamental states and entities as behaving in another way. Let's say that:

1. The less fundamental states and entities are those that *ABC*, while ...
2. ... the more fundamental states and entities are those that *XYZ*.

For example, pain leads people to ejaculate 'ow' and so on (at length), and hold the painful spot; while neuroscience describes the response of neurons (individually and in aggregate) to physical stimuli. That is, the states and entities are identified in terms of their functions according to the theories: i.e., *functionally*.[7] But suppose further that:

3. One comes to believe that the more fundamental states and entities (likely in aggregate) also *ABC*. Then ...
4. ... *as a matter of logic*, (aggregates of) the fundamental states and entities simply *are* less fundamental ones.

---

[6] We will have to say more about the notion of 'emergence' in later chapters, particularly in chapter 2 and in §6.3, where we compare our approach with similar ones that have been offered by Butterfield and Crowther.

[7] Sometimes this function is described as the 'causal role', but this is by no means an essential part of functionalism: For instance, one could characterize the functional role in terms of the Carnap sentence: 'the states and entities are the unique x that $ABC(x)$', for instance. Since causation is a spatiotemporal notion, it is important for the functionalist approach to emergent spacetime that causation not be essential.

The conclusion follows deductively because if a is identified as uniquely bearing some description, then also anything bearing that description must be a, and that is what functional identification involves.

So for example, if neuroscience can show that suitable aggregates of neurons can produce ejaculations of 'ow' and comforting grasping of painful spots, and so on (at length), then one would conclude that pain is in fact a suitable neuron aggregate. In the case of spacetime, we will follow the same program: functionally identify spacetime states and entities (as that which underwrites topology, locality, and geometry for instance); functionally identify QG states and entities (through the different theories of QG); show how the latter may perform the functions of the former; to the extent that one accepts that they do, conclude that they are identical with the former. We will discuss how this works (and how it differs from psychophysical functionalism) in the next chapter.

But we have said enough to see why some have objected to our use of 'emergence'; in philosophy of mind, functionalism is usually thought of as a form of reduction—opposed to a stronger kind of emergence! However, we generally choose not to speak of reduction because there are many notions of 'reduction', some of which are too strict (though we are happy with notions of 'explanation' or 'derivation', as will be seen). But we think that (weak) 'emergence' is justified even when spacetime is derived, because of the gulf between physical theories with and without spacetime.

1.3 The role of philosophy in physics

As we noted, the theories that we plan to investigate are all speculative at present, faced with considerable formal and empirical uncertainties. So what can we hope to learn from a philosophical enquiry into something that is at worst likely false, or at best a work in progress? In the first place, the theories we discuss are built from existing physics, especially GR and QFT, and our analysis will require novel contributions to the philosophy of those theories. But there is a far more substantial and important answer.

The situation of competing, partial theories just described is characteristic of emerging fundamental physics (and perhaps other sciences). The process of discovery takes place along various fronts: obviously, new empirical work constrains theory and requires explanation; also obviously, new mathematical formalisms are tried out and explored; less obviously, but just as importantly, conceptual analysis of the emerging theory is undertaken. In particular, we want to stress that this last kind of work is carried out concurrently with the empirical and theoretical. One should not view interpretation as something that merely happens after an uninterpreted formal structure is presented, but as an inextricable aspect of the process of discovery. As such, it is something that has to be carried out on inchoate theories, in order to help their development into a finished product.

10   OUT OF NOWHERE

We claim that this view is supported by the historical record: we will see this for Newtonian gravity in chapter 2, for example. But one can equally well point to the absolute-relative debate in the development of the concept of motion, or 19th century efforts to come to grips with the physical significance of non-Euclidean geometry. These debates did not wait until after a theory was developed to clarify its concepts; rather they had to be carried out simultaneously, as an integral part of the development of the theory (see DiSalle 2006). Of course we are hardly the first to realize that such philosophical issues have to be addressed together with the empirical and theoretical ones. Many of Kuhn's (1962) arguments illustrate this point, and it is a major theme of Friedman (2001). But while we agree with their focus on philosophical, conceptual analysis as an essential part of theory construction, we don't intend to get involved in issues involving incommensurability or the *a priori*, instead we want to emphasize the practical role for analysis in the development of QG.

In the search for a new fundamental theory, the goal is—as it was for Einstein and for Newton—a new formalism plus an interpretation that connects parts of the formalism to antecedently understood aspects of the physical world, especially to the empirical realm. That is, an interpretation of how the more fundamental plays the functional roles of the less fundamental. One never simply co-opts or invents formalism without some eye on the question of how it represents existing physics of interest; and as the formalism is developed it becomes possible to see more clearly how and what the new formalism represents. Addressing this question is of ongoing importance for finding the right formalism for the area under study. Moreover, constructing such a formalism does not typically proceed in a monolithic fashion; instead, different fragments of theory are proposed, investigated, developed, or abandoned. For example, think of the development of the standard model of QFT from the early days of QM. So the analysis of concepts of the new theory in terms of existing physics is often faced with a range of half-baked theories and models. All the same, lessons about how a more developed, less fragmented theory can be found depend on asking how the fragments represent known physics—the answers are potential clues to how the finished product could do so.

We believe that contemporary QG should be thought of in just this way—certainly the fragmentation is real! Our primary goal is to look at a range of the half-baked fragments and ask how they connect to spatiotemporal phenomena. Since they do not do so in a familiar way, in terms of a continuous manifold of points, the question becomes 'how does spacetime emerge from the underlying physics?' We hope, therefore, that we will be performing a service to physicists working in QG, by focusing their attention on what is already known—and reminding them that success depends on making it part of the search. Naturally, we do not expect to find solutions of the order of Newton or Einstein! Indeed, a lot of what we shall do is draw out answers already given by physicists;

we believe that careful philosophical analysis of these answers can help clarify them to reveal strengths and weaknesses, and hence aid progress. (Moreover, because we are focused on this quite narrow issue, we can survey a wider range of approaches than most physicists actively study, and so provide a helpful overview of the topic.) And hence we believe that in the examples we will consider there are important clues for the development of QG which philosophical analysis can reveal.

Let us emphasize that this understanding of philosophy in no way treats philosophy as a mere servant to science; philosophy is instead just as an essential activity within the scientific enterprise as formal and empirical work (though not as prominent). So if we can contribute useful philosophical analysis to the search for QG, we will thereby be contributing to philosophy. In fact we will go further: historically, many of the *most fertile and significant* moments in the history of philosophy (especially outside value theory) have come from such moments of scientific advance. To mention the most obvious and thoroughly studied: consider the deep interconnections of Aristotle's metaphysics and epistemology with the science of his day; or the work of Descartes, Locke, Leibniz, du Châtelet, and Hume, which is so deeply engaged with the scientific revolution that spanned their careers; or the nineteenth-century debates on empiricism and conventionalism spurred by advances in physics and mathematics; or of course the close connections of logical positivism and the quantum and relativistic revolutions then in progress; and so on. That is, what we view as some of the most important advances in philosophy were not mere responses to completed scientific theories, but were the fruit of the very philosophical participation in physics that we have described in this section. Of course, we don't expect to cast shadows as long as Aristotle or Hume, but we do claim that by engaging with the ongoing QG revolution we are solidly in the philosophical tradition, contributing (more modestly) to the philosophical stock of understanding in just the way that some of our most storied predecessors did.

1.4 The plan for the book

Thus, our primary aim is to see how spacetime disappears and re-emerges in several approaches to QG, and to show how this is not just a technical issue for physicists to solve, but instead elicits numerous foundational and philosophical problems. As we work through three such approaches—causal set theory (CST), loop quantum gravity (LQG), and string theory, which were all briefly introduced in §1.2—, we bring these philosophical issues to the fore and will concentrate our discussion on them.

As a preliminary, in chapter 2 we will articulate our functionalist approach more thoroughly, and explain how it addresses objections that might be made to the very idea of spacetime emergence: how could anything as physically and philosophically basic as space and time not be fundamental? Then we turn to applying our analysis to theories of QG.

It is common to divide approaches into those which start out from GR and attempt to convert it into a quantum theory of gravity in different ways and into those departing from the standard model of particle physics and aim to add gravity to the other three forces of the standard model. In the former approaches such as CST and LQG, we would not expect the resulting theories to fold in the physics of the standard model, whereas the latter, such as string theory, will presumably deliver more encompassing, unifying theories. It is clear that either way, a theory of QG needs to address how the geometrical degrees of freedom of spacetime interact with the matter degrees of freedom present in the world. But it is also clear that both kinds of degrees of freedom may well look very different from what we are used to from other theories.

The first two chapters after chapter 2 focus on CST. Chapter 3 introduces the basic kinematic axiom of the theory and shows how not only does space disappear rather radically from the fundamental ontology, but also that temporal aspects do not all survive. This raises the immediate question of the relationship between the fundamental ontology of causal sets and that of relativistic spacetimes, a question we start to address in chapter 3. Although some of the role played by space can tentatively be recovered, what is needed is a more systematic understanding of how causal sets generically give rise to worlds which appear to be spatiotemporal in ways described, to good approximation, by GR. The way in which this 'derivation' of spacetime is attempted in CST is sketched and discussed in chapter 4. In this chapter, we will discuss the role played by introducing a dynamics for the theory. We will argue that the emergence of spacetime in CST is closely tied to deeply philosophical questions regarding the metaphysics of space and time.

Chapters 5 and 6 turn to LQG, retracing the disappearance and emergence of spacetime in this approach. Just as CST, LQG builds a research program around what it takes to be GR's central lesson. In the case of LQG, this is the insight that GR postulates a truly dynamical spacetime, interacting with other fields. The demand is encoded in the theory's background independence. LQG seeks to articulate a theory of QG by delicately applying known quantization procedures to a Hamiltonian formulation of GR. Chapter 5 chronicles and discusses whether and, if so how, this approach leads to the disappearance of spacetime. Unlike CST, it is time whose existence is more threatened in LQG than space. Chapter 6 seeks to understand how relativistic spacetime then emerges from the fundamental theory, finding, again, close ties to philosophical questions.

Other approaches apply the strategies of perturbative QFT—so successful in understanding the other forces—to quantize gravity. The technique calls for starting a system in which the fields do not interact to build up a space of states: a lowest, vacuum state, and states of discrete, particle-like 'quanta'. Generally such a system is solved exactly, and the vacuum describes an obvious classical state. Then one introduces a small interaction, and uses approximation techniques to study the behavior of fields: especially the scattering of quanta. This approach was applied to gravity early on: Minkowski spacetime is a natural vacuum, and the gravitational field has quanta known as 'gravitons', very analogous to photons, the quanta of the electromagnetic field. Indeed, quite a lot is known about the quantized gravitational field through such methods, and this knowledge is taken as a constraint on a successful theory of QG. However, divergences prevent the theory from being generally applied; moreover, these divergence cannot be adequately resolved by 'renormalization' as they can for other QFTs.

String theory works within this approach, but with one important tweak: instead of quantized point like particles, it deals in quantized 1-dimensional, string-like objects. This, it appears, makes all the difference to the finiteness of the theory. Chapters 7–10 address the emergence of spacetime in string theory. Chapter 7 is a fairly technical introduction of the theory, aimed at philosophers of physics: it aims to be more intuitive, and more explicit about the conceptual and physical framework than physics textbooks usually are. For those who have some familiarity with classical and quantum field theory, it will tell you what you need to know about strings. Chapter 8 deals with string 'dualities': some fascinating and powerful symmetries that arise when space has an interesting topology (a cylinder, say). We argue that they are the kind of symmetries that are not merely observational, but 'go all the way down', showing that string theory does not possess, in its basic objects, familiar spacetime properties such as definite size or topology; it is for largely that reason that spacetime 'emerges'. Chapter 9 is again fairly technical, explaining and analyzing in some detail the derivation of the Einstein field equation for gravity, from string theory. This is a central part of emergence, for it derives the spacetime metric, giving empirical content to spacetime geometry, and gives rise to GR. Finally, chapter 10 draws on the material of the previous chapters to argue that indeed spacetime emerges in string theory, how this happens, and what 'principles of physical salience' are required.

In conclusion, chapter 11 draws on the results of the previous ones to return to the questions of this introduction. How does spacetime disappear, and just how does spacetime emergence by functional means work concretely?

Chapter 2
Spacetime functionalism

In the previous chapter we explained the imperative for a theory of QG, and why a crucial element of that search was philosophical, in particular to understand the disappearance and emergence of spacetime. In this chapter we consider in more detail the problems facing the attempt to explain spacetime emergence, and how—through case studies—they can be overcome, especially through our proposed approach, spacetime functionalism.

Before we proceed, we need to introduce some terminology to keep the discussion straight. The issue is that the theories of QG often contain some object referred to as 'space', even when they do not assume 'space' in the ordinary sense. For instance, there may be a 'Hilbert space', or a 'dual space', or 'Weyl space', or 'group space'. So we will refer to spacetime in the ordinary sense as 'classical', or 'relativistic', or sometimes just 'space' or 'spacetime' when the context makes matters clear. We eschew the phrase 'physical space', since the other 'spaces' may well be part of the fundamental physical furniture. We have previously used 'phenomenal space', to indicate that classical space is that of observable phenomena, according to the physicist's use of 'phenomenological'. However, this leads to confusion with the philosophical doctrine of 'phenomenalism', so we have dropped it.

By classical or relativistic spacetime, we mean that space theorized in QM (especially QFT) and relativity, approximated in non-relativistic mechanics, and ultimately implicated in our observations of the physical world. As stated, that is not an entirely homogeneous concept, so we will say more later (§2.4) about exactly what features of classical spacetime are emergent from our theories of QG.

Before that, in §2.1, we will review the challenges to the very idea that something as seemingly fundamental as spacetime could be derivative, from even more fundamental, yet non-spatiotemporal, physics. We then address these challenges, analyzing how they can be overcome, and illustrating this process in a historical case (§2.2) and in a short contemporary example (§2.3). This discussion encapsulates much of the work of the book: the chapters introduce, for philosophers, several proposals for a theory of QG, discuss the ways in which they eliminate spatiotemporal structures, and investigate the ways in which they are recovered as effective, apparent structures. What we propose is that the recovery of spacetime in each of these approaches is to be understood as implementing a form of functionalism, so in §2.4 we explain that position. We conclude in §2.5.

Out of Nowhere. Nick Huggett and Christian Wüthrich, Oxford University Press.
© Nick Huggett and Christian Wüthrich (2025). DOI: 10.1093/oso/9780198758501.003.0002

2.1 Challenges to spacetime emergence

Space and time are so basic to both our manifest and scientific images of the world that at first the mind boggles at the thought that they might be mere 'appearances' or 'phenomena' of some deeper, more fundamental, non-spatiotemporal reality. Is a physics without spacetime even intelligible? And if it is, is spacetime the kind of thing whose existence could be explained? At its core, this book seeks to address these questions: on the one hand explicating the worlds described by theories of QG, while on the other showing how spacetime can be derived from them. But to understand the nature and methodology of that project, it is important here to unpack the vertiginous panic about the very idea. Larry Sklar (1983) gave expression to this all too common sentiment among philosophers (and physicists) when he wrote[1]

> What could possibly constitute a more essential, a more ineliminable, component of our conceptual framework than that ordering of phenomena which places them in space and time? The spatiality and temporality of things is, we feel, the very condition of their existing at all and having other, less primordial, features. A world devoid of color, smell or taste we could, perhaps, imagine. Similarly a world stripped of what we take to be essential theoretical properties also seems conceivable to us. We could imagine a world without electrical charge, without the atomic constitution of matter, perhaps without matter at all. But a world not in time? A world not spatial? Except to some Platonists, I suppose, such a world seems devoid of real being altogether. (45)

According to Sklar, a non-spatiotemporal world is inconceivable, and thus presumably not even metaphysically possible, let alone physically. This monograph is concerned with establishing the possibility of a fundamentally non-spatiotemporal world, articulating the consequences of such a possibility, and defending the idea that spacetime may be merely emergent in a perfectly acceptable scientific explanation of the manifest world. So in this section we will discuss various more precise ways that one might doubt the possibility of deriving spacetime.

---

[1] A note on terminology: we take 'Platonists' to be committed to the existence of abstract entities, such as propositions, sets, love, and justice, but also to the existence of the concrete, physical, and spatiotemporal world. Those who maintain that our world is fundamentally mathematical in nature and thus entirely consists in ultimately abstract entities or structures are often labeled as 'Pythagoreans'. Since we are interested not in whether there exist abstracta, but in the possibility that all physical existence is grounded in non-spatiotemporal structures, we will refer to those who maintain that a fundamentally non-spatiotemporal physical world is not devoid of "real being"—no doubt historically inaccurately—as *Pythagoreans*. We take this Pythagoreanism to be Sklar's target—and the one of this monograph.

16 OUT OF NOWHERE

First, in Huggett and Wüthrich (2013a) we discussed the idea that a theory without spacetime might be 'empirically incoherent'. That is, any theory which entails that the observations apparently supporting it are impossible, cannot receive empirical support (Barrett 1996)—it undermines the very grounds for believing it. Since all observations ultimately involve events localized in spacetime, it might seem that theories without spacetime in their basic formulation are threatened with empirical incoherence; the confirmation of such a theory might be ruled out "a priori". However, it is clear that a conflation is involved. (More) fundamental theories of QG are non-spatiotemporal in the sense that spatiotemporal structure is missing in their furniture; but it is perfectly consistent to think that spacetime is present as an effective object, arising from the more fundamental ones. (And that observation events can be identified within effective spacetime.) That is, QG will not be empirically incoherent if the appearance of spacetime can be adequately explained, and of course that is exactly what our case studies aim to do.

Second, while this book concerns the idea of 'emergent' spacetime as it arises in QG, one of our key concerns has already arisen in discussions of a different kind of spacetime emergence. The quantum mechanical wavefunction of N particles is not a function in ordinary space, but of the positions of all the particles: $\Psi(x_1, y_1, z_1; x_2, y_2, z_2; \ldots; x_N, y_N, z_N)$. Thus Ψ lives in 'configuration' space, in which there are three dimensions for each particle. Albert (1996) has argued that we should take the wavefunction 'seriously' as the ontology of the theory, and conclude that configuration space is more fundamental than regular space—that the three dimensions of experience are mere appearances of the $3N$ dimensions of reality. Whatever the merits of that view, the general idea has been attacked by Tim Maudlin (2007a). In particular he argues as follows: one might

> derive a physical structure with the form of local beables[2] from a basic ontology that does not postulate them. This would allow the theory to make contact with evidence still at the level of local beables, but would also insist that, at a fundamental level, the local structure is not itself primitive. . . . This approach turns critically on what such a derivation of something isomorphic to local structure would look like, *where the derived structure deserves to be regarded as physically salient* (rather than merely mathematically definable). Until we know how to identify physically serious derivative structure, it is not clear how to implement this strategy. (3161)

We have italicized the key phrase here. Suppose that one managed to show formally that certain derivative quantities in a non-spatiotemporal theory took on values corresponding to the values of classical spatiotemporal quantities; one

---

[2] Our footnote: beables (in contrast to quantum observables) are "those elements which might correspond to elements of reality, to things which exist" (Bell 1987, 174). Local beables are those with spacetime locations.

would then be in a position to make predictions about derived space. However, according to the passage quoted, such a derivation (even if the predictions were correct) would not show that spacetime had been *explained*. In addition, we have to be assured that the formally derived structure is *physically salient*. We agree with Maudlin that physical salience is required of proper—one can say 'explanatory'— derivations: otherwise one simply has a formal, instrumental bookkeeping of the phenomena. Indeed, we agree with him that the issue is particularly pressing in theories of emergent spacetime. But we think that it can be addressed in QG: one of the goals of this book is to investigate the (novel) principles of physical salience for theories of QG, the principles whose satisfaction makes the derivations of spacetime physically salient. In the following chapters we will look in detail at the derivations, to make clear the assumptions and forms of reasoning that lie behind them. In the concluding chapter 11 we will analyze what we have learned, to start to explicate what makes a derivation of spacetime in QG physically salient.

But to explain that project—and its relevance to philosophy—we need to unpack the very notion of physical salience, as we understand it.[3]

2.2 Physical salience

There is a subtlety about the way that Maudlin makes the point, however (which we did not clearly address in Huggett and Wüthrich 2013a). For the target derived structure in itself is prima facie physically salient: it is the physical datum to which the more fundamental theory is answerable. So in that sense there is really no question of the physical salience of the 'derived structure'—in the sense of the structure *to be* derived. Perhaps a more fundamental theory will show that some less fundamental theory is profoundly confused; but more generally one expects that existing, well-confirmed theories have latched onto some genuine physical structures, and that new, better theories will simply explain how, by subsuming the old in some broad sense.

Rather, Maudlin is talking about a formal derivation within a proposed new theory, and the question of whether what is at present simply a mathematical structure, in numerical agreement with the target structure, in fact explains it, and is

---

[3] We are grateful to Maudlin for conversations on this topic. We believe that we capture the essence of his idea, even if we might differ in details; and especially regarding the depth of the problem in the case of spacetime emergence. We do, however, want to point out an important difference between the cases of emergence from QG and from configuration space: in the latter, but not the former, there is a way to formulate the theory in 3-space (as single particle wavefunctions with a tensor product). Thus in QM (but not QG) Maudlin can argue that the derivation isn't physically salient, because the formulation from which it is derived is unnecessary in the first place. Lam and Wüthrich (2021) have argued that spacetime functionalism has a different status in QM (and GR) versus QG, because there are alternatives in the former, while there are none in the latter.

18 OUT OF NOWHERE

not merely an instrument for generating predictions. We would break this question down into two interconnected parts (which will also help illuminate what is involved in explanation here). First, the question of whether and how the basal objects or structures of the more fundamental theory accurately represent physically salient objects and structures. As we shall see shortly, that question becomes far more pressing when none of the putative objects or structures are supposed to be in spacetime. Second, does the formal derivation of the phenomenal from the more fundamental make physical sense? That the derivation exists shows that it makes sense at the level of the formalism, and especially that the derivation is compatible with the mathematical laws. But, as Maudlin suggests, there is more to the question of physical salience than that. And the question is especially pointed when one wonders how the spatiotemporal could ever be 'made' of the non-spatiotemporal. We will illustrate these ideas with a homely (and idealized in many ways) example.[4]

The ideal gas law states that for a gas (in a box of fixed volume v) pressure is proportional to temperature. Ideal gas theory says nothing about the microscopic composition of gases, so these are (among) the primitive quantities of the theory, operationalized via pressure gauges (relying on forces measured via Hooke's law for springs), and thermometers (so relying on the linear expansion with temperature of some substance). This is the phenomenon to be explained by the more fundamental theory, the kinetic gas model, according to which the gas is composed of atoms with mass m, whose degrees of freedom are their positions and velocities. The latter can be expressed by a vector $\vec{V}$, with $3n$ components: for each of the n atoms that make up the gas, three components, to describe the speed with respect to each of the three dimensions of space. Each atom has a kinetic energy associated with its velocity; the average kinetic energy is simply their sum, divided by n: denote this quantity

$$T(\vec{V}) \equiv \overline{\frac{1}{2}m\vec{V}^2}. \tag{2.1}$$

Now one computes the atoms' momentum change (per second per unit area) resulting from their collisions with the sides of the box: assuming that the collisions are elastic, and that the atoms are distributed evenly throughout the box and with respect to their velocities, one formally derives that

$$P(\vec{V}) \equiv \frac{2n}{3v}\overline{\frac{1}{2}m\vec{V}^2}. \tag{2.2}$$

Clearly the two quantities are proportional:

$$P(\vec{V}) \propto T(\vec{V}), \tag{2.3}$$

---

[4] The following has also been discussed in Huggett (2022).

which has the form of the ideal gas law. However (and despite the suggestive names, P and T) we have so far said nothing to justify identifying the quantities with the pressure and temperature of the ideal gas law; we have only noted a formal proportionality.

From this example we can abstract the following schema:

If **(a)** fundamental quantities X can be 'aggregated' into $\alpha(X)$ and $\beta(X)$, such that **(b)** $f(\alpha(X)) = g(\beta(X))$ follows from fundamental laws, then the law $f(A) = g(B)$ relating less fundamental quantities A and B is *formally derived*.

The term 'aggregated' is supposed to be vague, in order to accommodate the many ways a derivation might proceed. But the underlying idea is that the more fundamental theory has (many) more degrees of freedom than the less fundamental, and somehow the more fundamental must be 'summarized' by the less, for example by averaging, or by coarse-graining.

Maudlin's claim is that formal derivability does not suffice to properly derive phenomena: in particular, 3-dimensional space can be formally derived from the full $3N$-dimensional configuration space, but for Maudlin that does not make it a plausible, more fundamental alternative to ordinary space. And more generally, one should worry that a merely formal condition does not distinguish instrumental calculi from serious physical accounts. And indeed, further analysis of the derivation of the ideal gas law shows that considerations of physical salience are at play.

In particular, $P(\vec{V})$ is derived by assuming that the atoms are striking the sides of the box and exerting a force there: so acting exactly at the place and in the way that would produce a reading on a pressure gauge. And $T(\vec{V})$ is (according to the randomness assumption) the amount of energy in any macroscopic region of the box, say the location of the bulb of a thermometer: and collisions with the bulb will transfer kinetic energy to the molecules of the thermometer, causing thermal expansion. Imagine if instead that $P(\vec{V})$ only referred to the center of the box, or if $T(\vec{V})$ referred to a single atom in the box. Then the formal derivation would not be convincing. Or suppose that instead of the atomic gas model we imagined that a gas was a continuous object, whose degrees of freedom were somehow described by $\vec{V}$, but not as the velocities of anything (certainly not atoms). Then the formal consequences of kinetic gas theory could still be taken to hold, but they would no longer have the interpretation that they do in the kinetic gas model; the whole derivation would go through, but its physical meaning would be obscure. In short, the reason, in addition to their proportionality, that we find $P(\vec{V})$ and $T(\vec{V})$ convincing as pressure and temperature, and not just quantities following a similar law, is that they are spatiotemporally coincident with those quantities, and involve processes capable of producing the phenomena associated with those quantities. The derivation is not merely formal, but also physically salient.

20 OUT OF NOWHERE

Continuing our schema:

A (non-instrumental) derivation of phenomena requires, in addition to a formal derivation, that **(c)** the derivation have *physical salience*.

Of course, this schema does not tell us what it is to be physically salient, but the example of the ideal gas above illustrates two very important aspects, spatiotemporal coincidence, and the action of a physically accepted mechanism. And this observation immediately reveals the problem for the emergence of spacetime, because such criteria simply cannot be satisfied by derivations from non-spatiotemporal theories, because they are explicitly spatiotemporal criteria. For instance, it makes no fundamental sense in such a theory to even ask *where* a structure is. So if such criteria are "a priori" constraints on science, then the QG program, to the extent that it involves non-spatiotemporal theories, is in some serious trouble. However, a second example indicates the contextuality of physical salience, and thereby the way in which QG can hope to achieve physical salience in its derivations.

Consider the competing Cartesian and Newtonian accounts of gravity, exemplified by illustrations from the *Principles of Philosophy* (Descartes 1644) and the *Mathematical Principles of Natural Philosophy* (Newton 1726), respectively: see figure 2.1. On the one hand, we have the vortices of Descartes, which aimed to provide a mechanical account of gravity, in terms of the motions and collisions of particles. On the other, there is Newtonian action at a distance, which allowed him (as in Proposition I.1) to formulate and use his mathematical principles. We pass over Newton's own ambiguous attitude toward the causes of gravitation (his refusal to 'feign hypotheses' on the one hand, but his speculations in the *Opticks* (Newton 1730) on the other). The point to which we draw attention is the controversy between the Newtonians and Cartesians regarding the need for mechanical explanation.[5] For the latter, Newton might have captured the effects of gravity in a formally accurate way, but offered no scientific explanation for the phenomena. For example, consider Leibniz's clear statement to Clarke:

If God would cause a body to move [round a] fixed center, without any [created thing] acting upon it ... it cannot be explained by the nature of bodies. For, a free body does naturally recede from a curve in the tangent. And therefore ... the attraction of bodies ... is a miraculous thing, since it cannot be explained by the nature of bodies. (*Leibniz-Clarke Correspondence* in Alexander 1956)

[5] Note especially that we strictly distort the logic of Newton's *Principia* here: as far as Proposition I.1 is concerned, the forces could be impulses directed toward the point *S*. However, though Newton's reader may not at that stage know the nature of the force, for Newton the figure represents the action of universal gravitation.

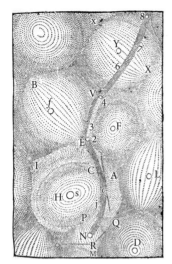 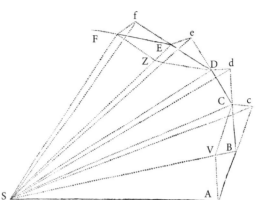

Figure 2.1 Descartes' and Newton's competing images of gravity. On the left is pictured Descartes' vortex model: each cell represents a ball of rotating matter, with lines to indicate the direction of rotation (e.g., those surrounding f, L, Y rotate about axes in the plane shown, while those surrounding D, F, S rotate about axes perpendicular to the plane). The bodies at the center of a cell represent suns: S is ours. On the right is the diagram from Newton's Proposition I.1 proof of Kepler's equal areas law for a central force (essentially, conservation of angular momentum). All that matters is the direction of the force (toward the point S), not any 'hypothesis' about its nature. Ultimately, Newton will apply the proposition to the case in which S is our sun.

We take Leibniz's complaint to be exactly that Newton's derivation of the phenomena lacks physical salience, because only mechanical causes are physically salient explanations of unnatural motions.

Of course, the Newtonians were ultimately victorious, and this Cartesian condition of physical salience was replaced by one that allows action at a distance, because of the success of universal gravity, and the failure of mechanical alternatives, such as Leibniz's. But that was not the end of the story: through the development and empirical success of electromagnetic theory, culminating in the development of special relativity, action at a distance was again rejected, with contact action replaced by the demand for local field interactions—and hence the replacement of Newtonian gravity with general relativity (GR). Again, we understand this demand as a criterion of physical salience, required for more than merely formal accounts. But even that is not the end of the story, for quantum mechanics experimentally conflicts with that concept of locality, and so quantum non-locality must be accommodated in some way.[6]

---

[6] Hesse (1961) is a classic telling of this tale.

22 OUT OF NOWHERE

By now, three points are indicated by this story: first, questions of physical salience, here in the form of the principles of locality, are genuine, controversial components of scientific enquiry. Second, such principles are historically contingent, changing in step with major advances in physics. Third, such changes are ultimately settled by, and epistemically justified by, empirical success: one of the things that we learn in a scientific revolution is a set of criteria of physical salience for explanation appropriate to the new domain of enquiry. Put this way, we see principles of physical salience as part of what Kuhn called the 'disciplinary matrix' in the *Postscript* to the second edition of (1962), or what Friedman (2001) refers to as the 'relative, constitutive a priori'. Though changes in the principles change wholesale what theories are even candidate explanations, we don't infer any catastrophic incommensurability here: as we said, innovations in physical salience are grounded in empirical success, like all other scientific knowledge.

So we have a general answer to the problem raised earlier. *How can a derivation of spacetime from a non-spatiotemporal theory ever be physically salient?* Well, it cannot satisfy the standards of physical salience that apply to theories with classical spacetime, but we should expect a non-spatiotemporal theory to require new standards. And so the real question is what are those new principles? Like Friedman, we see that question, and the development of such new principles as a foundational, interpretational, conceptual—hence *philosophical*—endeavor. We elaborated on how such a project is to be conducted in §1.3. We will see throughout the book how this endeavor is ineliminably philosophical in the different approaches to QG. For now we want to illustrate the problem with an example.

2.3 Non-commutative geometry

Of necessity, this section is somewhat more technical than the others, and could be skipped by those not requiring a concrete illustration of how interpretational considerations come into play in elevating a formal derivation into one (potentially) having physical salience. It elaborates an example of a non-spatiotemporal theory given in chapter 1, to show how one might come to view it as a theory from which spacetime emerges.

We start with familiar, commutative geometry, for which $xy = yx$, in a smooth manifold of points with Cartesian coordinates x and y; let it be 2-dimensional for simplicity.[7] Consider polynomials $P(x, y)$ of x and y. These are 'fields', meaning that they return a numerical value at each point (x, y). They form an algebra with respect to multiplication: this just means that when you multiply two polynomials together, the result is another polynomial. Moreover, because $xy = yx$ we have that

---

[7] This section is based on Huggett et al. (2021). See also Lizzi (2009) for a more mathematical survey.

$\mathcal{P}(x,y)\mathcal{Q}(x,y) = \mathcal{Q}(x,y)\mathcal{P}(x,y)$, so that the algebra is commutative. (Check with $\mathcal{P}(x,y) = xy$ and $\mathcal{Q}(x,y) = x^2 + y^2$ if you like.)

It may seem like a rather uninteresting structure, but in fact such algebraic relations alone contain geometric information about the space: in this case, that it is smooth, that it is 2-dimensional, and whether it is open or closed. This fact is shown by the important Gelfand-Naimark theorem (1943), which is the foundation of 'algebraic geometry'. Indeed, the whole structure of differential geometry can be recast in algebraic terms. (An interesting application is Geroch's 1972 formulation of GR as an 'Einstein algebra'; discussed by Earman 1989, §9.9 as a possible response to the hole argument.)

For a mathematician, the question of what happens when the algebra is 'deformed' so that it is no longer commutative is irresistible: so one sets $xy - yx = i\theta$ and sees what happens. (And similarly in spaces of any dimensions.) Surprisingly, one finds that the structure necessary to cast geometry in algebraic terms remains (at bottom, one can still define a derivative on the algebra, in terms of which the other structure is defined). Moreover, the Euler-Lagrange equation and Noether's theorem do not require commutativity, and so the structure of modern physics is preserved, even in such a 'non-commutative space'—in a purely algebraic formulation. (Recall our terminological clarification: just because we call this object 'space' does not at all imply that it is a classical space; indeed, we claim that it is incompatible with classical space!)

But suppose that such a physics were correct: how could it explain spacetime as it appears to us? Specifically, how are we to understand events localized in space in terms of an abstract algebra? When the algebra is commutative, a theorem due to Gelfand-Naimark lets us interpret the elements as fields, $\mathcal{P}(x,y)$ related to regions of space; but what about the non-commutative case? The question is just that which concerns us in this chapter (and indeed the whole book): how can we derive the appearance of classical spacetime from a non-spatiotemporal theory, in a physically salient way?

The obvious thing to try is to (i) interpret x and y not as elements of an abstract algebra, but as fields in an ordinary plane: taking the value of the x and y coordinates at any point (x,y). Then (ii) define a new binary operation, $\star$, such that $x \star y - y \star x = i\theta$. Then (iii) construct the algebra of polynomial fields, but with $\star$-multiplication instead of regular (point-wise) multiplication. Indeed, this is exactly how one typically proceeds in non-commutative geometry: in one formulation, the operation is 'Moyal-$\star$' multiplication,[8] and the fields form the 'Weyl representation' of the algebra.

The algebra of the fields with respect to $\star$ will be that of the abstract noncommutative algebra, and now we have referred that algebra to objects in an

---

[8] $\phi \star \psi \equiv \phi \cdot \psi + \sum_{n=1}^{\infty} (\frac{i}{2})^n \frac{1}{n!} \theta^{i_1 j_1} \dots \theta^{i_n j_n} \partial_{i_1} \dots \partial_{i_n} \phi \cdot \partial_{j_1} \dots \partial_{j_n} \psi.$

ordinary manifold. In particular, one could talk about the local region in which such-and-such a field has values < 1, say. Indeed, one might now wonder whether we should throw away the abstract algebra, and just treat physics in 'non-commutative geometry' as really physics in commutative geometry, but with an unfamiliar multiplication operation; in other words, wonder whether classical spacetime needs to be recovered at all?

Huggett, Lizzi, and Menon (2021) argue that indeed it must be, for the Weyl representation has formal representational structure that exceeds its meaningful, physical content. In particular, the concept of a region with an area smaller than θ—a fortiori that of a point—is undefinable in the theory. This can be seen in a couple of ways, but for instance the attempt to measure positions more accurately leads to unphysical results. The conclusion is that, although the Weyl representation contains points and arbitrarily small regions, they are purely formal, and do not represent anything real: non-commutative geometry—even the Weyl representation—is physically 'pointless'.

As a result, we cannot understand a point value of the Weyl fields as having any physical meaning. Rather we need to understand the fields as complete configurations: the unit of physical meaning for a field in non-commutative space is the *function* from each point to a value, $\mathcal{P} : (x, y) \to \mathbb{R}$; not its *values* $\mathcal{P}(x, y)$ at individual points (x, y). But the full configuration is equivalent to the place of the field in the abstract algebra, and so are back to the question of deriving locality.

Here is one way to proceed, using an ansatz proposed by Chaichian, Demichev, and Presnajder (2000) (discussed further in Huggett et al. 2021). They propose that an ordinary, commuting field—the kind observed in classical spacetime—be related to a Weyl field $W(x, y)$ by an operation of 'smearing'. One multiplies $W(x, y)$ by a θ-sized 'bell function' about (X, Y), and integrates over the Weyl space coordinates x and y.[9] This procedure is a form of 'aggregating', step **(a)** of the schema of §2.2. The formal result is a new field $\Omega(X, Y)$, living in a commutative space with points (X, Y), a partial realization of step **(b)**. Step **(c)** then would be to take this 'CDP ansatz' to have physical salience, to accept that the result of this smearing is to introduce physical, classical space into the theory: W lives in Weyl space, whose status, we argued, is only that of a formal representation of the fundamental algebra, while Ω should be interpreted as living in the physical space that we observe. We thus interpret smearing as relating a *function on one space* to a *value on another space*: it relates the non-commuting field W, represented as a function over Weyl space points (x, y), to the *value* of an observed, commuting field Ω at physical space point (X, Y). That it takes a function to a value is just mathematics; that it relates Weyl and physical spaces is a substantive physical postulate.

---

[9] $\Omega(X, Y) \propto \int \left(e^{-((X-x)^2 + (Y-y)^2)/\theta} \cdot W(x, y) \right) dxdy$.

However, it still makes no sense to consider Ω in regions smaller than θ: we have in fact erased the unphysical information at such scales by smearing W. So strictly a single coordinate pair (X, Y) does not label a physical point. Rather, the proposal is that these smeared fields are approximated by observable fields over regions greater than θ; thereby formally deriving the latter, spatially localized objects from the former, purely algebraic objects. (In this case, we have smearing as 'aggregating', in a very loose sense.) Of course, in our existing theories, fields live in a full commuting spacetime, but that is an extrapolation from our actual observations of fields, which are always over finite regions, to date larger than θ. On the proposed interpretation, then, any information contained in Ω about regions less than θ is not only unobserved, but unphysical, surplus representational 'fluff'.

Now, nothing in the theory forces this picture as a physical story—it is merely an interpretational postulate. (Though we claim that it is conceptually coherent.) However, it has empirical consequences: the dynamics magnifies the θ-scale non-commutativity to observable scales (e.g., Carroll et al. 2001). If those predictions are successful, then we have evidence that the underlying non-commutative field theory *and* the interpretational postulate are correct. Imagining that situation then, we claim that the situation is exactly analogous to that of the Newtonians regarding action at a distance. That is, we would be justified in accepting the CDP ansatz and our interpretational postulate as novel principles of physical salience: they regulate what constitutes a physically salient derivation in the theory. In both cases, the final ground is the empirical success of the theory.

So it should be clear how the example illustrates our points about physical salience and our scheme proposed above. In the first place we have argued that non-commutative geometry is non-spatial, in the sense that it is 'pointless', and so must be understood as a purely algebraic theory. Then we have explicated a possible formal derivation of localizable fields from this more fundamental theory. And finally, we have sketched a scenario in which such a derivation leads to successful predictions, and hence to the conclusion that the formal derivation is physically salient, in fact *explaining* the appearance of localized fields, and ultimately classical spacetime. Of course, we highlight that this discovery constitutes a *change* in what derivations 'deserve to be regarded as physically salient (rather than merely mathematically definable)', to paraphrase Maudlin. This is the pattern that we will see in more detail in the examples of the following chapters, and to which we will return in the conclusion.

2.4 Spacetime functionalism

The schema that we presented for a physically salient derivation relates closely (but not exactly) to David Lewis' (1972) account of functional identification. Since 'spacetime functionalism', of various varieties, has been recently discussed it is

26 OUT OF NOWHERE

worth drawing the comparison to better understand the nature of the proposed emergence. Suppose, in idealization, that a theory T is formulated as a postulate '$T[t]$'. t represents what were traditionally called the 'theoretical' terms, though we prefer 'troublesome' (Walsh and Button 2018, §3.1): the idea is that these are the new terms introduced by the theory, and the 'trouble' is the question of how they garner meaning. '$T[\cdot]$' also involves (traditionally) 'observational', or (according to Lewis) 'old', or (with Walsh and Button) 'okay' terms, and Lewis proposes that the t are defined in terms of them by

$$t = \imath x\, T\,[x]. \tag{2.4}$$

That is, 'the t are (if anything) the extant, unique things that satisfy the theory postulate'. The t are thus defined in terms of their nomic relations to one another and to the okay terms—i.e., in terms of their 'functional' relations—and so (2.4) is a functional definition.[10] As a result, the terms t are rendered semantically okay (though they may remain metaphysically problematic).

Suppose that we also hold a postulate $R[r]$, where r and t do not overlap, so that our acceptance of R does not depend on our views on the troublesome terms of T.[11] Further suppose that we come to believe $T[r]$. Now,

$$T[r],\ t = \imath x\, T[x] \ \vDash\ r = t, \tag{2.5}$$

so, by definition of the t, $T[r]$ deductively entails the identity of the objects of R and T. Lewis' point is the epistemic one that such functional identifications are thus not inductive: given a functional definition, once one accepts that the objects of R play the same roles as those of T, *logic and meaning alone commit one to accepting their identity*.

Lewis offered electromagnetic waves and light as an example of the scheme, but of course his point was that 'when' neuroscience showed that neural states played the functional roles of mental states, then they would—as a matter of logic and definition—be identified. The subject of this book also broadly fits Lewis' scheme: theory T is our spacetime theory assumed by QFT and relativity theory, while R is a theory of QG. Denote them ST and QG, respectively, and use ST to functionally define any troublesome spacetime terms. Then, according to the schema of §2.2, a physically salient derivation of ST from QG shows that 'aggregates' of

---

[10] Lewis proposes in passing that the actual definition be modified to allow for *approximate* satisfaction of $T[\cdot]$. In our opinion that is always going to be the case in actual theories, so this modification is not optional but required, and his discussion is a significant idealization (a point to which we return); in particular, the theoretical identifications will not be strict. The harder question of how exactly the modification is to be implemented is not carefully addressed by Lewis.

[11] Acceptance of a theory requires that it be meaningful, hence acceptance of $R[r]$ requires that the r be referential, and thus that any troublesome r can be functionally defined as $r = \imath x\, R[x]$. We accept this assumption for Lewis' cases, but we will see that things are more complex in the case of QG.

QG, described using its terms *q*, satisfy *ST* [·]—that they indeed play the functional roles of the objects of *ST*.[12] So the identity follows. However, there are differences to Lewis' functionalism, which we will explain presently.

Now, how to turn Lewis' scheme into a concrete plan for the functional reduction of spacetime? The spacetime functionalism recently introduced by Lam and Wüthrich (2018, 2021) is based on the general scheme in the spirit of Kim (2005, 101f) according to which a functional reduction of higher-level properties or entities to lower-level properties or entities consists in two necessary and jointly sufficient steps:[13]

(FR1) The higher-level entities/properties/states to be reduced are 'functionalized'; i.e., one specifies the causal roles that identify them, effectively making (2.4) explicit.

(FR2) An explanation is given of how the lower-level entities/properties/states fill this functional role, so that we come to accept *T* [*r*].

If these two steps are fulfilled, then it follows that the higher-level entities/properties/states are realized by the lower level ones.

Applying the template of functional reduction to the case of the emergence of spacetime in QG, the two steps above become:

(SF1) Spacetime entities/properties/states, *s*, are functionalized by specifying their identifying roles, such as spacetime localization, dimensionality, interval, etc. Effectively, one makes explicit $s = \imath x\, ST\,[x]$.

(SF2) An explanation is given of how the fundamental entities/properties/states, *q*, postulated by the theory of quantum gravity fill these roles, so that we come to accept *ST* [*q*].

Again, if these two steps are fulfilled, it follows that the (perhaps aggregated) QG entities/properties/states *are* the spacetime entities/properties/states. In the following chapters, after explaining each theory of QG and its conceptual foundations, we will follow this scheme in our discussions: on the one hand, describing the functional roles of spacetime entities/properties/states, and on the other showing how the theory of QG proposes that those roles are played. Of course, given the evidential state of QG, we do not claim that these proposals are correct: we only describe how the theories *may* functionally reduce spacetime, not—as far as we currently know—how they *do*.

---

[12] Yates (2020, §6.2.3) makes essentially the same proposal.

[13] Kim's model involves three steps, where the second is to identify the entities in the reduction base that perform the role at stake, and the third is to construct a theory explaining how these fundamental entities perform that role. We subsume these two steps in our second stage.

28 OUT OF NOWHERE

Several remarks are in order regarding the functionalist approach to spacetime. First, it should (again) be made clear that we take emergence and reduction to be compatible with one another, and hence functional *reduction* may serve as a template to explain the *emergence* of a higher-level feature, i.e., the fact that higher-level entities exhibit novel and robust behavior not encountered or anticipated at the more fundamental level.[14]

Second, there is a sense in which a functionalism about spacetime must start from a broader conception of functional reduction than is usual in the familiar functionalisms in the philosophy of mind or the philosophy of the special sciences. There, a mental or biological or other higher-level property is understood to be determined by—indeed, usually *identified with*—its *causal role* within the relevant network such as the network of mental or biological activities. If in spacetime functionalism the roles are still supposed to be *causal*, then a much broader notion of 'causal' must be at work, one that does not in any way depend on the prior existence of spacetime. As it is not clear what that would be, we have presented a functionalism devoid of any insistence that the functional roles be causal (it is especially clear that the Lewis-style semantic formulation makes no assumption of causal roles).

Third, the central claim of spacetime functionalism is that it is sufficient to establish only the functionally relevant aspects of spacetime. In particular, it is therefore not necessary to somehow derive relativistic spacetime in its full glory and in its every aspect in order to discharge the task. Naturally, this raises the question of what these functionally relevant aspects of spacetime are—the task of (SF1). As we will see in the following chapters, different approaches to QG take different stances on what functions are to be recovered, though broadly speaking, all aim to recover functions sufficient for the empirical significance of basic metrical and topological properties. Our stance will be that the list of functions cannot be determined "a priori" from conceptual analysis of classical spacetime theories, but by the twin demands of the empirical, and of the resources of the proposed reducing theory. In short, part of the work will be to identify the spacetime functions recovered in the different approaches, and indicate how they relate to observation.

Fourth, since we have drawn an analogy between spacetime and psychophysical functionalisms, is there a spacetime version of the 'explanatory gap' sometimes alleged to exist in the explanation the phenomenal character of mind? Are there some kind of spacetime qualia, essential qualitative features irreducible, inexplicable in functionalist terms? However, the case of spacetime is disanalogous to that of mind: we agree with Eleanor Knox (2014, 160) who states (regarding one attempt to cash out the idea) that where "the fan of qualia [in the philosophy of mind] has

---

[14] As restated many times in our earlier publications, and in agreement with what we take to be the consensus in philosophy of physics as stated, e.g., in Butterfield (2011a, 2011b) and Crowther (2016, §2).

introspection, the fan of the [spacetime] container has only metaphor"[15], and with Lam and Wüthrich (2018, 43f) who agree that the "nature and status of the evidence in favor of [mental] qualia may be equivocal, but the alleged ineliminable intrinsically spatiotemporal but ineffable quality of spacetime substance remains positively elusive". We conclude with them that the qualia worry in this form gets little if any traction in the spacetime case.[16]

Baptiste Le Bihan (2021) offers a somewhat opposed view. He claims that one can distinguish "hard" and "easy problems" of spacetime emergence, in analogy to that claimed in psychophysical functionalism, though in a thinner sense: the cognitive dissonance expressed earlier by Sklar is, he argues, evidence for an explanatory gap. What is it that requires explanation? Not a phenomenal 'what it is like' to be spacetime, but a physical 'what it is like' (S374): for instance, to possess "the metric field of GR, or the ordering of events observed through experience" (S382). Now, as we have indicated previously, we think metricity is likely one of the functions of spacetime, and are thus committed to a functionalist reduction of the metric (as indeed the following chapters aim to demonstrate). (The issue of the spacetime of experience is downstream from our targets, presumably to be explained in terms of GR and QFT.) In our view, a problem of reduction is 'hard' when, as for qualia, there are principled reasons to think that a functionalist account is impossible; but we do not think that is the case for the metric, or any of the other spacetime structures needed for physics. Of course, we do agree with Le Bihan about cognitive dissonance regarding spacetime emergence, but we explicitly address it with an appeal to (anticipated) new principles of physical salience; accepting them just is to accept that any explanatory gap is closed. In that sense, our disagreement is essentially over what makes a question of reduction 'hard' rather than 'easy', in the technical sense; Le Bihan's 'hard' problems are counted as 'easy' by us, since they have functional solutions, as we aim to demonstrate.[17]

Fifth, functionalism shows how the goal of reduction can be the scientific explanation of the functional roles of higher-level entities/properties/states by lower-level entities/properties/states (and to nothing more). But, it is debatable to what exactly spacetime functionalism is ontologically committed: substances,

---

[15] Or rather, attempts to render it non-metaphorical will be functionalist (Baker 2021, S293f), so not a spacetime quale.

[16] We also concur with Lam and Wüthrich (2018) in their rejection of the version of this concern articulated in Ney (2015), who worries that if the fundamental entities are not already appropriately (spatio)temporal in their nature, they cannot 'build up' or constitute spacetime as they are not the right kind of stuff (see also Hagar and Hemmo 2013). As diagnosed by Lam and Wüthrich, advocates of this worry seem to rely on an unreasonably narrow concept of constitution. We might also object that if we surrendered to this worry, there would be no principled reason to think that it would not also annihilate all other cases of presumed emergence and amount to an unyielding dualism.

[17] Chalmers (2021) also discusses spacetime emergence, but his focus is rather on the emergence of manifest, or as he calls it 'edenic', spacetime, rather than the emergence of the classical spacetime of GR and QFT from QG. On that latter issue he seems to agree that the kind of functionalism that we advocate will likely do the job.

30   OUT OF NOWHERE

relations, entities, states, structures, or something else. Along such lines, Le Bihan (2021, S378) poses the 'ontological problem'—'whether spacetime is real or not'—which (as he says) requires addressing just such questions (he also provides a useful taxonomy of possible answers). Such issues lie beyond the scope of this book, which is focused on the derivation of spacetime, the empirical content of the theories of QG, and the principles that make them physically salient: Le Bihan's 'scientific', 'empirical coherence', and 'hard' problems. We leave his fourth, onto-logical, problem for later investigation: the metaphysical nature of space and time is perennially contested, and we wish to remain as neutral as possible in our con-clusions. Thus we hope that the reader will forgive our switching between speaking of the spatiotemporal as if it were an entity, or a structure, or a state, or something different yet again. We simply aim to avoid torturing English more than neces-sary, and no deep philosophical commitment should, for instance, be read into our using 'spacetime' (or 'time' and 'space') as nouns.[18]

Sixth, we are far from the first to suggest functionally defining space or space-time. Robert DiSalle (2006, ch. 2) reads Newton's Scholium to the definition in much this way (though Huggett 2012 disagrees). Functionalist strategies have also become very visible in the philosophy of quantum mechanics, where it is deployed by David Wallace (2012) in his defense of an Everettian interpretation, and by David Albert (2015, ch. 6) in support of wavefunction monism. Such applications differ from ours because they are concerned with recovering three-dimensional physical space. In contrast, spacetime functionalism in QG is commissioned with functionally recovering four-dimensional spacetime, and so relates to work by Knox (2013, 2014, 2019) in the context of classical spacetime physics. For her, something 'plays spacetime's role' and thus *is* spacetime "just in case it describes the structure of inertial frames, and the coordinate systems associated with these" (2014, 15). In GR, the metric field performs spacetime's role in this sense and thus is identified with spacetime by her. As the metric may itself not be fundamental but instead emerges from the collective behavior of more fundamental degrees of freedom, she explicitly leaves open the possibility that the realizers of spacetime's functions may themselves not be fundamental (Knox 2013, 18). As the relationship between the fundamental degrees of freedom and the emergent spacetime realizer

---

[18] In fact, we have sympathies for the idea of the 'math-first' interpretation of physical theory recently advocated in Wallace (2022a) (developing suggestions by Simon Saunders). If successful, this approach would replace familiar 'language-first' interpretation in terms of objects and their (actual and possi-ble) relations (and their second order relations etc.), which underpins ontology in the familiar sense; instead one would most naturally speak of 'structures' in some sense. But as we say, we take no delib-erate stance on this question: we describe the reduction of spacetime to the non-spatiotemporal, and leave the question of whether it is an 'ontological' or 'structural' relation for further consideration. We note that Lewis' functionalism assumes a language-first approach, but we propose it only as low order approximation to the relationship between levels; we claim only that its outline and key features will remain in a math-first approach. We thank Eleanor Knox (who is skeptical of this claim in ongoing work), Henrique Gomes, and Lorenzo Lorenzetti (who are more sympathetic) for discussion of these points.

is left untouched by Knox's inertial frame functionalism, the latter also does not shed any light on it.[19]

Seventh and finally, there is an important but subtle difference in the application of Lewis' scheme to QG from that in the cases he has in mind. Suppose we accept a spacetime theory $ST[s]$, where whatever it is that performs the spacetime functions is denoted by the troublesome terms, s. These, following Lewis, we take to be defined by

$$s = \imath x\, ST[x]. \tag{2.6}$$

The okay terms appearing in $ST[\cdot]$ would refer to matter of various kinds, its relative motions and point-coincidences: so, for instance, the metric in GR might be defined locally in terms of its role in determining motions under gravity or scattering amplitudes. We think that this part of Lewis' picture—which corresponds to (SF1)—fits our cases well. But what about the second part of his scheme, involving $R[r]$? Although the result is still a functional identification, its significance has shifted somewhat, as we shall now explain.

Jeremy Butterfield and Henrique Gomes (2023; Gomes and Butterfield 2022) analyze recent proposals for spacetime functionalism in explicitly Lewisian terms. They emphasize, as we have, that in Lewis' scheme theoretical identification follows by definition alone (once the rs are known to play the roles of the ts), and that functional identification is a species of reduction. But they also show how various spacetime and temporal functionalisms follow the 'Canberra plan', according to which the troublesome ts are not only defined by T, but are also 'vindicated' by their functional identification as rs. For instance, as mental states, perhaps, turn out to be neural states so, in their examples, a temporal metric might be identified with purely spatial structure; then, if neural states or spatial structures are on a firm (or firmer) ontological footing than mental states or time, the identifications show that the latter are equally well grounded. They are, that is, vindicated against any metaphysical suspicions raised against them. That vindication is not by itself achieved by the functional definition (2.4) of the ts; that merely makes the terms referential, so that they can be meaningfully employed. Put another way, (FR1) alone does not vindicate the mental, for instance; (FR2) is also needed, to show how the mental is part of the physical.[20] Regarding these cases, we are in agreement with Butterfield and Gomes' emphasis of this important distinction, and its applicability to the cases that they discuss.

However, in our cases, for which R is some QG, the second step, while still involving a functional identification, does not follow the Canberra plan, because

---

[19] Cf. Lam and Wüthrich (2018, 40) and Lam and Wüthrich (2021, §3) for a more detailed discussion of inertial frame functionalism and how it relates to our project.

[20] Or put yet another way, the t are often troublesome both semantically and ontologically: the functional definition takes care of the first problem, while the functional identification takes care of the second. When we use 'troublesome' we always mean semantically.

32   OUT OF NOWHERE

the troublesome terms, q, of QG are non-spatiotemporal, and so on a *weaker*, not firmer, footing—the ontological and semantic correlate of empirical incoherence.[21] Ontologically, as we have discussed, our physical and metaphysical categories assume spatiotemporality, and so the natures of the q are mysterious. Semantically, we can expect an attempt to functionally define the qs as $q = \imath x\, QG[x]$ to fail. Lewis' scheme for functional definition requires that a theory have sufficient okay terms to *uniquely* define the troublesome ones: if many collections of terms satisfy the putative definition, then it fails to establish reference. But that is what one expects in a theory that breaks from established categories as radically as a non-spatiotemporal one; the terms that we take to be okay are systematically spatiotemporal in some way, and so are expected not to appear in QG. And indeed, we contend that the theoretical concepts of the theories we consider in this book cannot be defined without appeal to spatiotemporal concepts external to the basic formulation of the theory.

Given this situation, the significance of functional reduction is different from the way in which Lewis (and Kim) proposed. Rather than following the Canberra plan of vindicating spacetime objects by reduction, in our approach to QG things are *reversed*: the non-spatiotemporal objects of QG are vindicated via their identifications with spatiotemporal objects. Clearly this approach (one might call it the 'Geneva plan') only works to the extent that the spatiotemporal is itself on a firm ontological footing, which of course is a topic of endless debate. As we have said, to skirt such debates in this book we will remain as neutral as possible, and not take any stand on the metaphysical nature of spacetime features such as topology or metricity, so that our conclusions remain valid for anyone who accepts them under whatever interpretation.

Within Lewis' framework, the vindication of the q works as follows. Suppose that non-spatiotemporal $QG[q]$ has been proposed. As explained, the q are semantically troublesome and ontologically suspect. Moreover, until we accept that a derivation of (at least a fragment of) ST is physically salient, we have no empirical grounds for accepting QG. Such a derivation will provide not only grounds for believing QG, but also define and vindicate the q. Introducing the 'aggregate operator' $\alpha(\cdot)$, according to our schema, when we have a physically salient derivation of spacetime properties, then we accept

$$\alpha(q) = \imath x\, ST\,[x]. \tag{2.7}$$

In conjunction with $ST\,[s]$ this entails that

$$\alpha(q) = s \tag{2.8}$$

---

[21] Butterfield and Gomes do not claim otherwise, and indeed acknowledge that QG will look different (2022, 3).

more or less as for Lewis. However, the Genevese reversal of the Canberra plan makes several things different, particularly in regard to the critical role for physical salience.

First, semantics. As noted, the q were not antecedently defined, but now can be through their—or rather the $\alpha(q)$'s—role as spacetime entities/structures/states. In other words, (2.7) is in part definitional of the q: the troublesome non-spatiotemporal terms of QG can only be defined with reference to spatiotemporal terms not native in QG. Moreover, (2.7) only succeeds in defining the q if in physical fact they play the ascribed roles, and do not merely mimic them formally; something secured by the physical salience of the derivation.[22]

Second, ontology. The qs are placed on a firm ontological footing—are vindicated—when we accept that the $\alpha(q)$ are in physical fact those entities/structures/states that play the spacetime role.[23] Acceptance of the physical salience of the derivation secures just that.

Finally, epistemology. In Lewis' scheme, we have independently accepted theories of, say, neuronal and mental states, and *later* discover that they play the same functional roles, entailing that they are identical. In our case, the acceptance that QG's objects (or rather their aggregates) play the same roles as ST's objects, and hence are identical with them, is *simultaneous* with our acceptance of QG. In general terms, the evidence for $R[r]$ is no longer antecedent (or independent) of the evidence for $T[r]$, but rather the very same evidence. As such the epistemic calculus is different. In one case, observations of neuronal states can be made independently of mental states, and we only have to show that they perform the relevant functions: producing suitable behaviors, for instance. In the other, observations are not independent of spacetime states, and have to support both the truth of a theory of QG, and that its objects perform the right functions. To give evidence, that is, that the formal derivation of those functions is indeed physically salient. As a result, although the deductive logic is the same, the empirical inference to the premises of the identification is different, and indeed weaker. However, as we say, it is of the normal empirical kind, and we fully expect it to be made for a successful theory of QG. There is no special ground for skepticism.

2.5 Out of Nowhere

In this chapter we have dived much deeper into the philosophical framework for the emergence of spacetime, spacetime functionalism. Working backward, we have just seen that it is crucial to the vindication of QG according to the 'Geneva plan'

---

[22] (2.7) is not purely definitional, since it also involves an existential commitment that the qs exist. And it need not fully define q; we also still have that $q = \imath x\, QG[x]$ by definition.

[23] In Huggett and Wüthrich (2013a, 284), we described this approach to vindication as physical salience flowing down to the q 'from above'.

34   OUT OF NOWHERE

that the derivation of spacetime entities/structures/states be *physically salient*, tracking real physical identities, and not be a mere formal correspondence.[24] If the derivation is a mere formal correspondence then the terms of the theory will be neither well defined, nor vindicated as real physical objects; furthermore we will have inadequate empirical grounds to believe in them.

But in the first part of the chapter we posed the problem for physical salience posed by non-spatiotemporal theories: our conventional concepts of what is physically salient are deeply spatiotemporal. So how could our functionalist program ever be carried out? Our answer, supported by a historical case study (which could be multiplied many times over), is that criteria for physical salience are theory-dependent, changing over time, and supported by the evidence for new theories. Thus we expect that new principles of physical salience will come to be accepted for QG as it becomes a more mature, and more empirically confirmed, branch of physics. To demonstrate how this might come to be, we sketched an example: 'smearing' merely formal fields might explain how a non-spatiotemporal algebra behaves (above a certain spatial resolution) as a continuous field—if the derivation is taken physically seriously. And what else should one (if a scientific realist) conclude if such a theory is empirically well confirmed?

We thus have the template for the following chapters, which investigate the emergence of spacetime in more detail, in three of the central approaches to QG: introduce the approach, functionally identify the roles of spacetime to be recovered in the approach, explain how those roles can be formally played in the approach, then identify the new principles that are assumed if one takes the derivation to be physically salient. To do so for a complete, successful theory of QG would be to vindicate the significance and reality of the non-spatiotemporal elements of QG, according to the Geneva plan. Since we do not yet have such a theory (§1.1) our goals are slightly more modest: to describe what physicists have learned so far about how such a vindication might go according to the different programs. As we explained in chapter 1, we see such reconstruction—together with articulating the functionalist strategy for spacetime emergence—as crucial to the conceptual and philosophical development of QG, in tandem with formal and empirical development.

---

[24] Based on criticism we have received it seems that we did not sufficiently emphasize this crucial point in our Huggett and Wüthrich (2013a), though it was certainly part of our view there. We triple underline the point—in red felt-tip!—in this chapter.

Chapter 3
Spacetime from causality: causal set theory

As a first step toward its ultimate goal of formulating a quantum theory of gravity, causal set theory (CST) assumes that the fundamental structure is a discrete set of basal events partially ordered by causality. In other words, it extracts the causal structure that it takes to be essential for relativistic spacetimes, posits that structure as fundamental, imposes discreteness, and tries to establish that such spacetimes generically arise in the continuum limit. Precursors can be found in David Finkelstein (1969), Jan Myrheim (1978), and Gerard 't Hooft (1979), although the endeavor did not get started in earnest until 1987, when the seminal paper by Luca Bombelli, Joohan Lee, David Meyer, and Rafael Sorkin (Bombelli et al. 1987) hit the scene.

This chapter gives an introduction to the leading ideas of the program and offers a philosophical analysis of them. In §3.1, we introduce the theory by putting it into its historical context of the tradition of causal theories of time and of spacetime, and by showing how it grew out of concrete questions and results within that tradition. §3.2 presents and discusses the basic kinematic axiom of causal set theory. Finally, in §3.3, we will articulate what we will call the 'problem of space' in causal set theory and illustrate the work required toward solving it.[1] This will involve a consideration of what space is (§3.3.1), how it might naturally be identified in causal set theory (§3.3.2), and how the dimension (§3.3.3), topology (§3.3.4), and metrical properties such as distance (§3.3.5) of such spatial structures could be determined.

One might think that stipulating causal relations as fundamental might amount to a covert posit of spatiotemporal notions. The first key point of the present chapter, and in particular of §3.3, is to dispel that worry by showing how different causal sets are from relativistic spacetimes, in particular concerning spatial structure. The second key point is to exemplify how, by recovering aspects of spatial topology and geometry, physicists contribute to the spacetime functionalist program of establishing the emergence of spacetime from CST. In order for spacetime functionalism to succeed in this task, it will not be sufficient to present formal derivations; in addition, the derivations must be imbued with physical salience.

---

[1] This locution is intended to stand in analogy to the 'problem of time', which we will encounter in chapters 5 and 6, and is thus distinct from the use of the same term in the history of science to refer to the epistemological problem concerning our *knowledge* of the geometry of space as it has been discussed by Helmholtz, Poincaré, Weyl, and others. We take it that the issues are sufficiently distinct so that there is no risk of confusion.

Out of Nowhere. Nick Huggett and Christian Wüthrich, Oxford University Press.
© Nick Huggett and Christian Wüthrich (2025). DOI: 10.1093/oso/9780198758501.003.0003

36 OUT OF NOWHERE

A note before we start: CST as we will present it in this chapter and the next, and as it has to a large part been elaborated to date, is still a classical rather than a quantum theory.[2] The extension of CST to a proper quantum theory is of course the implicit goal of the research program. We proceed on the reasonable, but non-trivial, assumption that much of what can be learned from classical CST will offer valuable insights into a quantum CST.

3.1 Motivation and background

3.1.1 Historical prelude: causal theories of time and of spacetime

Starting in 1911, just a short few years after Hermann Minkowski articulated the geometry of the spacetime of special relativity (SR), Alfred A. Robb recognizes that this geometry could be captured by the causal structure among the events of Minkowski's spacetime. In fact, as Robb (1914, 1936) proves, the causal structure of Minkowski spacetime determines its topological and metrical structure.[3] In other words, the geometry of Minkowski spacetime can be fully reconstructed starting from the set of basal events and the binary relation of causal precedence in which they stand. As there are spacelike-related events, there exist pairs of events such that neither of these events precedes the other. This signifies a loss of 'comparability' and entails that the primitive causal relation imposes a *partial* order on the set of events, as is appropriate for a special-relativistic theory.[4] The spacelike relations between incomparable events can be defined in terms of combinations of fundamental causal relations: there are always events in Minkowski spacetime which are to the causal future of any two spacelike-related events (i.e., in the intersection of their causal futures). Thus, two spacelike-related events stand in an indirect causal relation.

The derivation of the full geometry of Minkowski spacetime from a few axioms mostly relying on a set of primitive events partially ordered by a primitive binary relation, and the early interpretation of this relation as *causal*, has led philosophers to causal theories of spacetime more generally, reinvigorating a venerable tradition of causal theories of time dating back at least to Gottfried Wilhelm Leibniz. This rejuvenated tradition starts off in Hans Reichenbach (1924, 1928, 1956) and

---

[2] For some suggestions as to how CST might be turned into a quantum theory, see for example Sorkin (1997a), Dowker et al. (2010), Gudder (2014), or Surya and Zalel (2020).

[3] It can be shown that the group of all automorphisms of the causal structure of Minkowski spacetime is generated by the (inhomogeneous) Lorentz group—and dilatations, of course. This result was independently proved by Zeeman (1964), and apparently also by A. D. Aleksandrov in 1949. For a detailed analysis, see Winnie (1977).

[4] Consult Huggett et al. (2013, §2.1) for details. For a systematic account of the various attempts to axiomatize the structure of Minkowski spacetime and an assessment of the characteristics of the resulting logical systems, see Lefever (2013).

continues in Henryk Mehlberg (1935, 1937) and Adolf Grünbaum (1963, 1967) as a causal theory of *time order*, rather than of spacetime structure. Arguing from an empiricist vantage point, the goal was to explicate temporal relations in terms of their physical, i.e., more directly empirically accessible, basis. In this spirit, Reichenbach postulates a set of events, merely structured by basic relations of a causal nature, where the relations of 'genidentity' and of causal connection play a central role.[5]

Reichenbach's early attempts to execute this program fail on grounds of circularity:[6] his theory makes ineliminable use of the asymmetry of causal connection to ground the asymmetry of temporal order, but his criterion to distinguish between cause and effect—often called the 'mark method'—relies, implicitly, on temporal order.[7] The circularity arises because the distinguishability of cause and effect is necessary in an approach that assumes as fundamental an asymmetric relation and thus cannot take recourse to a more fundamental description of the events that may ground the distinction. Reichenbach's early formulation of the causal theory of time order also uses spatiotemporal coincidence as primitive and thus fails to explicate temporal order entirely in nonspatiotemporal terms.[8] His later formulation (Reichenbach 1956) mends some of these deficiencies by attempting to explicate the causal asymmetry in terms of factual asymmetries in actual series of events ordered by temporal betweenness. This move does not get entirely rid of primitive temporal notions, but at least it provides an independent, 'physical' grounding of the asymmetry of the causal connection. Grünbaum (1963, 1967) adopts Reichenbach's basic strategy, but tries to overcome the difficulties that befell its predecessor. The result is mixed, and no full explication of spacetime in terms of physical relations is achieved.[9]

Bas van Fraassen (1970, ch. 6) offers what he argues is a significant simplification of the theory that does not rely on purely spatial or spatiotemporal notions (182). His causal theory of *spacetime* (as opposed to just *time*) takes as primitive the notion of an *event* and the binary relations of *genidentity* and *causal connectibility*. Genidentity is an equivalence relation and is used to define world lines of objects as equivalence classes, while the reflexive and symmetric relation of causal connectibility captures the causal structure of spacetime. Van Fraassen's final version of the theory—the details of which we will leave aside—dispenses with genidentity in favor of primitive persisting objects.

---

[5] Two events are *genidentical* just in case they are temporal stages of the same physical object. It is thus clear that genidentity also involves, apart from mereological considerations, a causal connection between the genidentical events.

[6] Speaking of circularity, he rules out closed causal chains *ab initio*—on empirical rather than logical grounds.

[7] As criticized by Mehlberg (1935, 1937) and Grünbaum (1963), among others.

[8] Cf. van Fraassen (1970, §6.2.a).

[9] Cf. also van Fraassen (1970, §6.3).

38   OUT OF NOWHERE

Evidently, at least in the context of SR, the (possibly directed) causal connectibility relation is equivalent to the spatiotemporal relation of 'spatiotemporal coincidence or timelike relatedness' (the first disjunct is necessary because of the reflexive convention chosen). This raises the worry—to which we will return below—whether the theory is trivial and the reconstruction hence futile in the sense that it just restates the spatiotemporal structure of Minkowski spacetime in different terms. Such a futile reconstruction would obviously fail to ground that structure in more fundamental, *non-spatiotemporal*, relations. Against this charge, van Fraassen, like his precursors, insists that the causal theory outstrips the relational theory of spacetime because, unlike blandly spatiotemporal relations, the fundamental causal relations are *physical* relations. This line of defense may incite the further worry that the causal theory attempts to illuminate the obscure with the truly impervious: analyses of causation are famously fraught with a tangle of apparently impenetrable problems. This concern is allayed, van Fraassen retorts, by the fact that no general definition of either causation or physical relations is required for the project to succeed; since the non-modal component of 'causal connectibility'—causal connection—gets explicated in terms of genidentity and 'signal connection', and since these are evidently physical and empirical relations, 'causally connectible' has a meaning which is not derivatively spatiotemporal. Furthermore, van Fraassen (1970, 195) notes, the causal theory does not insist that all spatiotemporal relations are in fact *reduced*, but merely that they are *reducible*, to causal connectibility. We will need to return to this point below when we consider whether the fundamental structures of causal set theory are spatiotemporal at all.

3.1.2 Earman's criticism of causal theories of spacetime

Shortly after van Fraassen publishes what was then the most sophisticated articulation of the causal theory of spacetime, John Earman (1972) airs devastating criticisms of the theory. Earman's objections trade on results in general relativity (GR) and establish that no causal theory of spacetime has the resources to deal with the full range of spacetime structures licensed by GR.

In order to appreciate Earman's points, and the story unfolding after his attack leading directly to CST, a few concepts need to be introduced. A *relativistic spacetime*, or simply 'spacetime', is an ordered pair $\langle \mathcal{M}, g_{ab} \rangle$ consisting of a 4-dimensional (Lorentzian) manifold $\mathcal{M}$ with a metric field g_{ab} defined everywhere on it. A *model of GR* then is a triple $\langle \mathcal{M}, g_{ab}, T_{ab} \rangle$ including a stress-energy tensor T_{ab} that satisfies Einstein's field equation (EFE),

$$R_{ab} - \frac{1}{2}Rg_{ab} + \Lambda g_{ab} = 8\pi T_{ab}, \tag{3.1}$$

given here in natural units $c = G = 1$, where R_{ab} is the Ricci tensor, R the Ricci scalar, and Λ the cosmological constant.

A spacetime is *temporally orientable* just in case a continuous choice of the future half of the light cone (as against the past half) can be made across $\mathcal{M}$. Since a temporally orientable spacetime always affords a smooth, everywhere non-vanishing timelike vector field t^a on $\mathcal{M}$,[10] we can take such a vector field to encode the temporal orientation of spacetime. In what follows, we will assume that all spacetimes are temporally orientable. If in fact the orientation is given, e.g., by a smooth timelike vector field, then we will call the spacetime temporally orient*ed*.

In order to get to the relevant causal structure, we need the notion of a timelike relation:

Definition 1 (Timelike relation). *Let $\langle \mathcal{M}, g_{ab} \rangle$ be a temporally oriented relativistic spacetime. The binary relation $\ll$ of timelike separation is then defined as follows: $\forall p, q \in \mathcal{M}, p \ll q$ if and only if there is a smooth, future-directed timelike curve that runs from p to q.*

The relation $\ll$, which is technically a relation of *timelike* separation in GR, will nevertheless be used to encode the *causal* structure. Let us grant, at least for the time being, that this relation is indeed fundamentally causal. We will return to this point below.

Since we are interested in the causal *structure*, we need a criterion to identify causal structures. Here is the relevant isomorphism:

Definition 2 ($\ll$-isomorphism). *Let $\langle \mathcal{M}, g_{ab} \rangle$ and $\langle \mathcal{M}', g'_{ab} \rangle$ be temporally oriented relativistic spacetimes. A bijection $\varphi : \mathcal{M} \to \mathcal{M}'$ is a $\ll$-isomorphism if, for all $p, q \in \mathcal{M}, p \ll q$ if and only if $\varphi(p) \ll \varphi(q)$.*

It is clear that two spacetimes have the same causal structure just in case there is a $\ll$-isomorphism between their manifolds. The causal structure is precisely what is preserved under these isomorphisms. Analogously, we can introduce the notion of a causal relation, which is only slightly more general than that of a timelike relation:

Definition 3 (Causal relation). *Let $\langle \mathcal{M}, g_{ab} \rangle$ be a temporally oriented relativistic spacetime. The binary relation $<$ of causal separation is then defined as follows: $\forall p, q \in \mathcal{M}, p < q$ if and only if there is a smooth, future-directed timelike **or null** curve that runs from p to q.*

---

[10] Wald (1984, lemma 8.1.1).

40   OUT OF NOWHERE

We have highlighted in bold font the relevant difference to definition 1. Causal relations $<$ also give rise to a corresponding isomorphism, denoted '$<$-isomorphism', in full analogy to the $\ll$-isomorphism as defined in definition 2. Note that the two relations $\ll$ and $<$ are generally not interdefinable.[11] But since their physical meaning is closely related, we will generally only state the results using $\ll$.

The essential goal for a causal theorist must be to take causal structure as fundamental and show how this structure grounds everything else about a spacetime. In particular, their goal is to show that its metric structure is determined by its causal structure. Thus, the success, or at least viability, of the causal program gets measured by the extent to which the causal structure of relativistic spacetimes determines their metric (and topological) structure.

It turns out this success cannot be complete, as Earman (1972) notes. Conceding, as he does, the causal connectibility relation to his explicit targets Grünbaum (1967) and van Fraassen (1970), Earman demonstrates that there are relativistic spacetimes for which there is no hope that the causal program succeeds, thus showing that the transition from a causal theory of time to a causal theory of relativistic spacetime is not trivial. To add insult to injury, Earman concludes that the doctrines of the causal theorist about spacetime "do not seem very interesting or very plausible" (75). However, subsequent work by David Malament and others and in CST has clearly nullified *that* appearance, as should become apparent below.

Earman starts out by precisifying the notion of causal connectibility. In order to do that, let us introduce the notions of the 'chronological future' and 'past':

Definition 4 (Chronological future and past). *For all points $p \in \mathcal{M}$, let $I^+(p)$ and $I^-(p)$ be the* chronological future *and the* chronological past, *respectively, as determined by:*

$$I^+(p) := \{q : p \ll q\},$$
$$I^-(p) := \{q : q \ll p\}.$$

The *causal* future and past sets, denoted J^+ and J^-, respectively, can be defined in complete analogy to definition 4, substituting throughout the causal relation $<$ for the timelike relation $\ll$. These notions permit the precise articulation (of one notion) of 'causal connectibility' of two events in a given spacetime as the physical possibility of a causal signal of nonnull (affine) length to connect them:

---

[11] More precisely, as Kronheimer and Penrose (1967) show, it generally takes two of the three relations, causal, chronological, and null, to (trivially) define the third. All three relations can be reconstructed from any one only under the imposition of appropriate restrictions.

Definition 5 (Causal connectibility). *An event $q \in \mathcal{M}$ is* causally connectible *to another event $p \in \mathcal{M}$ just in case $q \in J^+(p) \cup J^-(p)$.*

Earman then asks whether the causal theorist can supply a criterion of spatiotemporal coincidence that relies purely on the basal causal notions. Suppose we have a set of otherwise featureless 'events' that partake in relations of causal connectibility but that do not, fundamentally, stand in spatiotemporal relations. Van Fraassen (1970, 184) then defines pairs of events to be *spatiotemporally coincident* just in case they are causally connectible to exactly the same events, i.e., for every event r, r is causally connectible to the one if and only if it is causally connectible to the other. This criterion is only adequate for spacetimes which satisfy the following condition: for every pair of points $p, q \in \mathcal{M}$, if $J^+(p) = J^+(q)$ and $J^-(p) = J^-(q)$, then $p = q$. As Earman points out, since GR permits many spacetimes which violate this condition,[12] a causal theorist could never hope to reduce spatiotemporal coincidence to causal connectibility for *all* general-relativistic spacetimes. This condition is very closely related to another one:

Definition 6 (Future (past) distinguishing). *A spacetime $\langle \mathcal{M}, g_{ab} \rangle$ is* future distinguishing *iff, for all $p, q \in \mathcal{M}$,*

$$I^+(p) = I^+(q) \Rightarrow p = q$$

(and similarly for past distinguishing).

If a spacetime is both future and past distinguishing, then we simply call it *distinguishing*. As we will see momentarily, the conditions of future and past distinguishing mark an important limit of what sorts of spacetimes a causal theory of spacetime can hope to successfully capture. This is perhaps not too surprising given that a distinguishing spacetime cannot contain any closed timelike or null curves and that non-distinguishing spacetimes do at least contain non-spacelike curves which come arbitrarily close to being closed. Relying on causal structure as it does, the causal program remains incapable in cases where this structure is pathological. However, pathological causal structures are certainly possible in notoriously permissive GR (Smeenk and Wüthrich 2011). And for Earman (1972, 78), there are no good reasons to reject those spacetimes violating the condition as physically unreasonable, even though that is sometimes done, and not just by defenders of the causal program. He thus concludes that the causal theory

---

[12] This condition is crudely violated, e.g., in spacetimes containing closed timelike curves, which consist in numerically distinct points which nevertheless all have the same causal future and past. For a more subtle and interesting example for a past but not future distinguishing spacetime as defined in definition 6, see Hawking and Ellis (1979, fig. 37).

42   OUT OF NOWHERE

of spacetime does not command the resources necessary to offer a novel basis "for getting at the subtle and complex spatiotemporal relations which can obtain between events set in a relativistic space-time background" (Earman 1972, 79).

Some spacetimes violating Earman's condition are such that $\mathcal{M}$ can be covered by a family of non-intersecting spacelike hypersurfaces. These spacetimes are topologically closed 'timewise' and every event in $\mathcal{M}$ contains closed timelike curves. In this case, it is always possible to find an embedding space which is locally like $\mathcal{M}$ but does not contain closed timelike curves. Given that such embedding spaces are causally well behaved, the causal theorist might be tempted to use their causal structure to recover the spacetime structure. Earman notes that such an enterprise would be bound to fail, given that the original spacetime and the embedding space also differ in global properties: in the former, 'time' is closed, while in the latter it is not. Thus, the causal structure of the embedding space could not possibly render the correct verdict about such important global properties of the spacetime at stake. But note that the causal set theorist could respond to this by weakening the ambition to recovering merely the *local*, but not necessarily the *global*, spacetime structure.

In fact, this strategy could be more widely applied to those non-distinguishing spacetimes which can be partitioned into distinguishing 'patches' of spacetime. A patch $\langle S \subseteq \mathcal{M}, g_{ab|S} \rangle$ of a spacetime $\langle \mathcal{M}, g_{ab} \rangle$ would be distinguishing just in case it is with respect to the events in the patch, but not necessarily with respect to those outside of it. This covers many, though not all, spacetimes which violate Earman's condition. In this case, the causal theorist could claim to be able to recover the spatiotemporal structure of the patch from the patch's causal structure. In this sense, the program could then still be brought to fruition 'locally'. Thus, what Earman identifies as a problem for the causal theorist could be turned into a strategy to deal with many of the spacetimes ruled out of bounds by Earman. Of course, this strategy would only be acceptable if accompanied by a commensurate attenuation of the ambition of the program.[13]

Earman's criticism, though sound, may thus not spell the end of the program. The causal theorist can surely respond to Earman's conclusion that "taking events and their causal relations as a primitive basis is not sufficient for getting at the subtle and complex spatiotemporal relations which can obtain between events set in a relativistic space-time background" (79) with a dose of healthy revisionism. The causal theorist is bound to fail if her pretension was to produce an alternative foundation for full GR, as Earman conclusively establishes; however, if the aspiration is instead to offer a *distinct*, and perhaps more fundamental, theory, Earman's

---

[13] Moreover, it would still need to find a way to fix the conformal factor, which according to Malament's theorem (to be discussed in §3.1.3) causal structure does not fix. This means that metrical structure of a relativistic spacetime is not fully reducible to its causal structure even in otherwise serendipitous circumstances. In causal set theory, the conformal scale will be fixed by cardinality—a move unavailable in the continuum case.

objections can be bypassed.[14] The latter, of course, is precisely what CST aims to provide, as we will see shortly.

3.1.3 Malament's theorem

That the situation is not hopeless for the causal theorist, and, more specifically, that the causal structure often provides a powerful ground for the entire geometry of a relativistic spacetime, was established by the remarkable results of Stephen Hawking and collaborators (Hawking et al. 1976) and David Malament (1977), building on results regarding 'causal spaces' in the seminal paper of Kronheimer and Penrose (1967). It is these results by Kronheimer and Penrose, Hawking and collaborators, and Malament that motivate the causal set theory program. They can be thought of as an extension of Robb's reconstructive project from SR to GR. Mathematically, the basic idea is that we take a point set with its topology, its differential structure, and its conformal structure to extract the relation $\ll$ defined on the point set. Then we throw away everything except the point set and the relation $\ll$ and try to recover the entire spacetime geometry. This is accomplished—to the extent to which it is accomplished—in terms of implicit definitions using invariances under the relevant group of mappings. Metaphysically, the idea is of course that, fundamentally, there is only the point set structured by $\ll$ and every other aspect of spacetime geometry ontologically depends on that structure.

Using the definitions above, as summarized by Malament (1977, 1400), the relevant result of Hawking, King, and McCarthy is the following theorem:

Theorem 1 (Hawking, King, and McCarthy 1976). *Let ϕ be a $\ll$-isomorphism between two temporally oriented spacetimes $\langle \mathcal{M}, g_{ab} \rangle$ and $\langle \mathcal{M}', g'_{ab} \rangle$. If ϕ is a homeomorphism, then it is a smooth conformal isometry.*

Homeomorphisms, i.e., continuous mappings that also have a continuous inverse mapping, are maps between topological spaces that preserve topological properties. Furthermore,

Definition 7 (Conformal isometry). *A $\ll$-isomorphism ϕ is a conformal isometry just in case it is a diffeomorphism and there exists a (non-vanishing) conformal factor $\Omega : \mathcal{M}' \to \mathbb{R}$ such that $\phi_*(g_{ab}) = \Omega^2 g'_{ab}$ (where ϕ_* is the 'push forward' map which takes tensors defined on $\mathcal{M}$ to tensors defined on $\mathcal{M}'$).*

Thus, Hawking and collaborators showed that the following conditional holds: if two temporally oriented spacetimes have the same topology and causal structure,

---

[14] For a detailed assessment as to what extent a causal theory can succeed in grounding GR, see Winnie (1977).

44 OUT OF NOWHERE

then they have the same metric, up to a conformal factor. This leads to the natural question of under just what conditions do two temporally oriented spacetimes with the same causal structure have the same topology? Obviously, if we knew the answer then we could state under what conditions two temporally oriented spacetimes with the same causal structure have the same metric, up to a conformal factor. Malament's result gives a precise answer to this question: just in case they are distinguishing, i.e., just in case they are both future and past distinguishing. Here is the theorem:

Theorem 2 *(Malament 1977). Let ϕ be a $\ll$-isomorphism between two temporally oriented spacetimes $\langle \mathcal{M}, g_{ab} \rangle$ and $\langle \mathcal{M}', g'_{ab} \rangle$. If $\langle \mathcal{M}, g_{ab} \rangle$ and $\langle \mathcal{M}', g'_{ab} \rangle$ are distinguishing, then ϕ is a smooth conformal isometry.*

It is important to note that neither future nor past distinguishability alone is sufficient to clinch the consequent (Malament 1977, 1402). The theorem thus establishes that, for a large class of spacetimes, causal isomorphisms also preserve the topological, differential, and conformal structure, and hence the metrical structure up to a conformal factor. In this sense, we can say that the causal structure of a relativistic spacetime in this large class 'determines' its geometry up to a conformal factor. It is worth keeping in mind that this 'determination' is not underwritten by explicit definitions of the geometrical structure of these spacetimes in terms of their causal structures;[15] rather, the argument proceeds by showing that maps between spacetimes that leave the causal structure invariant will also leave the geometrical structure invariant (again, up to a conformal factor). Even though this may be too thin a basis on which to lay grand claims of ontological dependence, the results are sufficiently suggestive to motivate the entire program of causal set theory.

Before moving to causal set theory proper, let us make a few remarks regarding its immediate prehistory. Considerations concerning the relations between discrete and continuum spaces in philosophy, physics, and mathematics certainly predate causal set theory—in philosophy, for instance, by a couple of millennia. At least since the advent of quantum physics physicists have played with the idea of a discrete spacetime, which was seen as motivated by quantum theory.[16] It was recognized early on that a fundamentally discrete spacetime structure would create a tension with the invariance under continuous Lorentz transformations postulated by SR, and that solving this problem would be crucial to finding a quantum theory of gravity.[17] An early clear statement of something that starts to resemble the causal

---

[15] Whether or not such explicit definitions can be given remains an open problem (David Malament, personal communication, 24 April 2013).

[16] E.g., Ambarzumian and Iwanenko (1930).

[17] For the earliest attempt to reconcile Lorentz symmetry with a discrete spacetime that we know of, cf. Snyder (1947). We will return to this issue in the next chapter.

set program has been given by David Finkelstein (1969). Finkelstein argues that macroscopic spacetime with its causal structure may arise as a continuum limit from a 'causal quantum space'. He asserts in the abstract of the paper that "[i]t is known that the entire geometry of many relativistic space-times can be summed up in two concepts, a space-time measure μ and a space-time causal or chronological order relation C, defining a causal measure space" (1261). Unfortunately, no reference or proof is given in the text in support of this claim; in fact, the claim is essentially repeated, but with the qualification that this is true only of "many" spacetimes dropped:

> The causal order C determines the conformal structure of space-time, or nine of the ten components of the metric. The measure on spacetime fixes the tenth component. (1262)

That this is locally true is shown in Hawking and Ellis (1979, §3.2), and may have been known earlier. Similarly, in another pre-causal-set-theory paper, Jan Myrheim (1978) presumably invokes this local fact when he claims that

> [i]t is a well-known fact that in the standard, continuum space-time geometry the causal ordering alone contains enough information for reconstructing the metric, except for an undetermined local scale factor, which can be introduced for instance by means of the volume element dV. (4)

Of course, both Finkelstein's and Myrheim's statements are false if read as global statements, i.e., as applying to an entire spacetime. For the global scale, we need Malament's theorem with the relevant qualification to distinguishing spacetimes.

Myrheim's project can clearly be considered a precursor to causal set theory. He argues that the partial causal ordering can just as naturally be imposed on a discrete set as on a continuous one, and that the discrete set has the advantage of automatically providing a natural scale, unlike the continuous space. This natural scale, as Bernhard Riemann (1868, 135, 149) observed, results from the fact that a discrete space, but not a continuous space, possesses an 'intrinsic metric' given by *counting* the elements. The *number* of elementary parts of the discrete space determines the volume of any region of that space, which is of course not the case for continuous spaces. Myrheim's goal is to show how one can recover, in a statistical sense, not only the geometry, but also the vacuum EFE for the gravitational field from the discrete causal structure. While the project is suggestive, and the goal the same as in causal set theory, it does not succeed.[18]

---

[18] Although Myrheim is aware (cf. 3) that it is not in general true that the causal ordering of any relativistic spacetime is globally well defined, he proceeds as if it were; he also seems to assume that

46   OUT OF NOWHERE

Around the same time, Gerard 't Hooft (1979), in just a few short pages (338–344), sketches a quantum theory of gravity closely related to causal set theory. While the use of lattices in quantum field theory is generally considered a pretense introduced to simplify the mathematics or to control vicious infinities, 't Hooft takes the point of view that in the case of gravity, a kind of a lattice "really does describe the physical situation accurately" (338). He, too, uses the unqualified and hence probably local version of the statement that Malament proved in its qualified form (340)—he makes no reference to either Hawking et al. (1976) or Malament (1977). Starting out from a basic causal relation, for which he offers the temporal gloss as "is a point-event earlier than" (ibid.), and requiring transitivity, 't Hooft asserts four basic assumptions of the theory. First, the causal relation gives a partial ordering of events and defines a lattice, i.e., a discrete structure. Supposing that a continuum limit of this fundamental structure exists, he demands, secondly, that this lattice contains all the information to derive a curved Riemann space in this limit; that, thirdly, the existence of this limit constrains the details of the fundamental partially ordered structure; and that, finally, a curved Riemann space (with the appropriate signature) comprises all information on the entire history of the universe it represents, i.e., the entire spacetime. The second and fourth assumptions jointly entail that the fundamental structure contains all physically relevant information of the spacetime it describes. 't Hooft then sketches how the continuum limit could be taken and briefly mentions that a cosmological model could be obtained by simply adding the demand that there exists an event that precedes all others. The sketch omits any dynamics, which 't Hooft admits should be added such that it approximates the usual Einstein dynamics in GR and adds that we should expect such an action to be highly non-local.

't Hooft then closes his speculations with the following words:

> The above suggestions for a discrete gravity theory should not be taken for more than they are worth. The main message, and that is something I am certain of, is that it will not be sufficient to just improve our mathematical formalism of fields in a continuous Riemann space but that some more radical ideas are necessary and that totally new physics is to be expected in the region of the Planck length. (344)

Such a radically new start is precisely what the program of causal set theory seeks to offer.

the local uniqueness of geodesics can be extended globally (6), which is not in general true. There are other oddities in his analysis, for instance when he takes coordinates to be more fundamental than the metric (8).

3.2 The basic plot: kinematic causal set theory

Causal set theory as we now know it starts with the foundational paper by Luca Bombelli, Joohan Lee, David Meyer, and Rafael Sorkin (Bombelli et al. 1987), who, unlike Myrheim and 't Hooft, explicitly build on Hawking et al. (1976), Malament (1977), and Kronheimer and Penrose (1967). Advocates of causal set theory often insist that their theory consists of three parts, like any other physical theory: kinematics, dynamics, and phenomenology. Even though many physical theories may conform to this partition, there is, of course, nothing sacrosanct about this trinity. As long as the theory clearly articulates what it considers the physical possibilities it licenses to be and how this explains or accounts for aspects of our experience, it may have as many parts as it likes. Be this as it may, we will split the discussion of the theory into kinematics and dynamics, and deal with the former in this section and the latter in the next chapter, and particularly in §4.3.

3.2.1 The kinematic axiom

Causal set theory assumes a 4-dimensional point of view and, impressed with Malament's result, attempts to formulate a quantum theory of gravity *ab initio*. While it still takes GR as its vantage point, it does not apply canonical quantization as loop quantum gravity does (we will encounter loop quantum gravity in chapters 5 and 6). The 4-dimensional viewpoint regards spacetime as a fundamentally inseparable unity. In this sense, it stands in opposition to canonical approaches which divide 4-dimensional spacetime into 3-dimensional spacelike slices totally ordered by 1-dimensional time. Spacetime points are replaced in causal set theory with elementary 'events'. These events, just as points in the spacetime manifold, have no intrinsic identity; instead, they only acquire their identity through the relations in which they stand to other such events.[19] Consonant with Malament's result, and as reflected in its name, causal set theory takes the relations which sustain the identities of their relata to be *causal* relations. Finally, the fundamental structure is assumed to be 'atomic', and discrete. We are thus left with a fundamental structure of otherwise featureless elements partially ordered by a relation of causal precedence into a discrete structure:

Basic assumption. *The fundamental structure is a countable set of elementary events partially ordered by a relation of causal precedence. In short, it is a causal set.*

---

[19] The theory thus lends itself rather directly to a structuralist interpretation (Wüthrich 2012).

48   OUT OF NOWHERE

This assumption is too weak to articulate a sufficient condition delineating the models of the theory; rather, it should be regarded as a conceptually central necessary condition. More precisely, it can be stated as follows:

Axiom 1 (Kinematic axiom of causal set theory). *The fundamental structure is a* causal set *C, i.e. an ordered pair $\langle C, \preceq \rangle$ consisting of a set C of elementary events and a relation, denoted by the infix $\preceq$, defined on C satisfying the following conditions:*

1. *$\preceq$ induces a partial order on C, i.e., it is reflexive ($\forall x \in C, x \preceq x$), antisymmetric ($\forall x, y \in C,$ if $x \preceq y$ and $y \preceq x,$ then $x = y$), and transitive ($\forall x, y, z \in C,$ if $x \preceq y$ and $y \preceq z,$ then $x \preceq z$).*
2. *$\preceq$ is locally finite:[20] $\forall x, z \in C, |\{y \in C | x \preceq y \preceq z\}| < \aleph_0,$ where $|X|$ denotes the cardinality of the set X (sometimes called its 'size').*
3. *C is countable.*

A causal set, or 'causet', can be fully represented by a *directed acyclic graph*,[21] i.e., a set of nodes (usually represented by dots) connected by directed edges (often represented by arrows) without directed cycles (which are closed loops in which each node is traversed along the edges without changing direction). As is standard in causal set theory, we will often use so-called 'Hasse diagrams', which are directed acyclic graphs of the transitive reduction of finite partially ordered sets.[22] In Hasse diagrams, the direction of the edges is encoded in the relative positions of the connected vertices, rather than by arrows. The convention is that an element 'smaller' in the partial order is always drawn below a 'greater' element. Thus, an event x is drawn below a numerically distinct event y and there is a line connecting the two just in case x immediately causally precedes y. Figure 3.1 shows the Hasse diagram of a causet. Hasse diagrams are convenient because no two distinct causets can have the same Hasse diagram, so that a Hasse diagram uniquely determines the represented causet.[23]

---

[20] This condition may perhaps be more adequately called 'interval finiteness' (Dribus 2013, 19), where an *interval* $[x, z]$ in a causet C is a subcauset that contains all events $y \in C$ such that $x \preceq y \preceq z$. Intervals are sometimes also called 'Alexandrov sets'.

[21] At least if we gloss over the fact that acyclicity is often defined in a way that renders causets 'cyclic' because their fundamental relation contains trivial 'cycles' $x \preceq x$. If readers are bothered by our sloppiness here they are invited to substitute 'non-circular' for 'acyclic'.

[22] The *transitive reduction* of a binary relation R on a domain X is the smallest relation R' on X with the same transitive closure as R. The *transitive closure* of a binary relation R on X is a smallest transitive relation R' on X that contains R. Effectively, the transitive reduction of a graph is the graph with the fewest edges but the same 'reachability relations' as the original graph, or, in the case at hand, the transitive reduction of a causet is the causet with all those causal connections removed that are entailed by transitivity. At least in the finite case, the transitive reduction of a causet is unique.

[23] For a useful entry point onto the mathematics of partially ordered sets, see Brightwell and West (2000). For a mathematically motivated and hence much wider perspective on causets, see Dribus (2013, 2017).

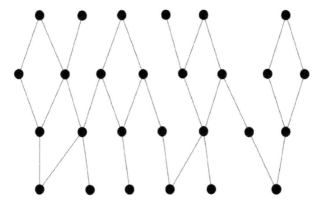

Figure 3.1 The Hasse diagram of a causet. The diagram is to be read from the bottom up, i.e., events (marked by dots) causally precede those events above just in case an edge is drawn between them. Relations of causal precedence between events implied by transitivity (or by reflexivity) are not indicated by a drawn edge.

3.2.2 Discussion of the basic assumptions

A few remarks on the basic assumption and the kinematic axiom. First, it is important that the ordering be merely partial, and not total. A *total order* induced by a binary relation R on a set X is a partial order such that every pair of distinct elements of X is *comparable*, i.e., $\forall x, y \in X$, either Rxy or Ryx. That the ordering imposed by the causal structure is partial rather than total is crucial to capture the causal structure of a relativistic spacetime, where two distinct events can be spacelike separated and hence not stand in a relation of causal precedence.

Second, it is just as important that the ordering be no weaker than partial; in particular, it matters that it is not a mere pre-order. A binary relation on a set induces a *pre-order* just in case it is reflexive and transitive, but not antisymmetric. The additional imposition of antisymmetry rules out the possibility of causal loops of the form of cycles involving distinct events $a, b \in C$ such that $a \preceq b \preceq a$. Two remarks concerning causal loops are in order. First, it should be noted that any events on such a loop would have the identical relational profile in the sense that any event that precedes one of them precedes all of them, and any event that is preceded by any one of them is preceded by all.[24] If the relational profile of an elementary event constitutes its identity, as the rather natural structuralist interpretation of the fundamental structure in causal set theory would have it, events on a causal loop would therefore not be distinct at all and the theory lacks the resources to distinguish between a single event and a causal loop. This may be a limitation of the theory, but it reflects the assumption that any two events which do not differ

---

[24] See Wüthrich (2012, 236f) for details.

50 OUT OF NOWHERE

in their causal profile do not differ physically and hence should be considered identical. Second, there are, of course, spacetimes in GR which contain causal loops (Smeenk and Wüthrich 2011). Given that axiom 1 prohibits causal loops at the fundamental level, it is not clear that causal set theory has the resources to show how relativistic spacetimes with causal loops can emerge from a fundamental causet without.[25] Many physicists consider causal loops 'unphysical' and hence do not mourn their absence from the fundamental level (and so perhaps from emergent levels), but it is a cost that the theory incurs and that needs to be kept in mind. The fact that causal loops appear in so many relativistic spacetimes, including in (extensions of) physically important ones, should make us wary to dismiss them too precipitously, as is argued in Smeenk and Wüthrich (2011) and elsewhere.[26]

Third, the fundamental structure is assumed to be locally finite, and hence discrete. This discreteness is *assumed* in causal set theory, not derived, as we shall see it to be in loop quantum gravity. A first set of justifications offered in the literature (e.g., in Henson 2009, 394) argues from the technical utility of assuming discreteness. A fundamentally discrete 'spacetime' can cure a theory from divergences in various quantities that may not be tamed by renormalization, as well as simplify the computational challenges faced by the physicist. A second set of justifications typically given trades on physical reasons for preferring a discrete structure. For instance, without a short-distance cut-off as supplied by a discrete 'spacetime', the semi-classical black hole entropy will not come out finite as desired (ibid.); or the discreteness may be an effective way of avoiding violations of the local conservation of energy in the form of photons with infinite energies (Reid 2001, 6). Perhaps most curiously of all, Henson (2009, 394) argues that the concurrence of many rather diverse approaches to quantum gravity on the discreteness of the fundamental structure adds support to the stipulation of such discreteness in causal set theory. It is questionable, however, how different programs with contradictory assumptions which presuppose, or infer, the same discreteness can be considered mutually supportive. The truth of such an alternative theory would be evidence against causal set theory as a whole, though not necessarily against all its parts. Conversely, if causal set theory is true, or at least on the right track to a true quantum theory of gravity, then this entails that its competitors are mistaken. False theories may of course rest on true assumptions or deliver true implications; but whether this is so in any particular case can only be argued by a careful analysis of the relation between the distinct theories. Such an analysis is at best

---

[25] But see Wüthrich (2021).

[26] See the references therein. This caution certainly also motivates Earman's earlier criticism of the causal theory of spacetime: if all spacetimes violating Earman's condition are dismissed as unphysical, the causal theory survives his criticism unscathed.

sketched in the literature; and of course, it may be that everyone is barking up the wrong tree.[27]

Discreteness may give us other advantages. Myrheim (1978, 1f) further points out that a fundamentally discrete 'spacetime' carries an intrinsic volume and would thus give us a natural fundamental scale. This is again Riemann's point so revered by Myrheim's successors in causal set theory. As Myrheim (1978, 1) also mentions, it may be that the discreteness has observable consequences. Although one might be worried that there could not be such observable consequences given the many ways in which a fundamentally discrete structure can be compatible with emergent continuous spacetime symmetries such as Lorentz symmetry, the discreteness scale of causal set theory has been used by Sorkin (1991, 1997b) to obtain a very rough estimate of the value of the cosmological constant Λ, with the result that it has a small positive value. Although Sorkin's estimate turned out to be too large by almost two orders of magnitude, it was a remarkable prediction of a non-zero Λ at a time when cosmologists commonly assumed that Λ was zero. Of course, this changed when astronomers observed the acceleration of the expansion of the universe in 1998 and a small positive value for Λ quickly came to be accepted. We will return to this remarkable prediction in §4.4.4.

In this spirit, discreteness may well be a central axiom of an empirically successful theory and could thus be vindicated indirectly and a posteriori. With Sorkin (1995), one may simply consider fundamental discreteness necessary to "express consistently the notion that topological fluctuations of *finite* complexity can 'average out' to produce an uncomplicated and smooth structure on larger scales" (173). This is an appealing thought, though the apparent necessity may of course evaporate on closer inspection. In sum, although all the reasons given in favor of discreteness, as defensible as they are, are clearly defeasible, and, as we shall see in the next chapter (§4.4), there is a sense in which the discreteness leads to a form of non-locality, the proof, ultimately, is in the pudding: a theory's success ultimately validates its assumptions.

Whatever its ultimate justification, it should be noted that the discreteness of causets has deep consequences. We will encounter some of these consequences over the course of this chapter and the next. One difference that matters in the present context can be seen by considering whether an analog of the program's basic motivating result—Malament's theorem (theorem 2)—holds in causal set theory. To state the obvious, this result is a theorem in *classical relativity theory*. There are other issues that would need to be settled for such a transliteration to be meaningful, but here it should be noted that an obvious causet analog of future and past distinguishability is easily violated in causal set theory. Call a causet *future (past) distinguishing* just in case for all pairs of events in the causet, if the set of

---

[27] See Wüthrich (2012, 228f) for a more detailed analysis of these arguments.

events that are causally preceded by (causally precede) them are identical, then the events are identical. It is obvious that a causet such as the one depicted in figure 3.2 violates this condition for the pair p and q—both events have the same causal future.

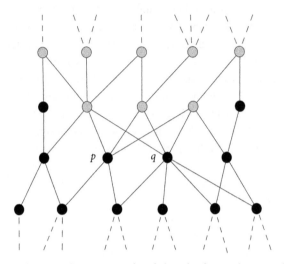

Figure 3.2 A past distinguishing causet that fails to be future distinguishing: the grayed out events are to the future of both p and q.

That such a structure can satisfy the causal conditions while still violating distinguishability is a direct consequence of its discreteness. A causet which is not distinguishing and thus contains at least one pair of events with either an identical future or an identical past contravenes the premises of Malament's theorem, but of course it is fully determined by its causal structure alone. (It should be noted that in general a non-distinguishing causet still uniquely identifies each event in terms of its structure alone: the causet in figure 3.2 is not future distinguishing, and hence not distinguishing, but the events p and q are structurally distinct in that they have different causal pasts.)

3.2.3 Causation?

There remains an important point to be recorded before we move on. By virtue of what is the fundamental relation that structures the causets a *causal* relation? One might object that such an austere relation cannot possibly earn this designation because 'causation'—whatever its precise metaphysics—refers to a much richer feature of our world. Why should one think that the partial ordering imposed on the fundamental and featureless events stands in any tangible connection to the causation that we attribute to the macroscopic and richly structured events that populate our world of experience? Of course, an answer to this question will require a detailed understanding of how an effective spacetime emerges from underlying causets. We will turn to this problem in chapter 4. But even granting—prematurely—a removal of these obstacles, there remain at least two worries.

First, it may be argued that the conditions listed in axiom 1 are not sufficient to make the relation at stake causal in the sense of relating causes and effects. A first insufficiency is that, as stated, these conditions are purely formal and will need to be imbued with physical meaning. In particular, they will acquire their meaning as causal by virtue of being appropriately related to relativistic spacetimes. As this will be the topic of the next chapter, we assume the conditions to have physical content.

Even so, they will not be sufficient. The worry is familiar from relativity, and surely from the causal theory of (space)time, to which pedants (ourselves included) have always interjected that the 'causal structure' of spacetimes merely captures a minimally necessary, but not sufficient, connection between events for them to be causally related as cause and effect. The objection is motivated by the observation that we do not attribute causal efficacy to all timelike or null relations; given an event, we take neither all events in its past lightcone to be its 'causes', nor all events in its future lightcone to be its 'effects'. But this interjection is neither very insightful, nor is it at all damaging to the causal set program. Those moved by it are hereby invited to mentally replace our talk of 'causation' with more precise, but also more awkward, substitutions in terms of (possibly directed) 'causal connectibility'.

Second, one may be worried that the conditions in axiom 1 are not necessary for a relation to qualify as *causal* generally speaking (as opposed to merely causally connectible). Note that in the thicker sense of 'causation', all of the conditions—reflexitivity, antisymmetry, and transitivity—may be deemed unnecessary. Questioning the demand for reflexivity is rather straightforward, as on many accounts an event itself is not considered either among its own causes or effects. However, the demand for reflexivity is the least central to causal set theory. In fact, a new theory can be formulated that is 'dual' to causal set theory except that the demand for reflexivity is replaced by one for irreflexivity. One sometimes finds the resulting irreflexive version of causal set theory, based on a strict partial order induced by an irreflexive relation $\prec$, in the literature.[28] Since the reflexive and

---

[28] A binary relation R imposes a *strict partial order* on its domain X just in case it is irreflexive (i.e., $\forall x \in X, \neg Rxx$) and transitive. A relation which induces a strict partial order is always antisymmetric, i.e., irreflexivity and transitivity imply antisymmetry. In fact, every such relation is asymmetric, i.e., irreflexivity and transitivity imply asymmetry. A binary relation R is *asymmetric* just in case $\forall x, y \in X, Rxy \rightarrow \neg Ryx$.

54 OUT OF NOWHERE

irreflexive versions of the theory are arguably empirically equivalent, and indeed dual, the choice between a reflexive and an irreflexive relation may be seen as conventional.[29] Thus, either a reflexive or irreflexive understanding of causation can arguably be reconciled with causal set theory. What could not be accommodated, however, would be a non-reflexive notion of causation.[30] Requiring causation to be antisymmetric may also be unnecessary, as an analysis of causation may want to leave room for cyclical causation as it can for instance be found in feedback loops. Finally, recent work on causation (cf. Paul and Hall, 2013, ch. 5) has found the requirement of transitivity unnecessary.

Taking a step back from the details of these objections, the causal relations of CST of course differ from those attributed to events in our ordinary lives. Nevertheless, given the tight relationship between the fundamental relation in causets and the causal structure of the spacetimes they give rise to, we take it to be legitimate to dub this relation 'causal'.

3.3 The problem of space

Having given and discussed the basic—kinematic—framework of CST, an important question arises: what is the relationship between causets and relativistic spacetimes? It cannot be one of identity. Mathematically, these structures are distinct; so distinct in fact that any physical interpretation of them which disregarded these differences would be metaphysically negligent. But even if strict identity is not an option, the claim that the fundamental relation is causal could be rejected on the grounds that it still remains essentially spatiotemporal. This objection is commonly raised to causal theories of (space)time, and is repeated in Sklar (1983).[31] The objection is intended to lead to the conclusion that the reduction of (space)time structure to causal structure has thus failed; at best, what one gets is the elimination of some, but not all, spatiotemporal relations from the fundamental ideology. However, to interpret the causal relation in causal set theory as spatiotemporal would be disingenuous. The guiding idea of CST is to offer a theory in which relativistic causation is fundamental and spacetime emerges. In that sense, all spatiotemporal structure is grounded in causal structure; spacetime

---

[29] Note that there is a bijection between 'reflexive' and 'irreflexive' causets: if $\leq$ is a non-strict partial ordering relation, then the corresponding strict partial ordering $\prec$ is the *reflexive reduction* given by $x \prec y$ just in case $x \leq y$ and $x \neq y$; conversely, if $\prec$ is a strict partial ordering relation, then the corresponding non-strict partial ordering relation is the *reflexive closure* given by $x \leq y$ just in case either $x \prec y$ or $x = y$, for all events x, y in the domain. The fundamental relations of the two dual theories, $\prec$ and $\leq$, thus merely differ by the identity.

[30] A relation is *non-reflexive* just in case it is neither reflexive nor irreflexive.

[31] Sklar claims that on some versions of the causal theory of (space)time at least, the fundamental causal relations such as 'genidentity' are based on what are at heart still spatiotemporal notions. CST does not, however, depend on the relations discussed by Sklar.

ontologically depends on causal structure, but not vice versa. So let's take this idea seriously and see how far we can run with it.

In order to appreciate that the objection that the causal relation is just spatiotemporal has no purchase in causal set theory—unlike in GR—let us consider just how different causal sets and spacetimes are. In the next chapter, we will see that causal sets are infested with a more virulent form of non-locality than we find in relativistic spacetimes. Furthermore, causal sets are discrete structures. This means that quite a bit of the geometric structure that we routinely attribute to space and time—and that is certainly available in GR—is simply missing in a causet. It is also important to note that the fundamental relation is *causal*, not *temporal*; temporal relations are supposed to emerge. Strictly speaking, thus, there is no time in CST— at least not at the kinematic level. Whether the dynamics re-introduces a form of time will be discussed in §4.3.

Having said that, however, despite the intended difference, the fundamental causal precedence relation $\preceq$ shares some properties with a generic relation of *temporal* precedence.[32] Just as with temporal precedence, we can define it as a reflexive or an irreflexive relation. If we further prohibit temporal loops and assume the transitivity of temporal precedence, both relations are at least antisymmetric (on the reflexive convention) or even asymmetric (on the irreflexive convention). Either way, any relation of temporal precedence consistent with relativity, just as causal precedence, orders its domain only partially, not totally. The causal precedence relation at work in causal set theory does not pick a 'now' or give rise to a 'flow'. But neither does temporal precedence on a B-theoretic metaphysics of time. Putting these elements together, then, the causal precedence of causal set theory though distinct has structural similarity with a B-theoretic version of a discrete form of (special-)relativistic time without metric relations such as durations.[33]

It should be noted that the remainder of the chapter concerns predominantly the presence (or absence, as the case may be) of spatial structures in causets in an attempt to grasp the connection between these structures to physical space. We will only be concerned with spacetime insofar as it may offer a path to space.

---

[32] The intended relation is more generic than the relation of 'chronological precedence' we find in the relativity literature, e.g., Kronheimer and Penrose (1967); in fact, the latter relation is arguably essentially *causal*, rather than the other way around.

[33] Dowker (2020) thinks that already the causal structure of relativistic spacetimes used as a vantage point for causal set theory really is a structure "of *precedence*, of *before and after*, not of *causation*" (147n) and that causal set theory should thus have more appropriately been called 'temporal set theory'. Unilluminating semantic debates aside, there is a clear sense in which the causal structure in GR is fundamentally and primarily *causal*, and only derivatively temporal, as it encodes relations of causal connectibility of events by light signals. This means, to repeat, that the fundamental relation of causal set theory is best interpreted as 'causal connectibility' or 'causal precedence'.

56 OUT OF NOWHERE

3.3.1 The 'essence' of space

This leaves us somewhat inconclusive as regards time in *kinematic* causal set theory. It should be noted that we will encounter a much deeper problem, the so-called 'problem of time' in canonical GR and quantum theories of gravity based on the canonical approach in the chapters on loop quantum gravity. In the remainder of this section, we will argue that there is a similar 'problem of space' in causal set theory, i.e., that 'space' is absent from fundamental causal set theory.[34] The next chapter will return to time in the context of causal set theory and offer a fuller analysis, which will include its dynamics.

In order to ascertain the absence of space, we need to state what kind of thing 'space' is. Of course it is notoriously problematic to proclaim what the nature of 'space'—or anything else for that matter—is, but we are going to do it anyway. Lest the readers mistake this for our being hubristic, we hasten to reassure them that we do not take this list to express any kind of deep and final truth about the essence of space. We merely offer it as a useful starting point to determine the absence of space in fundamental causal set theory. No particular item on the list will be necessary for this purpose, they may all be individually expendable; similarly, further items can be added to the list without threatening the purposes of this exercise—if anything, this would strengthen our case. The only thing that would undermine our argument for the absence of 'space' would be if 'space' had no nature at all—because that's precisely what we will find: 'space' in causal set theory has no structure at all and hence no natural properties whatsoever.

In his famed *La Science et l'Hypothèse*, Henri Poincaré (1905) confidently asserts at least some essential properties of physical space:

> In the first place, what are the properties of space properly so called? ... The following are some of the more essential:—
>
> 1st, it is continuous; 2nd, it is infinite; 3rd, it is of three dimensions; 4th, it is homogeneous—that is to say, all its points are identical one with another; 5th, it is isotropic. (52)[35]

An attentive metaphysician will have a few quibbles with this passage. First, essentiality is not generally regarded as admitting of degrees. Second, Poincaré seems to employ either an awfully strong condition of homogeneity (and, in the French original, of isotropy) when he paraphrases it as amounting to the identity of the points (or of all straight lines through a point), or else a notion of identity weaker

---

[34] We would like to remind the reader of our remark in footnote 1.

[35] Interestingly, the English translation omits the clause from the French original where the reader is informed that 'isotropic' is to say that all straight lines through one point are identical with one another.

than numerical identity. The former seems indefensible, as it would imply, among other things, that space cannot be extended—assuming that points themselves are not extended. However, points can be extended and cover entire 'spaces' (Ehrlich 2022). Assuming standard geometry, homogeneity should be taken to require that all points of space share their essential physical properties but remain numerically distinct.

Let us take Poincaré's lead though not his list and compile a list of essential properties of physical space, which seems better equipped to serve our purposes.[36] Space as we know and love it seems to have

- the structure of a differentiable manifold with a specified topology and dimension,
- affine structure,
- and metric structure.

'Space' as it is represented in various approaches to quantum gravity fails to have several or all of these features. In this sense, we have a 'problem of space'! This problem is particularly pronounced in causal set theory. But first two comments on the list. First, we do not mean to claim that 'manifest' space must have these properties, but only that GR assumes that space(time) does. So arguably, a precisification of 'manifest' space would yield a theory that ascribes these properties, or properties very much like these, to space. Second, although we will be concerned with spacetime as it figures in GR, the above list makes for perfectly natural assumptions about the nature of space in a much larger class of spacetime theories.

3.3.2 Space in a causal set

So how could one identify 'space' in a causet? What is the most natural conception of space 'at a time' in causal set theory? Given some vantage point, i.e., some particular basal event, how can one determine the set of events that are 'simultaneous' to it and jointly form 'physical space' at an instant? There are obvious worries here concerning the relativity of simultaneity, but let's leave those to the side for the moment. In GR, at least in its globally hyperbolic sector, we can foliate the 4-dimensional spacetime into totally ordered 3-dimensional spacelike hypersurfaces that the metaphysician might consider identifying with 'space at some time'. Of course, the choice of foliation of a relativistic spacetime is highly nonunique and in general not justifiable on physical grounds. Our best shot at a similar construction in causal set theory would be to partition the causal sets into totally

---

[36] See Hilbert and Huggett (2006) for a presentation of Poincaré's different purposes.

58   OUT OF NOWHERE

ordered maximal sets of pairwise 'spacelike' related events. Just as the foliation of globally hyperbolic spacetimes was non-unique, such partitions will in general not be unique in causal set theory. Furthermore, no two events in any such set can stand in the fundamental causal relation—if they did, one would causally precede the other and hence they could not be part of 'space' at the same 'time'. This means that the resulting subsets of events of the causet would, by necessity, be completely structureless—no two points in any subset could be related by the fundamental relation. Technically, this means that these subsets are 'antichains':

Definition 8 (Chain). *A chain γ in a causet $\langle C, \preceq \rangle$ is a sequence of events in C that are pairwise comparable, i.e., for any two events x, y in γ, either $x \preceq y$ or $y \preceq x$. This implies that a chain is a subset of C that is totally ordered by $\preceq$ (hence 'sequence' rather than just 'set').*

Definition 9 (Antichain). *An antichain α in a causet $\langle C, \preceq \rangle$ is a subset of events in C that are pairwise incomparable, i.e., for any two (numerically distinct) events x, y in α, $\neg(x \preceq y)$ and $\neg(y \preceq x)$. This implies that an antichain is a subset of C that remains completely unstructured by $\preceq$.*

Given that they are altogether unstructured, an antichain is completely characterized by its cardinality, at least intrinsically. Extrinsically, it is characterized by how its elements are embedded in the total structure of the causet. The extrinsic characterization of antichains matters for our purposes, since it is important to *partition*, i.e., to divide into non-empty and non-overlapping subsets without remainder the entire causet in order to obtain what can reasonably be considered a 'foliation'. Furthermore, the elements of the partition—the antichains—should be *inextendible*, i.e., they should be such that any basal elements of the causet not in the antichain is related to an element of the antichain by $\preceq$. The problem, to repeat, is that antichains have no structure, and hence the fundamental causets have no (intrinsic) *spatial* structure at all, no metrical structure, no affine and differentiable structure—there is no manifold—and, if any at all, very different dimensionality and topology than the continuous spacetime they are supposed to give rise to. Furthermore, there is no evident spatial ordering, such as we ordinarily find in terms of spatial proximity.

Before we have a closer look at some of these claims, let us state the important fact that at least for any finite partially ordered set, such a foliation-like partition into antichains is always possible.[37] We need some definitions before we can state the relevant theorem:

---

[37] A partition of any causet into antichains is always trivially possible: just take the antichains each consisting of a single event. That *such* a partition would not satisfy our present needs should be clear, though. The point of the following is to establish that for a large class of causets, we are guaranteed that partitions conducive to these needs exist.

Definition 10 (Height and width). *The height of a partially ordered set P is the cardinality of the largest chain in P. The width of P is the cardinality of the largest antichain in P.*

This allows us to formulate the theorem (Brightwell 1997, 55):

Theorem 3 *Let $P = \langle P, \leq \rangle$ be a finite partially ordered set with height h and width w. Then*

> *(a) the domain P can be partitioned into h antichains;*
> *(b) the domain P can be partitioned into w chains.*

As a corollary, this gives us an upper bound for the cardinality of P: $|P| \leq hw$. The part (b) is a major result, requiring a rather sophisticated proof, and is known as 'Dilworth's Theorem' (Brightwell 1997, 55). For our purposes, however, the first part (a) suffices to establish that for a finite causet, a partition into a sequence of putative 'nows' is always possible. That the result can straightforwardly be generalized to past-finite causets—a class of causets that will take center stage in the next chapter—is clear from the proof of (a) (Brightwell 1997, 56).

This proof proceeds by constructing antichains as follows. Define the height $h(x)$ of an event x in P as the cardinality of the longest chain in P with top element x, i.e., the maximal element in the chain. Then collect all the events of the same height into a set, i.e., a set of events of height 1, a set of events with height 2, etc. Every event has a height and will thus be an element of one of these sets. These sets turn out to be antichains, which completes the proof. Now the extension of theorem 3 to past-finite causets should be obvious: although the height of the causet will be infinite—and hence the partition will consist of infinitely many sets or 'layers'—it is still the case that every event has a height and that the resulting 'height sets' will be antichains. The same technique cannot be applied in the case of causets which are neither past- nor future-finite,[38] though at least some of these infinite causets admit a foliation. Let us remind the reader that these results only guarantee the existence, but not of course the *unique* existence of such a foliation; in fact, the foliations of a causet will in general be non-unique.

There is a *problem of space* in causal set theory: there is a clear sense in which causets have no spatial structure at all. Of course, this may be unsurprising given that in CST, the *only* fundamental structure is causal, which by concept requires 'temporal' thickness beyond purely 'spatial' structure at a 'moment of time. After all, spacelike hypersurfaces in relativistic spacetimes also have no causal structure in themselves. However, relativistic spacetimes have a richer structure than just their causal structure; in particular, a spacelike hypersurface, and so

---

[38] An obvious variant of the same technique will work for future-finite causets.

60   OUT OF NOWHERE

'space', has a rich metrical and topological structure we do not find in analogs of space in causets.

Given that the natural correlates of 'space' (at a time) in CST are altogether unstructured, inextendible antichains, can we nevertheless attribute some spatial structure to these antichains? Can we reconstruct spatial structure from a fuller structure? Could it be that perhaps a causal set induces some such structure on these antichains in a principled way? If so, then we would have completed an important step toward solving the problem of space in causal set theory. As it will turn out, this question cannot be addressed separately from the overarching problem of the emergence of spacetime from causets. So asking more generally, and in anticipation of the next chapter: is there a theoretically sound way of extracting geometrical information from the fundamental causets that could be used to relate them to the smooth spacetimes of GR?

Before we get to the full geometry and to how 'manifoldlike' causets are, let us consider the more basic notions of dimension and topology, with a particular eye toward trying to identify any 'spatial' structure in causets. This marks the start of the spacetime functionalist project of recovering aspects of spacetime from the fundamental causets.

3.3.3 The dimensionality of a causal set

The obvious proposal when ascribing a dimension to a causet is to import the general notion of the dimension of a partial order, though we shall soon see that this approach is unsatisfactory. The general idea is to identify some property of the causet itself that is indicative of the dimensionality of the 'smallest' manifold into which it can be comfortably embedded (if it can be so embedded). Following Brightwell (1997), a standard definition of the dimension of a partially ordered set starts out from what is called the 'coordinate order'. In order to illustrate this concept, consider the order structure imposed on the plane $\mathbb{R}^2$ by introducing an ordering relation $\leq$ on the elements of $\mathbb{R}^2$ such that $\langle x, y \rangle \leq \langle u, v \rangle$ just in case both $x \leq u$ and $y \leq v$, where we introduce Cartesian coordinates in the usual way such that a point $p \in \mathbb{R}^2$ is identified with an ordered pair of 'coordinates', and where '$\leq$' is interpreted as the usual 'less or equal' relation. $\leq$ induces a partial order on $\mathbb{R}^2$, as there are pairs of points in $\mathbb{R}^2$ such as $\langle 0, 1 \rangle$ and $\langle 1, 0 \rangle$ that are incomparable. The ordering can be straightforwardly generalized to yield what Brightwell (53) terms the *coordinate order* on $\mathbb{R}^n$ for any positive integer n. One can then define the dimension as follows (55):

Definition 11 (Dimension of partial order).   *The dimension $dim(P)$ of a partially ordered set $P = \langle P, \leq \rangle$ is the minimum integer n such that P can be embedded in $\mathbb{R}^n$ with the coordinate order.*

A common alternative, but equivalent, definition renders—of course—the same verdict. It defines the dimension of $\mathcal{P} = \langle P, \preceq \rangle$ as the smallest number of total ordering relations on P whose intersection is $\preceq$, i.e., $\forall x, y \in P, x \preceq y$ if and only if x is below y in each of the total orders. This definition goes back to Dushnik and Miller (1941) and is called 'combinatorial dimension' in Meyer (1988, 26), and 'order dimension' elsewhere.

What is the minimum condition on a mapping from P to $\mathbb{R}^n$ that it qualifies as an 'embedding'? In the general case at hand, the only substantive rule is that the embedding is 'order-preserving', i.e., it is a mapping f from P to $\mathbb{R}^n$ that preserves the ordering in that $\forall p, q \in P, p \preceq q$ if and only if $f(p) \le f(q)$, where '$\le$' is the coordinate order introduced above.

Applied to the case at hand, since any two elements are unrelated in an antichain, its dimension must be equal to its cardinality: an antichain of two elements requires an embedding into $\mathbb{R}^2$ such that both elements can be mapped in a such a way that they are not related by the coordinate order, an antichain of three elements needs an embedding into $\mathbb{R}^3$, etc.

What about the dimension for partial orders more generally? Dushnik and Miller (1941) not only show that the dimension is well defined for every partially ordered set, but also prove that the dimension of a partially ordered set of size n is finite if n is finite and no greater than n if n is transfinite (theorem 2.33). Hiraguchi (1951, 81) strengthens this to the statement that the dimension of a partially ordered set is no greater than its size n, for n finite or transfinite. He also shows (theorem 5.1) the stronger claim that the dimension of a partially ordered set increases by at most 1 with the addition of a single element to the partially ordered set.

While definition 11 captures the intuition behind what it is to be a 'dimension', and these results establish that 'dimension' is well defined and has an upper bound, the notions suffer from three major limitations that make it unusable for present purposes. First, even though the dimension is well defined, it is generally hard to compute it: for $n \ge 3$, the computation of whether a given partially ordered set has dimension n is what in computational complexity theory is called an 'NP-complete problem' (cf. Felsner et al. 2017). Second, it should be emphasized just how weak the established bounds are: for a universe like ours, i.e., the size of something like 10^{245} in Planck units, and hence with as many causet elements, we would like to strengthen the assertion that the fundamental causet has a dimension of at most 10^{245} by about 244 to 245 orders!

Third, and most importantly, definition 11 needs to be modified to suit the present context: here we are not concerned with embedding partially ordered sets into $\mathbb{R}^n$, but rather with embedding causets into *relativistic spacetimes*. Thus, while the first two limitations may be of a rather technical nature, this third point cuts directly to the heart of what we are interested in in this book: the emergence of spacetime from structures in quantum gravity. Therefore, the connection between

62   OUT OF NOWHERE

spacetimes and causets must be imbued with physical salience. Without it, there is no hope to advance the present project. For the purposes at hand, it is much more sensible to define the dimension of a causet in relation to relativistic spacetimes. We start by defining the relevant kind of embedding:

Definition 12 (Embedding).   *An embedding of a causet $\langle C, \preceq \rangle$ into a relativistic spacetime $\langle \mathcal{M}, g_{ab} \rangle$ is a injective map $f : C \to \mathcal{M}$ that preserves the causal structure, i.e., $\forall x, y \in C, x \preceq y \Leftrightarrow f(x) \in J^-(f(y))$.*

A causet that is embeddable, i.e., affords an embedding into a relativistic spacetime, could in this sense be regarded as a discrete approximation to this spacetime—their causal structures are consistent. Of course, systematically speaking, the fundamentality of the causet demands that this relationship runs in the other direction: relativistic spacetimes, at least to the extent to which they present physically reasonable models of the large-scale structure of the universe are low-energy approximations to an underlying fundamental causet. Given Malament's theorem 2 it is evident, furthermore, that an otherwise unrestricted embedding of a causet into a spacetime cannot fully recover all salient features of the spacetime; at most, it approximately determines the conformal structure of the spacetime. Additional assumptions will be necessary to fix the conformal factor, although the cardinality naturally suggests that the number of elements be regarded as a measure of the size. We will return to this problem in the next chapter; for the analysis of dimensionality it suffices to consider any embeddable causets.

Although quantum gravity theorists are ultimately interested in the regime of strong gravitational fields, the embeddability of causets into (subspaces of) Minkowski spacetime provides an important test case. It is natural, then, to follow the seminal Meyer (1988) and to study the dimension of causets in terms of their embeddability into Minkowski spacetime. Meyer defines the 'Minkowski dimension' as follows:

Definition 13 (Minkowski dimension).   *"The Minkowski dimension of a causal set is the dimension of the lowest dimensional Minkowski space into which it can be embedded (not necessarily faithfully)." (Meyer 1988, 16f)*[39]

Meyer (1988, 1993) proves that the Minkowski dimension of a partial order is identical to its dimension as defined in definition 11 in dimension two, but not in higher dimensions. It is natural to ask whether we can similarly find least upper bounds for dimensions higher than two. It turns out that it is not the case that a partially ordered set of size n has a Minkowski dimension of at most n, as there are some finite ones that cannot be embedded in any finite-dimensional Minkowski

---

[39] The condition of 'faithfulness' will be discussed in the next chapter.

SPACETIME FROM CAUSALITY: CAUSAL SET THEORY   63

spacetime (cf. Felsner et al. 1999). In the absence of analytical results with much traction, different methods to estimate the Minkowksi dimension of causets have been developed, with some encouraging numerical results particularly for high-density 'sprinklings'.[40]

To return to the problem of space, it should be noted that all the work on the dimension of partially ordered sets assumes a non-trivial relational structure on the set and, in the case of causets, considers entire causets, and not just some unstructured subsets. It is not obvious from such considerations what the Minkowski dimension of the antichains that are supposed to represent 'space' at a 'time' is, given that these were supposed to be interpreted purely spatially and Minkowski spacetime is defined by its causal structure, and thus is spatio*temporal*. Although the outcome will arguably not miss its mark by as much as the standard dimension for partially ordered sets, these difficulties illustrate just how unlike space an antichain is. If space is to be found in causal sets, it will require more structure.

3.3.4 The spatial topology of causal sets

This brings us to the next to final task for this chapter, continuing the project of spacetime functionalism: is it possible to induce a topology on the 'spatial' antichains which will endow them with a structure at least starting to look more 'spacelike'? Given that there is no structure in an antichain, we cannot hope to extract topological structure from it. It turns out that there is a way to *induce* a topology on the 'spatial' antichains of a causet, but it requires much or all of the structure of the entire causet.

A *topology* on a set X is a set $\mathcal{T}$ of subsets of X—the 'open sets'—which satisfies the following conditions:

1. $\emptyset \in \mathcal{T}$ and $X \in \mathcal{T}$;
2. the union of a collection of sets in $\mathcal{T}$ is in $\mathcal{T}$;
3. the intersection of any two (and hence any finite number of) sets in $\mathcal{T}$ is in $\mathcal{T}$.

It turns out that for a finite X, there exists a one-to-one correspondence between topologies on X and pre-orders on X. Apart from the fact that a spatial antichain may not be finite, this theorem does not help us to obtain a natural topology on the spatial antichain simply because there is no fundamental physical relation at all that obtains between the elements of the antichain, and a fortiori no pre-order. Case closed?

---

[40] Most prominent among them are the 'Myrheim-Meyer dimension' (Meyer 1988) and the 'midpoint-scaling dimension' (Bombelli 1987).

64   OUT OF NOWHERE

Let's not jump to conclusions; of course, one can easily *impose* topologies on an altogether unstructured set. For any such set X, take $\mathcal{T} = \{X, \emptyset\}$—the so-called *indiscrete topology* on X. But such a coarse topology will not further our goal in identifying some useful geometrical structure on a spatial antichain. At the other end of the spectrum, we find the *discrete topology* $\mathcal{T} = \mathcal{P}(X)$, where $\mathcal{P}(X)$ is the powerset of X, i.e., the set of all subsets of X, and hence the 'finest' possible topology on X. But two cautionary remarks expose the very limited appeal of the discrete topology for our purposes.

First, and as a first indication of the need to go beyond a 'spatial' antichain, imposing the indiscrete topology seems to presuppose a distinction between the set and the empty set. In other words, a presupposition of the indiscrete topology is that X is non-empty. That seems harmless enough. But by the same token, the discrete topology seems to assume not only that there are elements in X, but that there are *numerically distinct* elements. And that is decidedly less harmless. If basal events in causal set theory are indeed featureless and $\preceq$ is the only physical relation at the fundamental level, then any two events with the same relational profile, i.e., they constitute what is called a 'non-Hegelian pair', cannot be physically distinguished (cf. Wüthrich 2012; Wüthrich and Callender 2017). In an antichain, of course, elements cannot be so distinguished. This raises two challenges. The first is to metaphysically underwrite a difference between a singleton set and a set containing a plurality of events. If we accept primitive plurality even in cases the elements cannot be physically distinguished—as we may have to do for independent reasons (Wüthrich 2009)—then that problem can be circumvented (and the presupposition in the indiscrete case secured).

But there remains a second, related challenge. Merely to be able to assert that the antichain consists of n elements does not suffice to distinguish the various subsets of the antichain; even just for $n = 2$, there would have to be two distinct singleton sets in the discrete topology. But with nothing to distinguish the elements they contain, nothing can mark the distinctness of the sets. Of course we can assert that there must be *two* such singleton sets, but for a topology to be able to structure the set in a meaningful way, it seems necessary to be able to tell one of them apart from the other. This could be accomplished by insisting that the basal events of causal set theory command some haecceitistic identity; however, haecceities offend against the natural structuralist interpretation of causal set theory (Wüthrich 2012). Thus, some other means of ensuring the distinctness of elements will be required—arguably some form of primitive plurality.

It should be noted that if the basal elements in the antichain X enjoy numerical distinctness, then we can in fact define a distance function d such that $\langle X, d \rangle$ is a metric space. For instance, for any $x, y \in X$, define

$$d(x, y) = \begin{cases} 0, & \text{if } x = y, \\ 1, & \text{if } x \neq y. \end{cases} \tag{3.2}$$

This function straightforwardly satisfies the standard conditions demanded of a distance function (i.e., non-negativity, non-degeneracy, symmetry, and the triangle inequality), and $\langle X, d \rangle$ thus qualifies as a metric space (called a *discrete metric space*). Clearly, however, this d will not give rise to any 'space' that resembles physical space as found, e.g., in GR. And it remains opaque how a structure with such a distance function could be meaningfully embedded in a spacetime of low dimensionality. Furthermore, imposing topologies and metrics on a 'spatial' antichain seems to require a metaphysics of basal events that is anathema to the spirit of causal set theory.

Second, then, even granting the possibility of imposing a topology and indeed a metric of our 'spatial' antichains, no physically useful structure can be extracted from them. A topology on a 'spatial' slice should underwrite nearness relations. In order to successfully do that, it shouldn't be too coarse—as was the indiscrete topology—for then it would seem as if no events are 'nearby' one another. However, it also shouldn't be too fine—as was the discrete topology—for otherwise *all* events are near all other events. Either way, there are no physically useful nearness relations that appropriately discriminate some events to be nearer than others. So we will want a 'goldilocks' topology that finds the sweet spot in fineness. Furthermore, the topology of 'spatial' antichains should cohere with, and in fact give rise to, the nearness relations as we find them in spatial slices of the emerging relativistic spacetime. This means that it should be appropriately grounded in the fundamental physics in place. Thus, our best bet to impose some structure in general, and topological structure in particular, on 'spatial' slices of causal sets is to start out from the one physical relation present in a causal set—$\preceq$.

The imposition of a physically perspicuous topology based on $\preceq$ can be accomplished by considering how an inextendible antichain is embedded in the causet. Major, Rideout, and Surya (2006, 2007) offer just such a way of doing that. Their construction proceeds by 'thickening' the antichain and exploiting the causal structure gained by this 'thickening'. The idea is, roughly, as follows. For any subset X of the domain C of a causal set, define $F(X) := \{x \in C \mid \exists y \in X, y \preceq x\}$—the causal future of X—and $P(X) := \{x \in C \mid \exists y \in X, x \preceq y\}$—the causal past of X. Note that given the reflexive convention chosen for $\preceq$, $x \in F(x)$ and $x \in P(x)$ for any x in C. Next, define the *n-future thickening* of an (inextendible) antichain $A \subset C$ as

$$A_n^+ := \{x \in C \mid x \in F(A) \text{ and } |P(x) \setminus P(A)| \leq n\}, \tag{3.3}$$

where $\setminus$ means ordinary set-theoretic difference, $|X|$ denotes the cardinality of set X, and, mutatis mutandis, the *n-past thickening* A_n^- of A. Note that—again as a consequence of the reflexive convention—$A_0^+ = A_0^- = A$ for any antichain A. A_1^+ and A_1^- will contain the antichain together with all the immediate successors and predecessors of events in the antichain, respectively, etc. See figure 3.3 for an example of A_1^+. Finally, define an *n-thickening* A_n of A as $A_n^+ \cup A_n^-$.

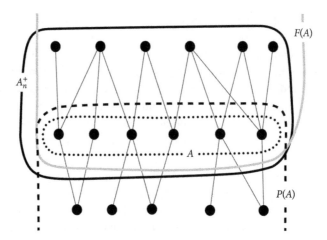

Figure 3.3 A 1-future thickening A_1^+ of an antichain A.

Next, identify the maximal or 'future-most' elements m_i of A_n^+ and form the sets $P_i := \{x | x \in (P(m_i) \cap F(A))\}$ of 'past lightcones' of the maximal elements truncated at the 'spatial' antichain A. Then $\mathcal{P} := \{P_i\}$ is a *covering* of A_n^+, i.e., $\cup_i P_i = A_n^+$. From these, we can construct the *shadow sets* $A_i := P_i \cap A$. See figure 3.4 for an example of a shadow set.

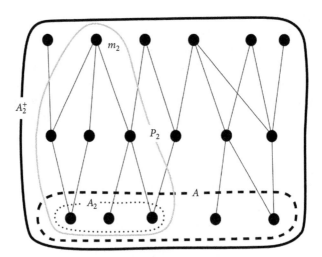

Figure 3.4 A maximal element m_2 of a 2-future thickening A_2^+ of an antichain A and the resulting shadow set A_2.

We find that $\mathcal{A} := \{A_i\}$ provides a covering of A and can be used as a vantage point to specify a topology on A. $\mathcal{A}$ itself will not, in general, be a topology because it may not satisfy any of the three conditions on topologies. Major et al. (2007)

SPACETIME FROM CAUSALITY: CAUSAL SET THEORY   67

use the shadow states to construct topological spaces called 'nerve simplicial complexes' and show that natural continuum analogs of these topological spaces are homotopic to (i.e., can be continuously deformed into) the globally hyperbolic spacetimes in which they are defined. This is a hopeful sign that the topologies constructed in this manner are indeed the physically salient ones and thus satisfy a central requirement of spacetime functionalism, assuming, of course, that the topologies of these globally hyperbolic spacetimes are. Furthermore, their results contribute toward establishing a relevant form of the so-called *hauptvermutung* and hence to the emergence of spacetime from causal sets. We will turn to these topics in the next chapter.

It should be noted that the topologies that will arise from just the maximal elements of an n-future thickening for some one particular n are generally going to be too coarse to be fully satisfactory. However, any such topology can be refined by adding shadow sets arising from thickenings with different n, and from the analogous construction based on past thickenings and 'minimal' elements. The finest topology obtainable on a 'spatial' antichain A will result from thickening in both causal directions and letting n range over all values $0, \ldots, N$, for an N such that $A_N^+ \cup A_N^-$ is more or less all of C. The finest topology for an antichain will thus be obtained from considering basically the full structure of the causet it inhabits. It should thus be noted that the spatial structure—in this case the topology of 'spatial' slices—asymmetrically depends on the fundamental causal structure.[41]

3.3.5 Finding distances

A next step for the spacetime functionalist project would be to recover metric structure from the causet. For metric relations between timelike related events, i.e., something like timelike distances or durations, there is a widely accepted recipe to obtain a straightforward proxy of continuum timelike distances. In analogy to relativistic spacetimes, one defines a geodesic in causal set theory as follows:

Definition 14 (Geodesic). *A geodesic between two elements $x \leq y \in C$ is the longest chain γ from x to y, i.e., the chain of the largest cardinality with past endpoint x and future endpoint y. If a geodesic γ has cardinality $n + 1$, i.e., $|\gamma| = n + 1$, then its* length *is n.*

---

[41] The constructions used to introduce topologies on 'spatial' slices can be used to introduce something akin a covariant sum-over-histories approach to dynamics, as articulated by Major et al. (2006). Informally, the picture resembles a three-layered cake with the middle bulk sandwiched between an 'initial' state of the universe consisting of the minimal elements of some A_n^- and a 'final' state constituted by the maximal elements of an A_m^+. This three-layered cake would then represent a 'spatial' slice that evolves from some 'initial' state to some 'final' state. Interestingly, it is possible to define physically meaningful 'transition amplitudes' from the 'initial' to the 'final' state as measures over the set of completed causets containing the fixed 'top' and 'bottom' layer of the relevant 'cake', but with generally differing 'interpolators' (i.e., middle layers) between them.

In general, a geodesic between two elements is not unique. That geodesics thus defined offer a natural analog of their continuous cousins was first conjectured by Myrheim (1978). Brightwell and Gregory (1991) show that for causal sets that are approximated (see next chapter) by a flat spacetime interval, for sufficiently large geodesics, the length of geodesics in those causets rapidly converges to (a multiple) of the proper time elapsed between the images of the endpoints in the spacetime. Though analytic results are only available for low dimensions and flat spacetimes, numerical studies (Ilie et al. 2006) suggest that the convergence holds also in dimensions up to four and in some curved spacetimes.[42]

Let us return to the topic of this section, 'space'. Rideout and Wallden (2009) offer a penetrating analysis of possible constructions of spatial distance for causets, again relying on the structure of the causet not contained in the antichain. For instance, an obvious, though doomed, first attempt (cf. figure 3.5) starts out from the continuum spacetime case and tries to generalize to the discrete case of causets. It defines the spatial distance between two spacelike separated events x and y in terms of an appropriate timelike distance, since we already know (from the previous two paragraphs) how to determine the timelike distance between two events in a causet. In the continuum case, let w be an event in the common past $J^-(x) \cap J^-(y)$ and z one in the common future $J^+(x) \cap J^+(y)$ such that the timelike distance between w and z is shortest for all such pairs. In other words, w is a maximal element of the common past and z is a minimal element of the common future (see figure 3.5).

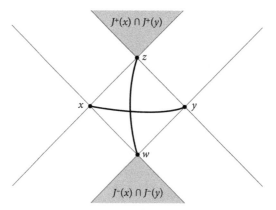

Figure 3.5 A first stab at determining the spatial distance.

Then the spatial distance between x and y is equal to the timelike distance between w and z in the continuum. For 2-dimensional Minkowski spacetime, the

---

[42] Cf. also Rideout and Wallden (2009, §1.3.2).

pair w and z is unique. In higher dimensions, however, this is not the case: in d-dimensional Minkowski spacetime, there is a $(d-2)$-dimensional submanifold of pairs of events that minimize the timelike distance (Rideout and Wallden 2009, 7).

This non-uniqueness thwarts the application of this simple recipe to the discrete case of causal sets that can only be embedded into higher-dimensional Minkowski spacetimes. Details of the embedding and the 'sprinkling' will be discussed in the next chapter, but the problem turns out to be that given the required kind of 'sprinkling' the events in the causet into Minkowski spacetime, there is a finite probability for each of the infinitely many minimizing pairs that the Alexandrov interval they encompass does not contain any of the images of events in the causet and is thus empty of 'sprinkled' events. Consequently, there will always be some pair $\langle w, z \rangle$ such that $w \leq x \leq z$ and $w \leq y \leq z$ are the longest chains between w and z (since $x \not\leq y$), both of which are of length 2. Hence, the timelike distance between w and z, and thus the spatial distance between x and y, will always be 2.

Although this would technically give us a distance function, it can hardly be considered physically adequate. Rideout and Wallden (2009) analyze a number of more involved approaches to extracting non-degenerate spatial distances from the fundamental structure of the causet. Their results are limited, but promising. As far as we can tell, they all resort to considering substantively larger parts of causets than merely unstructured antichains. Once again, this reflects the fundamentality of the causal structure over any spatial or temporal structure.

A different approach is taken in Eichhorn et al. (2019), who study the recovery of spatial distance from causal information in causets at 'mesoscales', i.e., scales small in comparison with the scale of the extrinsic curvature of spacelike hyper-surfaces in corresponding continuum spacetimes, but large in comparison with the discreteness scale of the causet. Numerical tests for simple spacetime regions in two and three spacetime dimensions support the hope that spatial distance functions can be defined which approximate the continuum counterparts in relativistic spacetimes.

3.3.6 Wrapping up

Relating causets to spacetimes via their 'spatial' and 'temporal' parts has thus thoroughly failed, in more radical ways even than in GR. In the next chapter, we will consider how causets and spacetimes might be related *in toto*, as wholes. For now, we can only state just how different causets are from spacetimes, lest we are inclined to see the causal relation at the core of the models of causal set theory as ultimately spatiotemporal. The geometric structure that we would normally attribute to space in particular—and that is certainly available in GR—such as topological, affine, differential, and metric structure is only very indirectly recoverable from the structure of causets, if at all. It is simply not built in at the fundamental level. In this sense, causal set theory offers a view of our world that is not ultimately spatiotemporal in the way that relativistic spacetimes are.

In this chapter, we have sketched the first steps of the spacetime functionalist project. In the next chapter, we will take this project much further by analyzing how relativistic spacetime could emerge from fundamental causal sets and by arguing how spacetime functionalism will helpfully delimit the to-do list to establish this emergence. In order for emergence to succeed, dynamical laws beyond the kinematics studied in this chapter will turn out to be called for. We will complete our discussion of causal set theory with two philosophically fruitful points that arise in connection to the dynamical aspects of the theory: the possibility of relativistic becoming and a form of non-locality that has nothing to do with quantum physics.

Chapter 4
The emergence of spacetime
from causal sets

In the last chapter we argued that according to causal set theory (CST), it is physically (and hence presumably metaphysically) possible that the world is not spatiotemporal. After having introduced CST in the previous chapter and having illustrated how at least space and arguably spacetime disappears in CST, and how some spatial structure may be recovered, we set out in this chapter to show how relativistic spacetime may emerge from a causal set, or 'causet'.

The immediate question before us concerns the relation between causets and spacetimes: how do relativistic spacetimes emerge from the fundamental structures, i.e., causets, in CST? Since CST, like any theory of quantum gravity, is supposed to be a theory of gravity more fundamental than GR, and since relative fundamentality in physics typically parallels scales of energy and size, we are looking to understand the low-energy, large-scale limit of CST. Thus, we are at the same time interested in the relationship between two theories—CST and GR in the present case—as well as between the entities or structures postulated by these two theories—presently, causets and relativistic spacetimes. Whenever there is no danger of conflating the two, we will often not explicitly indicate which is the present topic. As we will mostly be concerned with the relationship between causets and relativistic spacetimes, we will be asking, more specifically, *whether we can precisely state*, i.e., with full mathematical and philosophical rigor, *the necessary (§4.1) and sufficient (§4.2) conditions for generic causets to give rise to physically reasonable relativistic spacetimes*. Whatever the answer to this question, it better not merely deliver necessary and sufficient conditions for mathematical derivations, but supplement them with demonstrations of the 'physical salience' of these derivations. The argument in §4.2 for the sufficiency of the conditions listed in §4.1 relies on a defense of spacetime functionalism. What spacetime functionalism adds over and above mere derivation is precisely an account of the physical salience of the relevant derivations.

An important insight of this chapter will be that ensuring the physical salience of the relation between causets and spacetimes involves addressing foundational and even metaphysical issues. In particular, the philosophy of time will assume an important role in this undertaking. Consequently, this chapter contains sections on the implications of CST for the metaphysics of time (§4.3) and a (novel) form

Out of Nowhere. Nick Huggett and Christian Wüthrich, Oxford University Press.
© Nick Huggett and Christian Wüthrich (2025). DOI: 10.1093/oso/9780198758501.003.0004

72 OUT OF NOWHERE

of non-locality implied by the discreteness of the fundamental structure together with the demand for (at least emergent) Lorentz symmetry (§4.4). We summarize in §4.5.

4.1 Identifying necessary conditions

4.1.1 The basic setup

The restriction to *physically reasonable* spacetimes is supposed to leave room for CST to correct GR regarding which spacetimes are physically possible: GR is a notoriously permissive theory, and we would generally expect the more fundamental theory to rule out some relativistic spacetimes as physically impossible.[1] If this were to happen, the question of what it is for a relativistic spacetime to be physically reasonable would find some serious traction. Before some theory of quantum gravity is established though, answers to this question merely trade in human intuitions and the prejudices of physicists. Since these preconceptions are all trained in our manifestly spatiotemporal world, relying on them in the quest for a fundamental theory of gravity becomes deeply problematic.[2]

In this sense, we will use the qualification of 'physical reasonableness' merely as a reminder that we ought not to expect a quantum theory of gravity to return all sectors of Einstein's field equation (EFE), i.e., all spacetimes which are admissible by GR's lights. Thus, CST may only reproduce some, but not all, sectors of GR. For instance, as we have seen in chapter 3, the antisymmetry assumed in the partial ordering rules out causal loops at the fundamental level, and so may prohibit spacetimes with closed timelike curves.[3] Furthermore, models in GR with high energy or matter density may be eliminated by underlying 'quantum effects', some sectors of GR may violate energy conditions (i.e., additional conditions on the energy-matter content of the world) in ways inconsistent with CST. In this sense, it is the fundamental, 'correcting', theory which fixes what it is for a spacetime to be 'physically reasonable', not our intuitions.

Conversely, we seek to understand how spacetime arises from *generic* causets. On the one hand, it would be unreasonable to expect that all causets give rise to a spacetime, or even just nearly so. There will be causets without anything like an

---

[1] Could CST or another fundamental theory also give rise to *more* spacetimes than are permitted by GR? This will depend on what we take GR to be, exactly. In chapter 3, we defined a spacetime as a pair $\langle M, g_{ab} \rangle$, and to any such pair corresponds a spacetime model $\langle M, g_{ab}, T_{ab} \rangle$ satisfying EFE, if no further conditions are imposed. Thus, if GR is understood as to not include any further restrictions, then there are no non-GR spacetimes.

[2] For an argument that physicists' intuitions can mislead us about what it is for a spacetime to be physically reasonable, e.g., in the case of singular or non-globally hyperbolic spacetimes, see Smeenk and Wüthrich (2011), Manchak et al. (under contract), and Doboszewski (2017).

[3] But cf. Wüthrich (2021).

appropriately spatiotemporal structure. That such 'pathological' cases will also satisfy the demands of the theory on a causet, and so qualify as physically possible, is one of the main reasons to consider the theory, in general, as non-spatiotemporal. On the other hand, we would want 'most' causets, or at least 'many' of them to give rise to spacetimes, for otherwise we might ask ourselves why we were so incredibly lucky to be born in a spatiotemporal world when 'most' of them are not so.[4] Clearly, more will need to be said about this—and more will be said soon—but we hope that the intuition behind demanding genericity is reasonably clear for now.

Thus, we wish to relate generic causets to physically reasonable spacetimes. More specifically, we wish to see how the properties and structure of a part of a causet can give rise to the geometric and topological structure of a region of a relativistic spacetime. The basic idea is captured in figure 4.1. Importantly, as argued in §2.2, it needs to be demonstrated how the relationship between features of causets and properties of spacetimes is not just mathematically definable, but physically salient: figure 4.1 pictures not only a correspondence between abstract objects, but also an identity between physical entities. The method to accomplish this is not by staring at causets and ruminating over which of its parts are physically salient; instead one identifies the functionally relevant properties of relativistic spacetimes (in step SF1 of the spacetime functionalist program outlined in §2.4) and determines how physical salience percolates down to the fundamental structure: one tracks physical salience from the top down, in the direction opposite to that of derivation. But let us start with the more formal aspects of the relationship, so that the rest will then become more intelligible.

Under what circumstances do causets give rise to relativistic spacetimes? A first necessary condition is that a causet is 'sufficiently large' ("at least 10^{130}" elements, according to Bombelli et al. 1987). The reason for this demand is evident: if it is too small, the question of what approximates it at large scales does not make sense. Thus, let us eliminate all causets which are not sufficiently large. Unfortunately, there is no hope that 'sufficiently large' causets will generically give rise to spacetimes, due to the following problem:[5]

Problem 1 *Almost all 'sufficiently large' causets permitted by the kinematic axioms of chapter 3 are so-called 'Kleitman-Rothschild orders', i.e., discrete partial orders of only three 'layers' or 'generations' of elements.*

---

[4] Arguably, we *couldn't* have been born in a world best described by a non-spatiotemporal causet, and so a straightforward anthropic argument should eliminate any astonishment as to why our world is spatiotemporal, given our existence. However, this leaves bewildering the issue of why the world came to be spatiotemporal in the first place.

[5] This problem is what Smolin (2006, 211) calls the "inverse problem" and is often referred to as the "entropy problem" in the literature (see, for example, Brightwell et al. 2009; Dribus 2017, 188; Surya 2019, 50), in reference to the fact from statistical physics that the large-scale behavior of a physical system may be characterized in terms of the multiplicities of microscopic states sharing their macroscopic behavior.

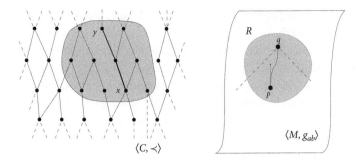

Figure 4.1 The relata: a causal set $\langle C, \prec \rangle$ (left) and a relativistic spacetime $\langle \mathcal{M}, g_{ab} \rangle$ (right). The 'volume' or 'size' of a region or part of a causet (such as the shaded region) is measured by the cardinality of the corresponding subset. The volume of a spacetime region R (shaded) is given by $\int_R d^n x \sqrt{-g}$, where g is the determinant of g_{ab} (and n the dimension). Generally, the relation between causet and spacetime should be such that their causal relations match in the sense that if for two events x, y in the causet we have $x \prec y$, then for their counterparts p, q in the spacetime, we will expect that $p \in \Gamma^-(q)$.

The notion of 'layers' can be made precise using the terminology of chains, antichains, heights and widths as defined in §3.3.2, but the main idea is just that we partition the partially ordered set into maximal antichains totally ordered by maximal chains. In the case of causets, we can think of these layers as being 'space' at a 'time'. 'Kleitman-Rothschild orders' or *KR orders* are partially ordered sets consisting in just three such layers of elements such that no chain consists in more than three elements, such that roughly half of the elements are in the middle layer and a quarter each in the top and bottom layers. Furthermore, each element of the middle layer is related by the order relation, on average, to half the elements of the top layer and to half the elements in the bottom layer.[6] More precisely, Kleitman and Rothschild (1975) prove the following theorem:

Theorem 4 (Kleitman and Rothschild 1975) *Let P_n denote the number of partial orders on a set of n elements. Let Q_n denote a special case in which n-element partially ordered sets are KR orders. Then*

$$P_n = \left(1 + \mathcal{O}\left(\frac{1}{n}\right)\right) Q_n.$$

In other words, in the limit of $n \to \infty$ of n-element partially ordered sets, almost all of them are KR orders. Why is this a problem? Because KR orders, if considered cosmological causets and under the standard assumption that a causet element

---

[6] For details, see Kleitman and Rothschild (1975).

THE EMERGENCE OF SPACETIME FROM CAUSAL SETS   75

gives rise, on average, to something like a Planck-sized volume of spacetime, would represent highly non-locally connected 'universes' which 'last' for a lousy three Planck times, i.e., roughly 10^{-43}s, during which they first double in size and then shrink by the same factor.[7] If causets generically have the form of KR orders, then CST cannot offer a satisfactory answer to how they could give rise to anything like our world. It seems obvious that the vast majority of even sufficiently large causets cannot give rise to anything like spacetime as we know it. But how to stem the KR flood? Clearly, if the theory is conceived of as only comprising the kinematics described in the last chapter, it is too weak to clear out the KR weeds. Thus, what is needed are additional, and sufficiently restrictive laws.

4.1.2 Classical sequential growth dynamics to the rescue

This is precisely the motivation behind introducing a 'dynamics':

Solution 1. *'Dynamical' rules should be appended to the kinematic axioms such that the KR catastrophe does not befall those causets that satisfy these additional axioms.*

Accordingly, we impose 'dynamical' laws in order to avoid the KR catastrophe and to restrict the vast set of kinematically possible causets to the physically reasonable models of the theory. By far the most popular proposal for causet dynamics is a classical, probabilistic law of sequential growth as introduced by Rideout and Sorkin (1999).[8] The central idea of a sequential growth dynamics is that a causet 'grows' by the sequential addition of newly 'born' events one by one to the future of already existing events. Thus, what grows is the number of elements, and it is assumed that the 'birthing' of new elements is a stochastic physical process in the following sense: the dynamics specifies transition probabilities for evolving from one causet in $\Omega(n)$ to another one in $\Omega(n + 1)$, where $\Omega(n)$ is the set of n-element causets. Since the growth always happens to the causal 'future' of 'already' existing events, the resulting causets are all finite toward the past or 'past-finite', in the sense that they possess at least one minimal element. Only causets which could have been grown by a process consistent with the dynamical laws of sequential growth are then considered physically admissible.

What dynamical laws ought to be postulated? They should capture our best guesses as to the conditions necessary to 'produce' spacetime. In particular, they should thus be 'natural'—i.e., physically salient—principles, not just arbitrary

---

[7] See figure 9 in Surya (2019).
[8] For a recent alternative approach, see Carlip et al. (2023).

76 OUT OF NOWHERE

mathematical rules. Thus, these natural conditions encode the physical requirements on an acceptable dynamics. Rideout and Sorkin (1999) impose four such
requirements, of which we give here their own gloss:[9]

Axiom 2 (Internal temporality). *"[E]ach element is born either to the future of, or
unrelated to, all existing elements; that is, no element can arise to the past of an
existing element." (5)*

Axiom 3 (Discrete general covariance). *"[T]he 'external time' in which the causal
set grows ... is not meant to carry any physical information. We interpret this in
the present context as being the condition that the net probability of forming any
particular n-element causet C is independent of the order of birth we attribute to
its elements." (5f)*

Axiom 4 (Bell causality). *"[E]vents occurring in some part of a causet C should be
influenced only by the portion of C lying in their past." (6)*

Axiom 5 (Markov sum rule). *"[T]he sum of the full set of transition probabilities
issuing from a given causet [is] unity." (7)*

Axioms 2 and 3 are intended to jointly underwrite the idea central to classical
sequential growth that although there is a birth order in the way the dynamics is
described, it and the 'time' in which it plays out has no physical significance. One
would expect axiom 4 to be violated in a quantum theory, which should naturally
be non-local in that it admits Bell correlations, i.e., correlations among spacelike-
related events. For a classical theory, it seems unproblematic, and indeed felicitous,
to postulate axiom 4, despite its being earmarked for being dropped in a quantum theory. Finally, axiom 5, required for any Markov process, just innocently
assumes that for any given finite causet there is exactly one way of possibly several in which it will in fact grow. Following Rideout and Sorkin (1999), let us call a
dynamical rule which complies with these four axioms a *classical sequential growth
dynamics.*

The four axioms can be thought to encapsulate the basic conditions any specific
dynamic law of classical sequential growth must obey. Obviously, many dynamical
laws specifying particular transition probabilities are conceivable. Yet a remarkable theorem by Rideout and Sorkin (1999) shows that if the classical dynamics
conforms to the four axioms 2–5, then the dynamics is sharply constrained. In
particular, it must come from a class of dynamics of sequential growth known as
'generalized percolation'. Since only the general properties of this class matter for

---

[9] The technically more precise statements can be found in Rideout and Sorkin (1999, §III).
To give just an example, internal temporality demands, as stated above, just that if $x \preceq y$, then label
$(x) \leq$ label (y).

our purposes, let us introduce the simplest model of classical sequential growth which satisfies the Rideout-Sorkin theorem—namely, 'transitive percolation'—a dynamics familiar in random graph theory.

A simple way to understand this dynamics is to imagine a sequential order of event births, labeled using positive integers $1, 2, 3, \ldots$ consistently with the causal order: i.e., if $x \preceq y$, then label $(x) \leq$ label (y). The reverse implication does not hold, because the dynamics at some label time may birth a spacelike-related event, not one for which $x \preceq y$. This is essentially the requirement of 'internal temporality' introduced above. We begin with the 'moment of creation', the singleton set. When the second event is birthed, there are two possibilities: either the second event (labeled '2') causally succeeds the first event (labeled '1'), or it does not, i.e., $1 \preceq 2$ or $\neg(1 \preceq 2)$. Transitive percolation assigns some probability p to the two events being causally linked and $1 - p$ to the two events not being causally linked. The same holds for the third event, which has probability p of being causally linked to 1 and $1 - p$ of not being causally linked to 1, and probability p that it is linked to 2 and $1 - p$ that it is not linked to 2.

In general, an alternative way to conceive of the dynamics is that when a new causet with $n + 1$ events comes into being, it chooses a previously existing causet of n events to be its ancestor with a certain probability. Thus, one can think of transitive percolation as involving two steps at each evolutionary stage from an n-element causet C_n to an $(n + 1)$-element causet C_{n+1}. First, select the subset of elements in C_n which will have a direct causal link to the newborn element. Second, add all ancestors implied by transitivity. In this way, the dynamics enforces transitive closure, so if, e.g., $1 \preceq 2$ and $2 \preceq 3$, then $1 \preceq 3$.

This second step means that the possible subsets of elements which can be ancestors is a subset of the power set of all elements: not all combinations of ancestors are generally permissible. Given an ancestor set, how is the probability determined? For each element in C_n, the probability of it being part of a given set of ancestors is p. This determines the probability of the transition from a given n-element causet to a given $(n + 1)-$ element causet. For instance, the transition from a 2-chain ($1 \preceq 2$) to a 3-chain ($1 \preceq 2 \preceq 3$) is the sum of the probability of picking both elements 1 and 2 of the 2-chain as ancestors and the probability of picking only the causally later element 2 as ancestor, since in this case, the second step (to ensure transitive closure), adds the first element 1 again. Thus, the probability of this transition is $p^2 + p(1 - p)$, which adds up to p.

A classical sequential growth dynamics can generally be thought of as a 'tree' of permissible transitions where each transition is assigned a probability consistent with the axioms above. The resulting tree is itself a partial order, where the ordering relation is not a causal relation intrinsic to a world, but instead a relation of permissible sequential growth. Figure 4.2 shows the first three stages with the transition probabilities exemplified by those of transitive percolation.

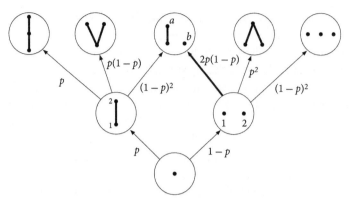

Figure 4.2 The first three stages of transitive percolation dynamics.

In the figure, the thicker arrow indicates that there are two 'ways' in which the transition from the 2-antichain to the 3-element causet with a chain of two elements and an isolated third element can occur: the third element can either be to the causal future of 1, or to the causal future of 2. Axiom 5 requires that the transition probability must count both these possibilities; hence the factor of 2 in the transition probability. In turn, this naturally suggests that the dynamics presupposes a non-structuralist metaphysics of elementary events according to which these events have a primitive identity—a 'haecceity', i.e., an identity which is independent of, and prior to, the causal relations they entertain to other events.[10]

That a careful consideration of the metaphysics of causets and their dynamics is needed can be directly seen when we seek reassurance that transitive percolation satisfies axioms 2 through 5. While axioms 2 and 4 are directly and unproblematically built in, it may appear as if axiom 3 and axiom 5 stand in tension. To repeat, the transition probability from the 2-antichain to the 3-element causet in the top middle of figure 4.2 must include the factor 2 for all three probabilities of the transitions emanating from the 2-antichain to sum to 1. However, axiom 3 demands that the product of the two transition probabilities from the ur-causet to the same 3-element causet along the two paths through the 2-chain and through the 2-antichain must be the same. Assuming $p \neq 0$, this cannot be the case if we maintain the factor 2. But the contradiction is only apparent: in order to check axiom 5, we must add up the probabilities of all possible ways in which the 2-antichain could evolve (including the two ways in which it can evolve to the 3-element causet of interest); in order to verify axiom 3, however, we only need to multiply the probability to get to that 3-element causet from the same ur-causet along one path.

---

[10] This raises the same metaphysical issue we have already encountered in the discussion of 'distinguishing' causets in §3.2.2, viz. whether elements are solely individuated by their structural position or whether they possess a haecceity beyond that. For further discussion of this point, see Wüthrich (2012).

Strictly speaking, there are three paths from the ur-causet to the 3-element causet at stake: one through the 2-chain, and two through the 2-antichain, and along all three paths, the probabilities must factor to the same product. And they do: they all factor to $p(1 - p)^2$.

There are further metaphysical assumptions built into this model of classical sequential growth dynamics. First, the probability of the 'big bang' causet, i.e, the minimal element of the partial order depicted in figure 4.2—the ur-causet—is assumed to be 1. In other words, the dynamics presupposes that there is something rather than 'nothing': the completely empty world containing no basal element at all is not a physical possibility. Second, as already mentioned, all causets which comply with this dynamics are past-finite and grow indefinitely into the future, at least as the transition probabilities are assumed to be non-zero.

Third, as Rideout and Sorkin (1999) declare right at the outset, and as has been reaffirmed by leading causet theorists since, it is their hope that this dynamical 'birthing' of events, this sequentially growing 'block universe', will turn out to underwrite and give rise to the "phenomenological passage of time", which "is taken to be a manifestation of this continuing growth of the causet" (2). This idea raises an immediate concern of conceptual inconsistency. On the one hand, physical time, and thus truly dynamical processes are supposed to only arise at the emergent level as an aspect of the emerging spacetime, and is at best only implicitly present in the causal structure of the causet. On the other hand, the rhetoric is infested by temporal locutions concerning the 'growth' of causets, the 'birthing' of events in 'sequential' order following a 'dynamical' law, suggesting the presence of time in a way that is metaphysically prior to the causet. Clearly, this tension has to be resolved, and the status of time will have to be settled. We will return to this and related metaphysical questions in section 4.3.

Let us add a final but crucial point concerning sequential growth dynamics as a physical dynamical law for CST. Although it involves stochastic transitions between stages, there is nothing quantum in nature about the sequential growth dynamics—hence, it is known as *classical* sequential growth dynamics. Just as discreteness postulated in the kinematics of the theory does not make the theory a quantum theory, neither does the stochasticity introduced in the dynamics. Alas, this is largely the state of the art as the philosopher studying CST finds it. Although there are incipient efforts to turn the theory into a full quantum theory, such as those based on quantum measure theory (Sorkin 1997a) or a quantization of classical sequential growth dynamics (Gudder 2014),[11] their incompleteness and inconclusiveness does not really permit a discussion of anything but the classical theory. Consequently, our conclusions are conditional on the proviso that this is just the classical theory so far, and so will have to remain tentative.

---

[11] See also the references in footnote 2 in chapter 3.

80   OUT OF NOWHERE

4.1.3 Manifoldlikeness

Transitive percolation or something like it ought to be able to turn the KR tide, at least as long as the probability p does not vanish. Is adding a viable and physically motivated dynamical law thus everything required to understand the emergence of spacetime from causets? A priori, no: resulting causets are generally still not 'manifoldlike' and so not yet candidates for giving rise to spacetime.

Problem 2. *The conditions stated so far may not guarantee that causets give rise to manifolds of reasonably low dimensionality (i.e., significantly lower than 10^{130}) or that it has Lorentzian signature or a reasonable causal structure.*

The discrete structures of causets and continuum Lorentzian spacetimes are mathematically rather different structures, as already noted. But in order for a causet to give rise to something that is at some scales well described by a relativistic spacetime, these structures cannot be too different. Thus, the second necessary condition after the requirement that candidate causets be 'sufficiently large' is that they are 'manifoldlike':

Solution 2. *Impose conditions such that, generically, causets $\langle C, \prec \rangle$ are 'manifold-like'.*

But what is it for a causet to be manifoldlike in the required sense? It means that it stands in an appropriate relation to spacetimes:

Definition 15. *A causet $\langle C, \prec \rangle$ is manifoldlike* just in case it is 'well approximated' *by a relativistic spacetime $\langle \mathcal{M}, g_{ab} \rangle$.*

If a causet is manifoldlike, we also say that the associated spacetime $\langle \mathcal{M}, g \rangle$ *approximates* the causet. Clearly, this definition, and hence solution 2, can only get any traction once we spell out what is meant by 'well approximated'. To provide a rigorous understanding of what it would be for a causet and a spacetime to stand in the required relation turns out to be challenging and is the subject of ongoing research. Although in the first place mathematical, such proposals, if we are to show that spacetime emerges from a causet, must also be understood as embodying (possibly competing) claims regarding physical salience. That is, the hope is not merely that there is the proposed *mathematical* correspondence between causets and spacetimes, but also that it expresses a *physical* identity between them, as physical objects. In this way, the proposals can be understood as hypothesized principles of physical salience, to be vindicated (perhaps) ultimately by the empirical success of CST; although at this stage, as we shall discuss, it remains even to vindicate them mathematically, as we shall now see.

There are at least two ways of defining 'well approximated'. First, one could understand both causets and (distinguishing) Lorentzian manifolds as examples of 'causal measure spaces' and use a so-called *Gromov-Hausdorff function* $d_{GH}(\cdot, \cdot)$ (or the 'Gromov-Wasserstein function' if we have a probability measure) to characterize the closeness or similarity between such spaces.[12] Such a distance function relating any two causets, any two Lorentzian spacetimes, or any causet and any Lorentzian spacetime in terms of their relevant similarity could provide a rigorous way to spell out the idea of 'approximation'.

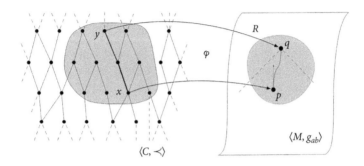

Figure 4.3 A faithful embedding φ of a causet into a spacetime.

But worked out in much more detail is a second approach, using the notion of a 'faithful embedding' of a causet into a spacetime. The task here is to find a map $\varphi : C \to \mathcal{M}$ such that the image looks like a relativistic spacetime, as shown in figure 4.3. More precisely, a spacetime approximates a causet if there exists a faithful embedding of the causet into the spacetime in the following sense:

Definition 16. *A causet $\langle C, \prec \rangle$ is said to be* approximated *just in case there exists a 'faithful embedding', i.e., a injective map $\varphi : C \to \mathcal{M}$ such that*

1. *the causal relations are preserved, i.e., $\forall a, b \in C, a \prec b$ iff $\phi(a) \in J^-(\phi(b))$;*
2. *$\varphi(C)$ is a 'uniformly distributed' set of points in $\mathcal{M}$;*
3. *$\langle \mathcal{M}, g_{ab} \rangle$ does not have 'structure' at scales below the mean point spacing.*

Let us discuss the three conditions in turn.

The first condition requires that the causal structures of the causet and the set of images of its events in the spacetime to which it gives rise are isomorphic. This requirement captures the idea that causal structure is fundamental and 'percolates'

---

[12] This line is pursued by Bombelli et al. (2012), although that paper has been withdrawn and never been followed up, as far as we can tell. Sam Fletcher gave talks about this in 2013 developing this ansatz, but has so far not published this material.

82 OUT OF NOWHERE

up the scales in a precise sense. This demand seems acceptable for a fundamentally classical theory, such as the version of CST discussed here. Once this theory is replaced by a quantum CST, we cannot reasonably uphold this demand as we should expect quantum 'fluctuations' of the causal structure rather than some determinate causal structure, and so no untainted emergence of that structure.

The second condition expresses the expectation that regions of causets of a similar number of elements give rise to spacetime regions of correspondingly similar volume. Thus, the mapping cannot be such that the image points of the causet elements are too dense or too sparse in the spacetime, resulting in a more or less uniform distribution. The intuition behind this condition is thus clear, although its articulation obviously needs to be precisified.[13]

The third condition, which will eventually also need a more rigorous formulation, and is strictly speaking not a restriction on ϕ, insists on the absence of non-trivial structure in the spacetime between the uniformly distributed image points, and so at scales so small that they could not possibly be captured by the causet. If this condition were violated, then the spacetime would have local features, such as a high or uneven curvature or a non-trivial topology in a very small region, which the fundamental causet would simply lack the expressive power to include. Suppose a given causet can be mapped into two distinct spacetimes with the first two conditions satisfied, such that one of the spacetimes contains non-trivial small-scale structure, while the other does not. Thus, the mapping into the second spacetime is faithful, but the one into the first is not. Since the point of the manifoldlikeness condition is to determine how we can conceive of the (mathematical and physical) relation between causets and spacetime as one of emergence, and since there is no reason to think that a causet can physically give rise to a spacetime with small-scale structure completely absent in the fundamental causet, there is no reason to validate the first mapping.

A prototypical example to illustrate the relation between manifoldlike causets and relativistic spacetimes is the 'Poisson sprinkling' of elements into 2-dimensional (or higher-dimensional) Minkowski spacetime.[14] Such a sprinkling selects events in Minkowski spacetime at the required density uniformly and at random and then imposes the unique partial ordering among them which is induced by the causal structure of Minkowski spacetime. Thus, Minkowski spacetime's causal structure is inherited by the causet and the first condition of a faithful embedding is satisfied by construction, as are the second and third.[15]

---

[13] One way in which it needs to be precisified is that the required uniformity must be with respect to the spacetime volume measure of $\langle \mathcal{M}, g \rangle$.

[14] The image points of the causets in the spacetime cannot stand in a 'regular', lattice-like structure since this would violate Lorentz invariance (see §4.4). The required 'randomness' is naturally (though not uniquely) captured by a Poisson distribution.

[15] For a useful illustration of Poisson sprinkling, see Dowker (2013, figure 1) and Surya (2019, §3, particularly figure 6).

Note that the relationship between fundamental causets and emergent space-times assumed in this 'sprinkling' is really the wrong way around: we started out with the spacetime given, and then sprinkled events into it, and connected those which were causally related as given by the causal structure of the space-time. Surely, this is putting the cart before the horse: the fundamental causet is prior to the relativistic spacetime, which is merely derivative. In this sense, what we ultimately want to understand is how a generic causet can give rise to the sorts of relativistic spacetimes we believe to be physically realistic.

It remains unclear to date what conditions need to be imposed in solution 2 in order for the abiding causets to be manifoldlike. In CST, it is hoped that the dynamical conditions imposed to solve problem 1 double up to also solve problem 2; that a minimal set of natural conditions or physically meaningful axioms suffices. Computer simulations fuel the hope that this may indeed be the case by providing evidence, e.g., that a significant class of dynamical laws closely related to transitive percolation yields a repeatedly collapsing and re-expanding universe in which each era contains phases of exponential expansion approximated by de Sitter spacetime.[16]

Note finally that a sprinkling of a causet 'into' a spacetime maps subsets of causet points into spacelike hypersurfaces, bypassing the 'problem of space' discussed in §3.3. In light of that discussion, we emphasize that the structure of a spacelike hypersurface does not supervene on the structure of the causet points it 'contains'; points in any such set are spacelike, so alike in having only trivial structure, while the corresponding hypersurfaces contain rich metrical and topological structures, which can vary between them. Rather, the spatiotemporal structures of the hypersurfaces supervene (approximately) on the causet as a whole.

4.1.4 The Hauptvermutung

The restriction to manifoldlike causets is surely a necessary condition for there to be any way in which a causet lends itself to the emergence of spacetime. But this in itself does not solve the challenge: GR is a notoriously permissive theory in that the number of spacetimes which satisfy the basic requirements of physical possibility imposed by GR, such as that the spacetime be Lorentzian and satisfies the EFE, is overwhelming. Many of these physically possible spacetimes are intuitively 'unphysical' and so sometimes ruled out by physicists. This may be because they feature a pathological causal structure (containing, perhaps, closed timelike

---

[16] See Ahmed and Rideout (2010). The class of dynamical laws is that of so-called 'originary percolation', which is transitive percolation where the possibility of an element being born unrelated to any prior element is excluded.

84 OUT OF NOWHERE

curves), an implausible topology, or a dubious curvature—or so we judge. In other words, we face the next problem:

Problem 3. *Given just how many 'unphysical' spacetimes are physically possible according to GR, causets may generically give rise to 'unphysical' relativistic spacetimes.*

One might argue that this is not really a problem of CST, or indeed of any quantum theory of gravity, since it is already at the level of GR that we are faced with rampant unphysical solutions admitted by the theory. Although this is undisputedly the case, we could, and perhaps should, invert this reasoning: it is precisely *because* GR is too permissive that we are looking to the more fundamental theory for guidance, and hope that it will rule out many or most of these unphysical solutions. So this problem would be solved by the following solution:

Solution 3. *Impose conditions such that qualifying causets are approximated by 'physically reasonable' spacetimes.*

Evidently, this solution could only be implemented once we specify what a 'physically reasonable' spacetime is, a locution on which we certainly have an intuitive grasp, but which turns out to resist systematic and satisfactory treatment.[17] Although a full explication is unlikely to emerge, we might hope to make some progress. For instance, one might think that the demand encoded in the kinematic axiom that the causal order be antisymmetric (cf. chapter 3) ensures that closed timelike curves cannot arise in the emerging spacetime. However, this implication only holds under some interpretations of the theory and must ultimately await its full articulation (Wüthrich 2021). Consequently, although the problem is real and the solution appears promising, its further elaboration is the task for another day.

Just as for problem 2, it is hoped that no *additional* conditions are needed and that the axioms postulated above would also suffice to generically deliver physically reasonable spacetime, whatever that is supposed to mean. But there is another reason why we would not want to impose any additional axioms just because they enforce our intuitions as to which spacetimes we take to be physically reasonable: part of the point of formulating a theory more fundamental than GR is for us to learn from it in which ways GR is false and needs to be corrected. In particular, this may apply to the range of physical possibilities which a fundamental theory would curtail much more restrictively than notoriously permissive GR. The point, however, would be to learn this from the fundamental theory, rather than to constrain that theory by the brute force of our antecedent intuitions. Thus, we take it

---

[17] See Doboszewski (2017) and Manchak (2021b) for further discussion of the notion of 'physically reasonable' spacetimes.

that we have excellent reasons to reject problem 3 as something we need to attend to at this point, all the while keeping in mind the issue as deserving revisiting once the theory is more fully understood.

In other words, the project is to determine, and to some extent to *guess*, what the fundamental physics is that will produce structures which are generically approximated by relativistic spacetimes.

But we are not quite done. Even if we thus arrive at the point at which judiciously chosen causets generically give rise to physically reasonable relativistic spacetimes, one further necessary condition must be imposed. A necessary condition for the spacetime to emerge is surely that it supervenes on the fundamental structure, to use the jargon of philosophers. In other words, there cannot be a difference at the level of the spacetime without there also being a difference at the level of the causet. The required covariance is asymmetric, as there can in general be differences at the level of causets that need not generate any difference for the emerging spacetime. We would want to think that the emergent spacetime asymmetrically depends on the causet. We can formulate this also as a problem:

Problem 4. *A given causet might be approximated by multiple relativistic spacetimes.*

This would indeed be a problem, since in this case, the fundamental structure would not uniquely determine the emergent large-scale structure. As Sorkin (2005, 313) puts it: "Implicit in the idea of a manifold approximating a causet is that the former is relatively unique; for if any two very different manifolds could approximate the same [causet] C, we'd have no objective way to understand why we observe one particular spacetime and not some very different one." The implicit condition then is something like this:

Solution 4. *If a causet is approximated by a spacetime, then the approximating spacetime is 'approximately unique'.*

Following Sorkin, this has become known as the 'Hauptvermutung', or principal conjecture. Of course, this is but a template to fill in the specifics of the solution. In particular, what it means for the spacetime to be 'approximately unique' will have to be specified. It should also be pointed out that it is hoped, or conjectured, that the satisfaction of the earlier conditions guarantees this solution, so that no *additional* condition need be imposed. In other words, the hope is that the Hauptvermutung can be established as a theorem of the theory as articulated so far and need not be added as an independent stipulation. In fact, to show that this is the case is the overarching goal of the research program (at least at the classical level).

86   OUT OF NOWHERE

To date, however, the Hauptvermutung remains an unproven conjecture. In order to have any hope of proving it, a more precisely formulated proposition will be needed. Just how to accomplish that will depend on the chosen notion of 'approximation'. Above, we identified two paths: one based on the Gromov-Hausdorff measure of similarity, and another in terms of a faithful embedding of causets into spacetimes. On the first research programme, the Hauptvermutung could be precisified as follows:

Solution 5 (Hauptvermutung, Gromov-Hausdorff version). *If there exist two spacetimes $\langle \mathcal{M}, g_{ab}\rangle$ and $\langle \mathcal{M}', g'_{ab}\rangle$ such that $d_{GH}(C, \mathcal{M}) < \epsilon$ and $d_{GH}(C, \mathcal{M}') < \epsilon$ for some causet C, then $d_{GH}(\mathcal{M}, \mathcal{M}') < 2\epsilon$.*

Although the choice of some particular $\epsilon \in \mathbb{R}^+$ is rather arbitrary, something like the above condition could encapsulate the idea that a causet is approximated by a spacetime in an 'approximately unique' sense. If, on the other hand, we followed the more standard path of articulating the relevant sense of approximation in terms of faithful embeddings, then the Hauptvermutung might be expressed differently:

Solution 6 (Hauptvermutung, standard version). *If there exist two spacetimes $\langle \mathcal{M}, g_{ab}\rangle$ and $\langle \mathcal{M}', g'_{ab}\rangle$, which approximate a given causet C in that there exist faithful embeddings $\varphi : C \to \mathcal{M}$ and $\varphi' : C \to \mathcal{M}'$, then they are 'approximately isometric'.*

It might appear as if not much has been accomplished by replacing solution 4 by solution 6, since we basically replaced the problematically vague notion of 'approximate uniqueness' with an apparently equally problematically vague notion of 'approximate isometry'. But we have made real progress: the first stab at solving problem 4 just offered a general template of what, in rather general terms, would be needed to solve the problem, whereas the much more precise solution 6 reduces the problem to one of defining and defending a measure of approximate isometry between spacetimes. This is by no means a trivial problem, as is witnessed by the fact that it has so far resisted resolution.[18] Such a resolution would presumably require that we identify the salient geometric properties of spacetime, introduce measures of similarity along the dimensions of the selected properties, and compound these measures into a score for approximate isometry. (Exact) isometry does not require any of the above; however, if the isometry is no longer 'exact',

---

[18] This is not trivial, since Malament's theorem is a theorem of *GR*, but not of *CST* (we have seen in chapter 3 that an analogous statement is false in CST). Consequently, we cannot expect that 'having the same causal structure' is sufficient to have spacetime emerge. In other words, we cannot just assume that causal structure and discreteness suffice to fix everything about spacetime. This is the significant work that remains to be done.

and thus the metrics compared are no longer identical in the relevant sense, the notion of (exact) isometry gets no traction, and the salient geometric features with respect to which the equivalence is supposed to be determined must be selected.[19]

We shall leave it at that, as the point of this section is to sketch, only in general terms, what would be required in order to establish the emergence of spacetime in CST. Now suppose that the causet research programme offered satisfactory accounts of how to fill in the details in the above sketch and could thus be said to have fulfilled all necessary conditions listed. Of course, merely satisfying some necessary conditions gives us no guarantee that the work is jointly sufficient. And at the end of the day, we will want the assurance that the challenge of the emergence of spacetime is fully met, and this guarantee is only forthcoming if we convince ourselves that the necessary conditions we have discharged are also jointly sufficient.

4.2 The case for spacetime functionalism

4.2.1 The road to spacetime functionalism

Before we enter the fray of the debate regarding the sufficiency of these conditions, let us pause to consider what would have been accomplished had we successfully discharged all necessary conditions. In fact, in this case, we could generically 'derive'—i.e., systematically relate with requisite mathematical rigor—relativistic spacetimes from causets, including an account of the physical salience of these derivations. No mean feat. But would this accomplishment conclude the project? Would we be done? Not according to the *résistance*, which believes that the conditions listed in the previous section are merely necessary but not jointly sufficient for the emergence of relativistic spacetimes from fundamental causets.[20]

The résistance has different cells, which may be differently motivated. A first cell, inspired by Bell (2004), consists of *primitive ontologists* who insist on fundamental local (and so a fortiori *localizable*) beables in spacetime. According to this view, for a theory to qualify as a candidate description of our physical world, thereby accounting for some of our empirical data or, more broadly, for aspects of human experience, the theory must postulate an ontology of entities populating regions of spacetime. This condition is sometimes motivated by the following thought. If the entities posited by a theory are not localizable, we lose a compelling

---

[19] Two spacetimes $\langle \mathcal{M}, g_{ab} \rangle$ and $\langle \mathcal{M}', g'_{ab} \rangle$ are (exactly) *isometric* just in case there exists a diffeomorphism $\phi : \mathcal{M} \to \mathcal{M}'$ such that $\phi^*(g'_{ab}) = g_{ab}$, where ϕ^* is the pull back map which takes tensors defined on $\mathcal{M}'$ to tensors defined on $\mathcal{M}$. This definition delivers the identity "in the relevant sense" invoked in the main text.

[20] This résistance is somewhat hypothetical, reconstructed and elaborated from discussions with various people.

88   OUT OF NOWHERE

way of parsing that which exists into a plurality of distinct entities: location offers a straightforward criterion for individuation. Although the loss of this criterion is already adumbrated by the non-locality of quantum physics, without spacetime at our side, how are we to dissect that which exists fundamentally into distinct and separate entities? Clearly, there are alternative ways of distinguishing entities, and so the criterion of spatiotemporal location commands no necessity. Admittedly though, for these alternative criteria to get traction at the fundamental level of a theory of quantum gravity may be tricky, and so a structuralist or monist stance may most naturally fit these cases.[21] Even so, fundamental spatiotemporality is not necessary for parsing an ontology. Although the realist motivation of this camp is laudable, the particular form of realism demanded thus comes with an unduly narrow conception of fundamental ontology—one that quantum theories of gravity will be hard pressed to satisfy.

Perhaps the concern is better expressed by a second cell, the *anti-Pythagoreans* who fear a loss of the venerable distinction between physical stuff and abstracta, and object to the idea that the physical world is fundamentally mathematical or abstract by its nature. The distinction between concrete physical entities and abstract entities is thought to be threatened by the presumed non-spatiotemporality of a putative ontology of quantum gravity. On a standard demarcation, the concrete entities are taken to be those in spacetime, and the abstract ones those which are not. This criterion of demarcating the concrete from the abstract seems to suggest that the fundamental, non-spatiotemporal structures which make up our natural world must be abstract, perhaps mathematical. Alternatively, concrete physical entities have been characterized as those engaging in the causal commerce of the world. Worries of circularity concerning this criterion aside, attempts to explicate a notion of causal efficacy in the absence of spacetime may be thwarted by insurmountable difficulties (Lam and Esfeld 2013, §4.2). Consequently, at least the most common ways of demarcating the physical from the mathematical are not apt in the present non-spatiotemporal context, and it may be feared that the concerned theories thus vindicate the unwanted Pythagorean conclusion that our physical world is ultimately mathematical in nature.

A third—related—cell is formed by those worried about the empirical coherence of theories without fundamental spacetime. All data ever collected for the empirical confirmation of physical theories seem, at their core, to be inextricably spatiotemporal: the coincidence of the tip of a pointer with a mark on a scale, the flashing of a particular light bulb, or the printout of a measurement outcome all involve physical objects or processes to be located in space and time. Thus, it seems as if the existence of spacetime is a precondition for empirical science. It appears

---

[21] Cf. Wüthrich (2012), who argues for a structuralist interpretation of CST, and discusses the challenges of such an approach.

as if theories of quantum gravity *sans* spacetime deny the very conditions necessary for their empirical confirmation, and in this sense be 'empirically incoherent' (Huggett and Wüthrich 2013a). Although empirical incoherence does not amount to any kind of impossibility, it would be most unfortunate for empirical science if we lived in a world in which the necessary conditions for such an enterprise could not be set in place.

We believe that all these cells of the résistance can be put to rest if it could be established that spacetime emerges in the appropriate limit or at the requisite scales, along the lines outlined in chapter 2. Showing how spacetime emerges would involve demonstrating, among other things, how macroscopic (or indeed most microscopic) objects have location and are thus localizable. In this sense, to the extent to which the demand of primitive ontologists was reasonable, it would be met. Pythagoreanism would be averted, because although the (more) fundamental ontology would neither be spatiotemporal nor directly causally efficacious, the fundamental structures would have been shown to realize the spatiotemporal structures we associate with physical being. As we have stressed, such a demonstration requires not just a formal derivation, but acceptance, justified by the empirical success of the result, of principles according to which the derivation is physically salient: for instance, that the causal structures of suitable causets *are* the causal structures of classical spacetimes. As a result, the emergent spacetime would be shown to depend ontologically on such fundamental structures, thus endowing the latter with concrete physicality in a 'top-down' manner (Huggett and Wüthrich 2013a, 283f). Finally, the only condition necessary to evade the threat of empirical incoherence is that the conditions for empirical confirmation are set in place at the scales of the human scientist, not at the level of fundamental ontology. Establishing that at human scales the world is indeed spatiotemporal to a very good approximation circumvents the menace of empirical incoherence, at least insofar it was motivated by the apparent absence of spacetime.

In order to establish the emergence of spacetime, it would evidently not be enough to satisfy merely necessary, but jointly insufficient conditions. But the three resisting concerns listed above may guide us in what it would take to appease them and thus arguably arrive at jointly sufficient conditions. Collectively, they suggest that in order to accomplish this, we need to focus on the functions spacetime plays in structuring our ontology, distinguish its elements from abstracta, and enabling empirical confirmation, among others. Thus, spacetime should be analyzed in terms of the functions it performs to other ends, rather than for its own sake. Spacetime is as spacetime does, to invoke the motif in Lam and Wüthrich (2018), although this need not be interpreted in ontologically thick ways. In this spirit, establishing the emergence of spacetime involves showing how the fundamental existents may instantiate these functions or roles in favorable circumstances. This 'spacetime functionalism' we have already encountered in chapter 2 asserts that once the functional roles of spacetime have been identified

90 OUT OF NOWHERE

and the fundamental physics shown to fulfill these roles, the emergence of spacetime has been fully established with full physical salience and no work remains to be completed in this respect.

4.2.2 Spacetime functionalism in causal set theory

How would one implement a functionalist programme for CST? The first step (SF1) of spacetime functionalism as articulated in §2.4 will require the specification of the 'spatiotemporally salient' features of spacetime that need accounting for. We will not attempt an exhaustive and conclusive list here (arguably there is none), but it should be clear that these features will include both topological aspects such as the dimensionality of space or spacetime and something approaching a topology of (non-metrical) nearness relations, as well as geometrical, i.e., metrical, relations such as (spatial) distances and (temporal) durations. Nearness and metric relations are both involved in spacetime's central role of (relative) spatiotemporal localization. SF1 establishes the targets for the demonstrations of physical salience.[22]

In the second step prescribed by spacetime functionalism, (SF2), it must be shown how causets can fulfill these roles. This involves precisely the kind of work presented in §3.3 in the previous chapter, showing how the dimensionality (§3.3.3), spatial topologies (§3.3.4), and distance relations (§3.3.5) can emerge from the fundamental causet if the fundamental degrees of freedom come together in the right way. Centrally, this work includes tracking the physically salient features through the derivations. In the last chapter, we used this work to show that space in itself cannot be usefully thought to be an inherent part or substructure of causets but that in order to recover the salient features of space, we must generally take into account the entire causet. Here, we want to recall these reconstructions because they exemplify what SF2 will look like. To repeat, such reconstructions will not be required for each and every property of relativistic spacetimes: spacetime functionalism does not demand the derivation of full relativistic spacetime, lock, stock, and barrel.

What precisely the functions are which must be recovered from the fundamental ontology will be strongly confined by its epistemology: the crucial spatiotemporal functions to be recovered connect to ways in which we can come to know of these structures, at least in principle, and to what it takes, more generally, for the world to manifest itself to us as it does. There are reasonable and important debates to be had about what these functions are, and there may well be no unique basic set

---

[22] Given the empirical significance of such features, it is not surprising that versions of this list appear in our discussions of spacetime emergence from loop quantum gravity (chapter 6) and strings (chapter 10); these discussions contain elaborations of the physical significance—i.e., *functions*—of the features most relevant for the different theories.

(Baker 2021). But the constraints imposed by the epistemology arguably makes the debate more controlled than in the philosophy of mind, where the debate turns on the status of qualia. The project, then, is to explain how the structures postulated in QG play these functions crucial to the epistemology and the phenomenology, and not to articulate what the essence of the things performing these roles is.

Although the two steps to the functionalist emergence of spacetime are sketched but in outline here, and most work remains to be done by researchers in the various programs, we claim that the approach offers a full and satisfactory resolution of the disappearance of salient features of spacetime in what appears to be structures rather different from what we are used to in GR. We freely admit that the recovery of empirically vital functions of spacetime from causets is far from accomplished. What we reject is the misguided insistence that even were all relevant spacetime roles to be demonstrated to be fulfilled by the fundamental degrees of freedom, it would furthermore be necessary to identify the nature of spacetime in the fundamental structure, which must therefore be, by necessity, of an ultimately spatiotemporal character. Quite the opposite: we see no in principle obstacle that the fundamental structure can be wildly non-spatiotemporal—as long as it is demonstrated that it adequately plays the requisite spacetime roles at macroscopic scales. On a functionalist understanding, spacetime just is what it contributes to the overall scientific account of the world as it manifests itself to us in its myriad ways. Obviously, a quantum theory of gravity and the ontology it posits will have to assume its role in the wider context of this account. At the end of the day, it will earn its keep by pulling its weight in this bigger project.

4.3 Causal set theory and relativistic becoming

Now that we have sketched CST, illustrated how spacetime disappears in it, and sketched how it re-emerges, let us finish the discussion by addressing two points of considerable philosophical interest which come up in CST and in its conception of the emergence of spacetime: the possibility of a relativistically invariant passage of time or becoming, and the appearance of highly non-local behavior in this classical, relativistic, but discrete theory. We save the second point for the next section (§4.4) and turn to the first issue, the metaphysics of time based on CST.[23] Both points aptly illustrate the philosophical and conceptual issues that arise in developing a new quantum theory of gravity, and how spacetime emerges from it. The emergence of spacetime in quantum gravity is inextricably entangled with deep, and unavoidable, philosophical questions.

---

[23] This section is based on work previously published in Wüthrich and Callender (2017).

92 OUT OF NOWHERE

4.3.1 The basic dilemma of relativistic becoming

Contemporary physics is notoriously hostile to what philosophers have come to call 'A-theoretic' metaphysics of time, i.e., a metaphysics of time which fundamentally includes an element of becoming or dynamical passage, an aspect of time captured in our language by the use of tenses.[24] An important example of an A-theoretic metaphysics is *presentism*, according to which only present entities, objects, or events exist, but in a dynamically updated way, so that we arrive at a metaphysics of a dynamical succession of 'nows'. Another example is the *growing block view*, according to which only past and present entities, objects, or events exist so that the sum total of existence continually grows by including ever more slices of existence.[25]

CST apparently promises to reverse that verdict against A-theories: its advocates have argued that their framework is consistent with a fundamental notion of 'becoming'. Relevantly, in the dynamical 'growth' of causets we introduced in §4.1.2, the 'birthing' process of new elements is said to unfold in a 'generally covariant' manner and hence in a way that is perfectly compatible with relativity. Here is Sorkin (2006) advertising the philosophical pay-off:

> One often hears that the principle of general covariance [...] forces us to abandon 'becoming' [...]. To this claim, the [classical sequential growth] dynamics provides a counterexample. It refutes the claim because it offers us an active process of growth in which 'things really happen', but at the same time it honors general covariance. In doing so, it shows how the 'Now' might be restored to physics without paying the price of a return to the absolute simultaneity of pre-relativistic days.

The claim is that CST, augmented with a dynamics such as classical sequential growth dynamics, rescues temporal becoming and our 'intuitive' notion of time from relativity. The claim is routinely made in the pertinent physics literature, and has found its way into popular science magazines.[26] One might not believe that our intuitive notion of time needs or deserves rescuing, but there is no denying that if this claim is correct it would have significant consequences for the philosophy of time. Specifically, it may underwrite a 'growing block' model of the metaphysics of time, as John Earman (2008) has speculated.

---

[24] As opposed to a 'B-theoretic' metaphysics of time, which does not fundamentally include an ontologically privileged present and has no such elements of substantive becoming, but instead seeks to explain the latter as emergent or illusory phenomena (see, e.g., Callender 2017). On the question of which is in fact more 'intuitive', see Graziani et al. (2023).

[25] We will further discuss these positions in §6.5.

[26] Cf. e.g., Dowker (2003, 38). For other important expressions of the claim, see Sorkin (2007) and Dowker (2014, 2020).

As argued in Wüthrich and Callender (2017), we believe that the introduction of CST and its dynamics does not ultimately change the fundamental dilemma any fan of becoming or passage confronts when facing relativistic physics, even though some novel aspects arise, mainly due to the discreteness of causets. The dilemma is the following: any metaphysics of time including a fundamental sense of becoming or passage either answers to their fan's explanatory demands or is compatible with relativistic physics, *but not both.*[27]

To illustrate this with an example, suppose one thought, as does the presentist, that in order for there to be becoming or passage, there needs to be a (dynamically changing) present, a 'now', identified in the fundamental structures of the world. In the context of special relativity, with its relativity of simultaneity, one might introduce a foliation of spacetime into spacelike hypersurfaces totally ordered by 'time'. Presumably, that would answer to the presentist's notion of a (spatially extended) present and of becoming, but at the price of introducing structure not invariant under automorphisms of Minkowski spacetime and hence arguably violating special relativity. Conversely, the present can be identified with invariant structures such as a single event or the surface of an event's past lightcone, and successive presents as a set of events on a worldline or as a set of past lightcones totally ordered by inclusion, respectively, but such structures will have radically different properties from those ordinarily attributed to the present by those seeking to save it (see Callender 2000 and Wüthrich 2013).

Returning to CST, a natural transposition of the idea of a foliation of spacetime into spacelike hypersurfaces is to partition a causet into maximal antichains, as considered in §3.3, and particularly in §3.3.2. Although such a partition always exists (unlike a complete foliation for relativistic spacetimes), it fails to answer to the explanatory demands of an advocate of becoming, as Wüthrich and Callender (2017, §3) argue in more detail. They conclude that, at least at the kinematical level, CST embraces the dilemma and in fact makes it more rigorous. But that may not be all that surprising; after all, the heart of the idea that CST rescues becoming involves taking sequential growth seriously: becoming is embodied in the 'birthing' of new elements, and so in the theory's dynamics.

Although we are interested in becoming, we should immediately remark that sequential growth is certainly compatible with a tenseless or block picture of time. In mathematics a stochastic process is characterized as a collection of random variables defined on a triad of a sample space, a sigma algebra on that space, and a probability measure whose domain is the sigma algebra.[28] The transition probabilities of the sequential growth dynamics are viewed merely as defining the elements of the triad. In the case at hand, the sample space is the set $\Omega = \Omega(\infty)$ of past-finite

---

[27] For earlier articulations of the same dillemma, see Callender (2000) and Wüthrich (2013).

[28] The sample space Ω is the set of all possible outcomes of the stochastic process. The sigma algebra is the set of events which form a space characterized by the algebra and the probability measure assigns a real number between 0 and 1 to all events in the sigma algebra.

94 OUT OF NOWHERE

and future-infinite labeled causets that have been 'run to infinity'. The events are the possible sequence of births. The 'dynamics' is given by the probability measure constructed from the transition probabilities; for details, see Brightwell et al. (2003). On this picture, the theory consists simply of a space of complete histories with a probability measure over them.[29]

4.3.2 Taking growth seriously

However, let's take growth seriously. There are different extents to which this can be done. At a more modest level, and consistent with explicit pronouncements by advocates of causet becoming, we can articulate a localized, observer-dependent form of becoming. Here, the idea is that becoming occurs not in an objective, global manner, but instead with respect to observers situated within the world that becomes. The only facts of the matter concerning becoming are local, and are experienced by individual observers as they inch toward the future. In Sorkin's words, which are worth quoting in full,

> [o]ur 'now' is (approximately) local and if we ask whether a distant event space-like to us has or has not happened yet, this question lacks intuitive sense. But the 'opponents of becoming' seem not to content themselves with the experience of a 'situated observer'. They want to imagine themselves as a 'super observer', who would take in all of existence at a glance. The supposition of such an observer *would* lead to a distinguished 'slicing' of the causet, contradicting the principle that such a slicing lacks objective meaning ('covariance'). (2007, 158)

According to Sorkin, instead of "super observers", we have an "asynchronous multiplicity of 'nows'". It seems fairly straightforward that a perfectly analogous kind of becoming can be had in the context of Minkowski spacetime. Indeed, 'past lightcone becoming', in the sense of Stein (1991), and 'worldline becoming', as articulated by Clifton and Hogarth (1995), both fit the bill.[30] Furthermore, past lightcone becoming and worldline becoming are also available in general-relativistic spacetimes, as they do not depend on the spacetime admitting a foliation.

Although Sorkin himself remains uncommitted as to whether the analogy holds, Fay Dowker (2014, 2020) rejects it, arguing that 'asynchronous becoming' is not compatible with GR, but only with a dynamics like the one provided by the classical sequential growth. She discusses 'hypersurface becoming' in GR—which does

---

[29] This interpretation corresponds to Huggett's first option (2014, 16), which is fully B-theoretic. When we consider 'taking growth seriously', we mean to essentially follow the second route he offers: augmenting the causal structure with an additional, but gauge-invariant, dynamics.
[30] Cf. also Arageorgis (2016) who makes a similar point.

of course depend on the spacetime admitting a foliation—and rejects it for its obvious violation of general covariance. She is clear that hypersurface becoming is but one way to implement becoming in the context of GR, but does unfortunately not discuss alternatives. In particular, she does not discuss worldline becoming or past lightcone becoming, which both seem much more promising analogs of the asynchronous becoming identified by Sorkin in CST.

She maintains that what is needed for there to be becoming is not the mere *existence* of events, but a process she terms their *occurrence* (2014, 22). She claims that spacetime events do not "occur" in GR in this special sense, and so there is no genuine form of becoming possible in GR. While in GR spacetime events exist, they are never "happening" as the result of a dynamical process of "occurrence", they are never "born" like the events in a dynamical understanding of CST. Thus, whereas the births of events in CST deliver a true form of becoming, in GR we just find static being.

Against this, we note firstly that (an important subsector of) general relativity certainly can be described in a 'dynamical' manner via its many '3+1' formulations.[31] To make her objection, Dowker would first need to elaborate the reasons why a 3+1 dynamics does not provide the 'occurrence' she desires. Of course, we admit that our retort here lacks full generality.

Furthermore, we note here a possible tension. If occurrence is simply a label for some events from the perspective of other events, then there is no problem—but then we note that such labels can be given consistently in GR too. But if occurrence implies something metaphysically substantial, such as existence or determinateness—then there is a possible tension between occurrence and the local becoming envisioned by Sorkin and Dowker. If spacetime events that are spacelike related do not exist for each other, for instance, then that is a radical fragmentation of reality.[32] Not only would that be a high cost to introduce becoming, but it is also one that, again, could be introduced in GR.[33]

Our present interest is to determine whether a more ambitious, objective, global, observer-independent form of becoming is compatible with classical-sequential-growth-cum-dynamics in a way that does not violate the strictures of relativity. In other words, does Sorkin's assertion in the last sentence of the indented quote above hold up to scrutiny? We will argue that it does not, but that there is a weak sense in which a fully objective kind of becoming with relativistic credentials can be had.

The central problem with taking the primitive growth of classical sequential growth dynamics as vindicating becoming is that this dynamics uniformly treats

---

[31] Cf. Wald (1984, ch. 10).

[32] In Pooley's view (2013, 358n), dynamical classical sequential growth should best be interpreted as a "non-standard *A* Theory" in the sense of Fine (2005), i.e., as giving up "the idea that there are absolute facts of the matter about the way the world is." (2013, 334)

[33] For a more detailed discussion of Dowker's position, see Wüthrich (2024).

96 OUT OF NOWHERE

the labeling time as 'fictitious'. It is only 'real' to the extent to which it respects the causal order (in the sense of axiom 2). A choice of label would be tantamount to picking a time coordinate x_0 in a relativistic spacetime. Any dynamics distinguishing a particular label order will be non-relativistic. Not wanting the dynamics to distinguish a particular label ('coordinatization'), the authors impose *discrete general covariance* (axiom 3) on the dynamics. This is a form of label invariance. As stated above, the idea is that the probability of any particular causet arising should be independent of the path to get to that causet. In fact, returning to figure 4.2, there could be no physical fact of the matter whether the 3-element causet in the middle of the top row grew via a stage consisting in a 2-chain or a 2-antichain. In the 3-element causet, the event with another event in its past (labeled 'a' in figure 4.2) and the causally isolated event (labeled 'b') are spacelike related. Consequently, there is no fact of the matter which of the two came into being first and which one second. To say which one happened 'first' is to invoke non-relativistic concepts. It is therefore hard to understand how there can be growth happening in time.

Seeing the difficulty here, John Earman (2008) suggests a kind of philosophical addition to causets, one where we imagine that 'actuality' does take one path or another. With such a hidden variable moving up the causet, we do regain a notion of becoming. But as Aristidis Arageorgis (2016) rightly points out, such a move really flies in the face of the normal interpretation of these labels as pure gauge.[34] The natural suggestion, espoused by (almost all?) philosophers of physics, is then that the above non-dynamic interpretation in terms of a block universe is best because it does not ask us to imagine that one event came first.

Perhaps the sensible reaction to this problem is to abandon the hope that the classical sequential growth dynamics does produce a novel form of becoming. Still, we are tempted to press on. The intuition motivating us is as follows. True, the dynamics is written in terms of a choice of label, but we know that a consistent gauge invariant dynamics exists 'beneath' this dynamics. In fact, rewriting the theory in terms of a probability measure space, as indicated above, one can quotient out over relabellings to arrive at a label-invariant measure space (for construction and details, see Brightwell et al. 2003). And one thing that we know is gauge invariant is the number of elements in any causet. Focusing just on these and ignoring any labeling, we do have transitions from C_n to C_{n+1} and so on. There is gauge-invariant growth.

The problem is that we are generally prohibited from saying exactly what elements exist at any stage of growth. Take the case of the spacelike-related events above. The world grows from C_1 to C_2 to C_3. In figure 4.2, C_1 would correspond to the one-element stage at the bottom and C_2 and C_3 to the second and third lines from the bottom. The causet cardinality at each stage is gauge invariant. We just

---

[34] Cf. also Butterfield (2007, 859f).

cannot say—not due to ignorance, but because there is no fact of the matter—whether C_2 is the 2-chain or the 2-antichain, i.e., whether the growth has taken the path through the left or right causet of the second line from the bottom in figure 4.2. Causet reality does not contain this information. There simply is no determinate fact as to whether C_2 is the 2-chain or the 2-antichain; but there is a determinate fact that it contains (exactly) one of them. If it is coherent, therefore, to speak of a causet having a certain number of elements without saying what those elements are, then classical sequential growth dynamics does permit a new kind of—admittedly radical and bizarre—temporal becoming.

Whether this notion of becoming is coherent depends on the identity conditions one has for events. If to be an event, one has to be a particular type of event with a certain character, then the idea is not coherent. The C_2 world determinately has the 2-chain or the 2-antichain in it, but it does not have determinately the 2-chain or determinately the 2-antichain: 'determinately' cannot penetrate inside the disjunction. This feature is a hallmark of vagueness or of metaphysical indeterminacy more generally. Without going into any details of the vast literature, let us note that there is a lively dispute over whether there can be ontological vagueness. The causet program, interpreted as we have here, supplies a possible model of a world that is ontologically vague. Further discussion of this model seems to us worthwhile.

We would like to point out that Ted Sider (2003) has supplied arguments that existence cannot be vague. In fact, he asserts (2001, 135) that anyone who accepts the premise that existence cannot be vague is committed to 4-dimensionalism, the thesis that objects persist by having temporal parts. To the extent to which many advocates of becoming reject 4-dimensionalism anyway, they would thus be open to embrace ontological indeterminacy even if Sider's arguments of 2001 and 2003 were successful. And they may well not be successful: one of them, for instance, is based on the claim that it cannot be vague how many things there are in a finite world (2001, 136f). Obviously, a defender of observer-independent becoming in CST may agree that it is at no moment vague how many events there exist, but nevertheless disagree that existence cannot be vague. Thus, we may have ontological indeterminacy without vagueness in the cardinality of the (finite) set of all existing objects.

4.3.3 Some bizarre consequences

One may be worried that on this notion of becoming in CST, no event in a future-infinite causet may ever be determinate until future infinity is reached, at which 'point' everything snaps into determinate existence. This worry is particularly pressing as realistic causets are often taken to be future-infinite. So is any event determinate at any finite stage of becoming? In general, yes. One way to see this is by way of example. As it turns out, causets based on transitive percolation in general have many 'posts', where a *post* is an event that is comparable to every other event, i.e., an event that either is causally preceded by or causally precedes every other event in the causet. Rideout and Sorkin interpret the resulting cosmological model as one in which "the universe cycles endlessly through phases of expansion, stasis, and contraction [...] back down to a single element." (1999, 024002-4)[35] Consider the situation as depicted in figure 4.4. There is a post, p, such that N events causally precede p, while all the others—potentially infinitely many—are causally preceded by p. At stage $N - 1$, shown on the left, there exist $N - 1$ events. At this stage, all the 'ancestors' of p except those three events which immediately precede p, shown in black, must have determinately come to be. Of the three immediate predecessors, shown in gray to indicate their indeterminate status, two must exist; however, it is indeterminate which two of the three exist. At the prior stage $N - 2$, the gray set of events existing indeterminately would have extended one 'generation' further back, as it could be that two comparable events are the last ones to come to be before the post becomes. At the next stage, stage N, N events exist and it is determinate that all ancestors of p exist. There is no ontological indeterminacy at this stage. Event p has not yet come to be at either stage and is thus shown in white. At stage $N + 1$, not shown in figure 4.4, event p determinately comes into existence. At stage $N + 2$, one of the two immediate successor to p exists, but it is indeterminate which one. And so on.

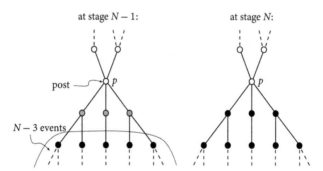

Figure 4.4 Becoming at post p (Figure adapted from Wüthrich and Callender 2017).

One may object that this interpretation of the dynamics of a future-infinite causet presupposes a given final state toward which the causet evolves, and thus involve a teleological element. Even though everything in the preceding paragraph is true under the supposition that the final causet is the one represented in figure 4.4, the objection goes, at stage N it is not yet determined *that p* is a post, as there could have been other events spacelike-related to p. Given that it is thus indeterminate whether p is indeed a post, and since this is the case for all events at

---

[35] Cf. also Bollobás and Brightwell (1997).

finite stages, no events can thus snap into determinate existence at any finite stage of the dynamical growth process.

First, it should be noted that even if this objection succeeds, it is still the case that it is objectively and determinately the case that at each stage, one event comes into being and that thus the cardinality of the sum total of existence grows. Although the ontological indeterminacy remains maximal, there is a weak sense in which there is objective, observer-independent becoming. Second, if the causet does indeed not 'tend' to some particular future-infinite causet, then all existence would always be altogether indeterminate (except for the cardinality). There would be no fact of the matter, ever, i.e., at any finite stage, not only of how the future will be, but also of how anything ever is. If this is the right way to think about the metaphysics of the dynamics of CST, we are left with a wildly indeterminate picture. Third, it should be noted that the mathematics of the dynamics is only well-defined in the infinite limit; in particular, for there to be a well-defined probability measure on Ω, we must take $\Omega = \Omega(\infty)$ (Sorkin 2007, 160n; Arageorgis 2016, §3), which can be interpreted to mean that the future-infinite 'end state' is metaphysically prior to the stochastic dynamics that grows the causet to that 'state'.

We close this section with a discussion of some of the strange features of this metaphysics. First, note that many philosophers, from Aristotle to contemporary ones, have thought that the future is indeterminate (see, e.g., Øhrstrøm and Hasle 2020 and references therein). According to some versions of this view, it is determinately true that tomorrow's coin flip will result in either a head or a tail, but it is not determinate yet which result obtains. Vagueness infects the future. We note that the above causet vagueness is quite similar, but with one big difference: on the causet picture, the past too can be indeterminate! In our toy causet of the previous subsection and its growth as shown in figure 4.2, it is not true at C_3 that C_2 determinately is one way rather than the other.

Second, note that as a causet grows, events that were once spacelike to the causet might acquire timelike links to future events. If we regard the growth of a new timelike link to a spacelike event as making the spacelike event determinate, modulo the above type of vagueness, then this is a way future becoming can make events past. That is, there is a literal sense in which one can say that 'the past isn't what it used to be'. Having said that, there is a relativistic analog in the growth of past lightcones, which also come to include formerly spacelike-related events as one moves 'up' along a worldline.

Finally, although we do not have space to discuss it here, note that despite appearances transitive percolation is time reversal invariant. This allows the construction of an even more exotic temporal metaphysics. If we relax the assumption that events can only be born to the future of existing events, then it is possible to have percolation—and hence becoming—going to both the future and past. Choose a here-now as the original point. Then it is possible to modify the theory so that the world becomes in both directions, future and past. Of course, similarly, we could have a causet that is future-finite and only grows into the past, and thus is past-infinite.

In sum, then, does CST rescue temporal becoming? At the kinematical level (not much discussed here, but see Wüthrich and Callender 2017), CST does offer new twists in dealing with time and relativity, but the basic contours of the relativistic challenge remain. Serious constraints also threaten becoming if we take the time in CST's dynamics seriously too. Here, however, if one is open to the costs of a sufficiently radical metaphysics, we maintain that there is a novel and exotic type of objective, observer-independent temporal becoming possible. It should be noted, again, that all this remains purely classical and so subject to potentially radical change in a future quantum theory of causets. So far, it is really just a classical, discrete, and dynamically stochastic theory.

Working out the implications of CST for the metaphysics of time is not only of intrinsic philosophical interest, but constitutes a valuable component of establishing the physical salience of features of causets by unpacking their potential contributions to the temporality of the world. Tracking these contributions through the emergence of spacetime from causets helps to fix the physical meaning of elements of CST. Similarly, understanding the apparently highly non-local behavior of discrete structures which are required to give rise to approximate Lorentz symmetry adds an important piece to the emergence of spacetime and so to identifying the physical salience of aspects of causets. It is to this behavior we now turn.

4.4 Lorentz symmetry and non-locality

The discreteness of its structure has another unusual consequence: a classical kind of 'non-locality', as the literature has it, of a quite different character to quantum non-locality. As we have discussed in chapter 3, the fundamental discreteness of causets has certain conceptual and technical advantages, as it promises to eliminate the nasty ultraviolet divergencies of quantum field theory (QFT) and the singularities of GR. However, it seems to also come at a price: it is incompatible with the demand for local Lorentz symmetry and the expectation of a reasonably local physics. It seems as if one can have any two of these three features—discreteness, Lorentz symmetry, and locality—in a theory, but not all three, as was already noted by Moore (1988) within half a year of the publication of the founding document of the causet program (Bombelli et al. 1987). As we will see, the discrete case differs from the Lorentz invariance of Minkowski spacetime in that events have arbitrarily many immediate predecessors (in the causal order) at arbitrarily large 'distances' away (rather than no immediate predecessors).

There is of course a sense in which no discrete structure can be Lorentz-invariant, as Lorentz transformations are only defined for continuum spacetimes. However, even if the fundamental structure is not Lorentz-invariant, one would not want to give up local Lorentz symmetry at the emergent level, at least to an extremely close approximation. This places a very strong constraint on both the spacetime structures and the dynamics of matter that can emerge from causets. As discreteness is built into the DNA of the causet program, it seems as if CST is committed to a form of non-locality, at least at the fundamental level. But our world seems manifestly local at all observed scales—apart from quantum non-locality of course—and so local physics better emerge from the fundamental physics of causets. Physics is local in that it is to offer a recipe to determine the physical state at an event as a function of the state at other 'nearby' events. Again, the non-locality we encounter in CST does not manifest itself as in the form of Bell correlations in entangled states, but is an entirely classical form of non-locality. Let's see how it arises.[36]

4.4.1 Lorentz symmetry and discreteness entail non-locality

How should we understand the claim that the demand for (local) Lorentz symmetry and the discreteness of causets entail a form of non-locality of the physics? The discreteness of causets in built in from the start—in axiom 1 in §3.2.1. Lorentz symmetry is the demand common in relativistic physics that the dynamics be invariant under Lorentz transformations. In a special-relativistic theory, such as the standard model of elementary particle physics, this demand is encoded in the invariance of the theory's action under Lorentz transformations and in the spacetime symmetries of Minkowski spacetime. In GR, spacetimes in general no longer have global symmetries like Lorentz symmetry, but Lorentz symmetry still holds 'locally', i.e., at sufficiently small scales. Consequently, causets must give rise to spacetimes which are locally Lorentz symmetric. In this sense, Lorentz symmetry must at least hold at some mesoscale: it may be violated at the fundamental level, and it may not hold as a global symmetry of emergent spacetime, but there must be a robust range of scales in which it holds to an extremely good approximation. No violation has ever been detected,[37] and so the demand is on empirically rather secure grounds.

In order to recognize the non-locality of a Lorentz-invariant, discrete structure, let us return to the standard way of conceiving of the relation between causets and spacetime as depicted in figure 4.3, prototypically captured by a Poisson sprinkling of elements into Minkowski spacetime (as in §4.1.3). The outcome of the sprinkling must be random, or erratic, in that it cannot be too regular. If it had even just

---

[36] The best references explaining this resulting non-locality in the context of CST are Sorkin (2009) and Dowker (2011).

[37] In the sense that the bounds for violating Lorentz symmetry are extremely tight (Will 2014, §2.1.2).

102   OUT OF NOWHERE

a somewhat regular lattice structure, then it would involve preferred directions because a distribution of elements in Minkowski spacetime will look different from its Lorentz-transformed cousins. Thus, the demand for Lorentz symmetry translates into a requirement that the sprinkling be random (Bombelli et al. 2009). In a Poisson sprinkling for instance, the probability of sprinkling n elements into a spacetime region of spacetime volume V depends only on the density ρ of the sprinkling and V:

$$P(n) = \frac{(\rho V)^n e^{-\rho V}}{n!}. \tag{4.1}$$

Since ρ is fixed, this probability is manifestly invariant under all transformations preserving spacetime volume, including Lorentz transformations (Henson 2012, §12.2.1).[38]

The density ρ of the sprinkling is a fundamental constant of the theory, but is naturally set at something like the Planck scale. As noted by Moore (1988), the spacetime volume of a hyperboloid shell of any finite thickness (defined spatiotemporally by an invariant interval Δs) at a given spatiotemporal distance in the past from any event p is infinite. The grayed out region in figure 4.5 represents such a hyperboloid shell to the past of an event p. Thus, whatever the fixed, constant density ρ is, the number of elements in the hyperboloid shell diverges. If the hyperboloid shell is, for example, at a Planck interval to the past of p, this means that p will have infinitely many predecessor elements a Planck distance in its past. In fact, whatever the spacetime interval chosen as the scale, and whatever the (finite) thickness of the shell, this will invariably be the case. Moreover, the spacetime volume of the region in p's past lightcone which consists of points *less* than the chosen interval in the past (i.e., the past lightcone of p minus the hyperboloid shell and *its* past) is also infinite and so contains infinitely many sprinkled elements. Thus, within any fixed spacetime interval to the past of an element of a causet being approximated by a Minkowski spacetime, there are necessarily an infinite number of elements. In this sense, there is no nearest neighbor to the past of any given event in a sprinkling of Minkowski spacetime.[39] For any event sprinkled close to p in p's past, there is always another event sprinkled to the past of p, which is even closer to p (Bombelli et al. 2009). The same holds for immediate neighbors to the future.[40] This is a form of 'non-locality' in discrete spaces in which in any inertial frame events arbitrarily far away spatially from p (in that frame) will be arbitrarily close to p in terms of the invariant spacetime interval.

---

[38] Of course there remains a finite chance that any finite sprinkling is (somewhat) regular, but this chance decreases with the size of the causet.

[39] 'Nearest' as measured in terms of the invariant spacetime interval Δs, i.e., smallest non-zero interval between timelike related events.

[40] We are not interested in spacelike relations between events, even though the same applies mutatis mutandis for spacelike-related neighbors.

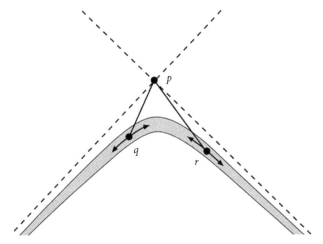

Figure 4.5 There are infinitely many sprinkled events in a thickened past hyperboloid of event p, such as q and r.

Inevitably, it thus seems as if any element of any causet being approximated by Minkowski spacetime must have an infinite number of 'immediate' predecessors and an infinite number of 'immediate' successors. What is an 'immediate' or 'nearest' neighbor? In the fundamental theory, it is defined as being directly related, i.e., the longest path between the pair has length one. In Minkowski spacetime, the sprinkled (timelike related) events are 'immediate' or 'nearest' neighbors if the Alexandrov interval defined by the two events does not contain another event. The Alexandrov interval is defined as the non-empty intersection of the causal future of one event and the causal past of the other. As Moore (1988, 655) concludes, it seems as if discrete 'spacetime' "has the nasty property that every point is influenced by an infinity of 'nearest neighbors' which, in a given frame, are arbitrarily far back in time." Analogously, infinitely many nearest neighbors are arbitrarily far away in space. Note that this form of non-locality does not arise in the continuous case, such as in Minkowski spacetime, where an event does not have any immediate predecessors, only an open sequence of ever smaller spheres of events.

In their reply, Bombelli et al. (1988) offer two lines of defense, both of which are sound in our view. First, they point out that this is really the natural discrete analog of the fact that an event in a Lorentz manifold has neighborhoods which converge to its lightcones as well, and so *should* be reflected by causets. In this sense, in a continuum spacetime, events spatially arbitrarily far away from p in a given frame are also arbitrarily close to p as measured by the spacetime interval. Perhaps the non-locality inherent in causets is really not all that troubling after all. Second, they point out that (general-)relativistic spacetimes are only locally Lorentz-invariant

104   OUT OF NOWHERE

and that in important cases such as the Friedmann-Lemaître-Robertson-Walker spacetimes, the past lightcones of each event has a finite volume and so a sprinkling would suggest only a finite number of events to the past in a discrete analog. In fact, they assert, this will be so in the general situation using a 'sum-over-histories' approach, which they prefer.

This latter point deserves some unpacking. In general, Lorentz invariance will not turn out to be a global symmetry of emergent spacetime. In this sense, space-time curvature will limit Lorentz symmetry, restricting the number of nearest neighbors of an element in an underlying discrete structure to a finite number. However, unless the spacetime curvature is rather large, i.e., unless the radius of curvature is comparable to the Planck length, the number of nearest neighbors of any element will still be very large. Hence, fundamental physics according to CST cannot be 'local' in the sense that the physics at an element of the causet depends only on that of a neatly confined, finite (and somewhat 'small') set of elements.

A remark before we proceed. This non-locality becomes apparent only once we consider embeddings of causets into Minkowski spacetime and countenance the symmetries of that spacetime. One might thus be tempted to think that the non-locality is somehow an artifact of this conception of the relation between causets and spacetimes, an artifact which might disappear if we replace the Poisson sprinkling of events into Minkowski spacetime with an appropriate form of relating the two. But this would be too quick: the argument shows, quite generally, that a causet whose elements do not have a very large number of immediate neighbors cannot possibly give rise to a spacetime with at least approximate local Lorentz symmetry. Thus, the high valency of its elements should be expected of any physically realistic causet.

What then to make of this non-locality? As Benincasa and Dowker (2010) explain, it is both a blessing and a curse vis-à-vis two central challenges of any approach to quantum gravity trading in fundamentally discrete structures. The first goal of quantum gravity is to explain how the dynamics manages to drive these discrete kinematic structures towardthose that are well approximated by Lorentzian manifolds. The worry, recall, is that the sheer number of non-Lorentzian (approximated) causets grows so fast with the number of nodes that they are overwhelmingly probable. The aim is then to show that the dynamics favors the Lorentzian causets. In the statistical approach that Benincasa and Dowker adopt, the probability of a causet therefore depends on the action: it is proportional to e^{-S}. Now consider a finite spacetime region, containing n sprinkled points; if the interaction is local, then broadly speaking points in the region interact with no more than n neighbors, and the action $S \sim \alpha n$ (at most), where α is some constant. Even if this weight is far greater than that for a non-Lorenzian causet, then the latter will still be far more probable if, as

THE EMERGENCE OF SPACETIME FROM CAUSAL SETS   105

seems plausible given our earlier discussion, their cardinality grows faster than exponentially in n. The only hope, that is, is that in Lorenzian causets a node interacts with far more nodes than those in the region considered—that they interact non-locally in just the way that Moore pointed out. In this sense, non-locality may be a blessing. Although this is of course not a conclusive argument, Benincasa and Dowker thus do not expect a local dynamics to be able to account for the manifoldlikeness and thus not be a solution in the sense of §4.1.2. The second challenge of a discrete approach to quantum gravity according to Benincasa and Dowker is to explain why the geometry of the emergent spacetime should be a solution to the EFE. With regard to this challenge, the non-locality turns out to be more curse than blessing, as local dynamics and the demand for general covariance "pretty much guarantee" (Benincasa and Dowker 2010, 2) that the EFE be satisfied. In the absence of a local dynamics, there is no such guarantee, and it remains an open question whether CST or a similarly non-local discrete approach to quantum gravity can explain how the geometry of emergent spacetime conforms to GR. They fear that "if causets were incorrigibly non-local, this would be fatal" (ibid.) for the project. But fortunately, there is hope.

4.4.2 Implications: hierarchy of scales and phenomenological signatures

This hope will be fulfilled if there exists an intermediate scale at which effective physics is approximately local. In the previous subsection, we have seen that the physics of CST must be non-local at the fundamental level. Due to the presence of spacetime curvature, the global structure of spacetime will not be Minkowskian, and hence there is no requirement that the physics be Lorentz-invariant at cosmological scales, and it may well be local. However, to repeat, physics must definitely be Lorentz-invariant across a wide range between the fundamental and the cosmological scales. Thus, whatever the fundamental physics, and in particular the fundamental dynamics, it must give rise to quasi-local physics, which is Lorentz-invariant to an extremely close approximation. What exacerbates the situation, argues Sorkin (2009, 28), is that as the non-local couplings by far outnumber the local ones, the non-locality cannot be entirely restricted to the fundamental level.

However, as Sorkin goes on to explain, there is promising evidence that the emerging physics is reasonably local. He considers (in §3.1) the limited but paradigmatic example of a massless, classical scalar field ϕ in two dimensions with the standard equation of motion, i.e., $\Box\phi = 0$, where '$\Box$' is the d'Alembertian operator. It is the significant achievement of this paper to propose a 'discretized' version of a 2-dimensional d'Alembertian in terms native to the fundamental causet ("fully

106 OUT OF NOWHERE

intrinsic" in Sorkin's words 2009, 33).[41] This d'Alembertian is thought to exemplify the sort of tools needed to construct a 'field' theory on the basis of CST, which will give rise, ultimately, to standard QFT. Sorkin's d'Alembertian includes non-local couplings at the fundamental level, which, however, will be suppressed at larger scales, or so he argues. Sorkin thus outlines a research program for developing the tools of recovering established physics from fundamentally discrete and thus highly non-local physics. Benincasa and Dowker (2010) add to this evidence by formulating a more general non-local operator, which approximates the local d'Alembertian for a scalar field in 4-dimensional flat spacetime.[42] Their operator is effectively local as well, courtesy of the non-local contributions to magically cancel out, leaving just 'local' contributions. Here, 'local' refers to the frame defined by ϕ itself, i.e., to the frame in which ϕ varies 'slowly'. It turns out that the operator is effectively local in this frame, with non-local contributions suppressed.[43]

Although it must be clearly stated that none of this suffices for anything like a definitive verdict on the matter, these preliminary results amount to a 'proof of concept', as Benincasa and Dowker (2010, 4) insist, in that they establish "the mutual compatibility of Lorentz invariance, fundamental spacetime discreteness and approximate locality".

One upshot of this incipient research is also that the fundamental non-locality in CST cannot be completely tamed. Nor should it be: the proposed physics should ultimately have empirically detectable consequences, after all. As the theory delivers Lorentz-invariant mesoscopic physics by construction, no violation of Lorentz symmetry should be expected in the approach. Thus, the novel phenomenology must come from elsewhere, and the non-local discreteness seems a promising source. We will close the chapter by discussing two such potential effects: swerving particles and the cosmological constant.

---

[41] Assuming that $\square$ acts linearly on the field ϕ, as Sorkin does, the d'Alembertian for a causet $\langle C, \prec \rangle$ can be expressed as a suitable matrix B_{xy}, where x and y range over the elements of the causet. Sorkin also requires that B is 'retarded' or 'causal' in that $B_{xy} = 0$ in case y is spacelike to, or precedes, x. Using a series of informed guesses based on methods in theoretical physics and computer simulations, Sorkin (2009, 33) proposes that (ignoring a constant scale factor)

$$B_{xy} = \begin{cases} -\frac{1}{2} & : x = y \\ 1, -2, 1 & : x \neq y \text{ for } n(x,y) = 0, 1, 2, \text{ respectively,} \\ 0 & : \text{otherwise,} \end{cases}$$

where $n(x, y)$ is the cardinality of the order interval $\langle y, x \rangle = \{z \in C | y \prec z \prec x\}$. This d'Alembertian is local in that the physics at events removed by more than two 'ticks' has no influence. However, it is still non-local in that events which are immediately causally connected will be arbitrarily far away spatially in some frames.

[42] Glaser and Surya (2013) propose an order-theoretic characterization of local neighborhoods in manifoldlike causets, thus providing a frame for this as well as further evidence.

[43] We thank Fay Dowker for private correspondence on this point.

4.4.3 Swerving particles: putting matter on causets

An issue wide open for CST is how causets interact with matter, e.g., with particles or non-gravitational fields. The d'Alembertian constructed in Benincasa and Dowker (2010) and related research efforts paint a way in which one might see fields propagate on causets. How should we put matter 'on' causets even kinematically, prior to giving it the proper dynamics? In the simple case of a (real-valued) classical scalar field, one just sticks a number on each element of the causet. More precisely, a scalar field on a causet is represented by a map from the causet to the field $\mathbb{R}$ of real numbers. The 'dynamics' then gives us a rule restricting the relations among these numbers. Complex-valued fields and quantum fields—although more complicated—will intuitively be constructed along similar lines, mutatis mutandis.

What about particles? Restricting ourselves to the simpler case of point particles, a particle will presumably trace out a timelike path on a causet. In other words, its worldline will consist in a set of pairwise causally related elements of the causet, i.e., on a chain of causet elements. Any chain in a causet is a kinematically possible worldline for a point particle. Surely, we will want to restrict the physically possible trajectories to some subset of such worldlines. Thus, we also need to formulate dynamical rules which determine, at the emergent level at least, some standard of inertial or geodesic versus forced motion. This is a challenging task, which remains to be solved in full generality.

However, Dowker et al. (2004) have proposed a simple model of a point particle moving on a causet sprinkled into Minkowski spacetime. This allows the use of the background Minkowski spacetime as a standard for inertial or accelerated motion. In the simple model, the task is to determine the continuation of the particle's trajectory given an initial portion of that trajectory. A central assumption of the model is that the determination is quasi-local in time: the particle's past trajectory within a certain proper time to the past fully determines its future continuation. Dowker et al. then propose an approximately Markovian, Lorentz-invariant dynamical rule, which, they argue, is the best discrete analog of geodesic motion in Minkowski spacetime. Without going into the details here,[44] the basic idea is that the continuation which best preserves linear momentum (in the frame in which the three-momentum going 'into' the event vanishes) is the dynamically chosen one. In general, the three-momentum going 'forward', although minimized, will not be zero. In this sense, the particle appears to 'swerve' away from the geodesic, apparently undergoing a random acceleration. Since this swerving diverges from the expectation of strictly geodesic motion, it would be in principle an empirically detectable signature of CST.

---

[44] A useful illustration can be found in figure 1 in Dowker (2011). See also Henson (2009, §21.3.2), as well as Dowker et al. (2004) of course.

108   OUT OF NOWHERE

That the effect size would be rather small compared to observationally accessible scales is just a practical problem. There are also more principled reasons to be critical of the model. First, the particles are not modeled in a credible manner: realistic particles will not be point particles, and be quantum in nature, so that the classical, deterministic rule above seems ill-fitted. Second, the model makes ineliminable use of Minkowski spacetime to deliver a standard of inertial motion. Surely, the details of the particle motion should not depend on the embedding of the causet into a spacetime, but be intrinsically defined in the causet. Third, and perhaps most damning, there is no sense in which the dynamics includes a back-reaction of the matter content on the causet, as we would expect from GR. There are ways to overcome some of these difficulties (see e.g., Philpott et al. 2009), but a full solution remains an open question.

4.4.4 The ever-present Λ

Let us end with a brief discussion of CST's most suggestive 'prediction', viz., that the cosmological constant Λ has a small, but non-zero positive value of the order of 10^{-120} in natural units. The 'prediction' earns its scare quotes due to its being a heuristic, and certainly defeasible argument, rather than a tight quantitative calculation. But it clearly is a *prediction* in that it was made prior to the relevant observations, and not a mere ex post facto construction of a just-so justification. And this makes it rather intriguing. The prediction is due to Rafael Sorkin and appears to date from the late 1980s, roughly a decade before the discovery of the accelerating of the universe's expansion in 1998 by Saul Perlmutter, Adam Riess, and Brian Schmidt and their teams. The first brief published version of the argument can be found in the last paragraph of Sorkin (1991), a published version of the talk given a year earlier, in 1990. We here mostly follow Sorkin (1997b, §7.2).

The central idea is that the non-vanishing cosmological constant arises not from a form of 'dark matter', but instead is just due to fluctuations in the volume V of spacetime. Thus, it offers an explanatory paradigm for a small positive Λ (or for the accelerated expansion), which differs interestingly from the standard account invoking dark matter.

CST uses, as its only measure of 'size', the number N of causet elements. Earlier (§4.1.2), we have seen that N is a (in fact, *the*) generally covariant quantity in classical sequential growth dynamics. Thus, N is the measure of the spacetime volume V of emergent spacetime. As a second ingredient, the argument uses the idea (borrowed from unimodular gravity) that the cosmological constant Λ is in some sense conjugate to V. In a loose analogy to the energy-time uncertainty, with Λ an 'energy' and V a 'time', we have $\Delta V \Delta \Lambda \sim 1$ (in natural units).

The third input required is the recognition that, returning to the paradigm of a Lorentz-invariant sprinkling of causets into the approximating spacetime, in CST

THE EMERGENCE OF SPACETIME FROM CAUSAL SETS   109

although N is a measure of V, the latter suffers from Poisson fluctuations in N, with a typical magnitude of $\sqrt{N}$. Consequently, for a fixed N in the fundamental causet, V will only be fixed up to fluctuations of magnitude $\sqrt{N} \sim \sqrt{V}$. Thus, the resulting uncertainty of V is given as $\Delta V \sim \sqrt{V}$. Putting this together, we obtain for the uncertainty of Λ:

$$\Delta\Lambda \sim \frac{1}{\sqrt{V}}. \tag{4.2}$$

Assuming that Λ fluctuates about zero (by whatever mechanism), then what we observe today is only this fluctuation, i.e., $\Lambda \sim V^{-1/2}$. Further, assuming that V is a 4-dimensional 'Hubble sphere', we replace V by the fourth power of the Hubble radius H^{-1}, where H is the Hubble constant, and thus come to expect

$$\Lambda \sim H^2, \tag{4.3}$$

which turns out to be of the order of the 10^{-120} in natural units mentioned above for present-day values. This is remarkably close to the value 10^{-122} derived from astronomical observations (see Barrow and Shaw 2011).

A few remarks in closing. First, it should be noted that (4.3) holds for any cosmic epoch. Unlike for other approaches where Λ can vanish during some epochs, it is thus 'ever-present' here. Second, given the immense advances in observational cosmology, the heuristic argument presented above, although suggestive, will need to be supplemented by physically more rigorous and quantitatively more precise models based on CST in order to fulfill the explanatory promise issued by the above argument. Ahmed et al. (2004) offer the beginning of a more sophisticated analysis; other models exist. Third, just as we should be used to by now, both the heuristic argument reproduced here and the more sophisticated model in Ahmed et al. (2004) and elsewhere derive the effect from purely classical fluctuations. Of course, in order to deliver a fully quantum theory of gravity, these considerations cannot be but first steps toward the full theory. Nevertheless, the prediction of the ever-present Λ on the basis of such simple and straightforward ideas connected to CST surely is intriguing, particularly also in light of the explanatory plight of standard approaches to account for the acceleration of the universe's expansion.

4.5 Summing up

We could have cast the issue of the emergence of spacetime in CST in terms of Pythagoreanism (in terms of footnote 1 in §2.1), the view according to which all being is mathematical, formal, or abstract. The worry would be that if our physical world is not fundamentally spatiotemporal, then the Pythagoreans were right and there could not exist something physical, material, or concrete. But this implication does not hold: the world according to CST is quite clearly non-spatiotemporal in its fundamental nature (as we have argued in chapter 3), and yet, space, time, and perhaps even material objects of our manifest world emerge, or at least *can* emerge, as we hope to have sketched in this chapter. We have articulated what we take the task of establishing this to consist in, and we have invoked spacetime functionalism in order to reject further (and impossible) tasks such as deriving the qualitative nature of spacetime from mathematical structures. With spacetime functionalists, we maintain that a functional reduction of spacetime and its roles in other theories and in their empirical confirmation exhausts the task to be completed. It is by means of this functional reduction, and by means of it alone, that the mathematical derivations involved in the task obtain physical salience.

Apart from this central task of establishing the emergence of spacetime as sketched in §4.1 and §4.2, we have also addressed foundational and philosophical issues that arise in the context of the emergence of spacetime in CST. We discussed the consequences of CST and its dynamics needed to regain spacetime for the metaphysics of time (§4.3) and the non-locality inherent in CST and the related questions of how to incorporate matter into the framework and thereby identify its empirical signatures (§4.4). Both of these issues have put the difference and the commonalities between (classical) CST and (continuum) relativistic spacetimes into sharper contrast. But more interestingly still, they show how inextricably connected physics, mathematics, and philosophy are in the pursuit of a theory of QG.

Chapter 5
The road to loop quantum gravity

5.1 Learning from general relativity for quantum gravity

General relativity (GR) is an enormously successful theory: from precision measurements confirming the bending of light rays near massive bodies and the slowing of clocks in gravitational wells, to its prediction of the existence of black holes and gravitational waves, to its applications in terrestrial navigation. To date, there is no accepted empirical disconfirmation of GR, even though there are of course accepted limits of its applicability. Given this situation, it is arguably desirable to try to preserve as much of GR as possible. It is certainly a clear condition for the acceptability of any future quantum theory of gravity that it be able to reproduce the successes of GR—and, ideally, to explain its possible failures and limitations. In order to accomplish these goals, it is not necessary that full GR be recovered in some appropriate reductive 'limit', as we will argue in the next chapter. However, it may well pay to seek to articulate a theory of quantum gravity which in some sense embraces the main lessons of GR. This is by no means a necessary condition (and string theory is not motivated by a strategy seeking to preserve those lessons), and may well fail to deliver success. In fact, Newton's theory of universal gravitation has been replaced by GR, and even though much of Newton's theory can be recovered in the limit of weak gravitational fields, the two theories postulate significantly different ontologies. A similar situation could arise in quantum gravity.

This chapter and the next are concerned with loop quantum gravity (LQG) (and with canonical quantum gravity more generally), which starts out from a heuristic principle that we ought to be as 'conservative' as possible when formulating a quantum theory of gravity. Consequently, canonical approaches attempt to transpose what they take to be GR's central innovation, viz. its lesson that spacetime is not a fixed 'background' determining inertial 'forces' acting on the content of the universe, but instead a dynamical structure interacting with that content. It seems as if canonical quantization offers a recipe which honors the heuristic principle of 'conservativeness' in this sense.

This strategy of 'quantizing' GR is opposed to the approach chosen by causal set theory (the topic of the previous two chapters), even though causal set theory also starts out from GR and tries to preserve what it takes to be GR's main lessons. The difference is that approaches such as causal set theory abstract away from GR quite a bit and rather radically rebuild a theory mostly from scratch, while the

Out of Nowhere. Nick Huggett and Christian Wüthrich, Oxford University Press.
© Nick Huggett and Christian Wüthrich (2025). DOI: 10.1093/oso/9780198758501.003.0005

112   OUT OF NOWHERE

former approaches attempt to keep GR as much as possible and employ established techniques such as known procedures of 'quantization', even though these techniques may only be used as heuristics to find a quantum theory whose classical limit is GR (or something close to it). The difference is one of degree, however, rather than in kind.[1] Although causal set theory is more radical in this sense, its concepts, ideas, and techniques tend to be clearer and simpler than those of LQG. For this pedagogical reason, we discussed it first.

The quantization procedure of interest here is so-called 'canonical quantization' (Dirac 1925), which is at the heart of LQG, but also of other approaches in 'canonical quantum gravity', such as 'quantum geometrodynamics' (Kiefer 2004, ch. 5). The method starts out from a classical theory cast in a Hamiltonian formalism and proceeds by converting the canonical Poisson brackets, which form an algebra, into an algebra of canonical commutation relations among basic operators acting on an appropriately chosen Hilbert space. In this sense, it tries to formulate a quantum theory which closely tracks the formal structure of the classical theory, and so can be expected to preserve its basic features. Canonical quantization offers a powerful recipe for churning out quantum theories based on classical ones, as for instance the success of Dirac's construction of quantum electrodynamics testifies. Given the prominent success of the scheme, it is only natural to try and apply it to GR. We will see in the next section in more detail how this can be done.

Before we get there, we need to get clear on two points. First, quite generally, why would one seek to preserve GR's central lessons in a theory that will obviously supplant it? Second, what are, more specifically, the main lessons and the basic principles of GR that physicists who work on this approach consider to be worthy of preservation into a quantum theory of gravity?

As to the first point, it is clear that not all principles, basic tenets, lessons, and ideas of GR can be preserved into quantum gravity, for otherwise we will just end up with GR, and not quantum gravity. Something has to give. As a matter of logic, therefore, any of the principles of GR, even apparently central ones, may have to be given up in quantum gravity. Although fallible, it seems reasonable to first stick to what one takes to be those lessons and innovations of GR responsible for its past success in the domain in which it worked so well, and to let go of those parts which appear more peripheral. Moreover, we plausibly assume that among the aspects of GR which will not survive into quantum gravity will be those that are typically missing when we move from a classical theory to a corresponding quantum

---

[1] This is also distinct from string theory—to be discussed in chapters to come—which attempts to deliver a unified theory of all interactions covered in the standard model as well as gravity. Thus, the vantage point of string theory is QFT, not GR. In the framework treated in this chapter and the next, one does not care much either about these other interactions or about the constitution of matter. Instead, the primary focus lies on gravity and so on spacetime geometry. If this is a mistake, then the framework presented here is unlikely to succeed.

theory; slightly more strongly, we may also suppose that it is only those aspects of GR which will not survive. We take it that there is really no stronger methodological imperative than this plausible, but clearly uncertain, heuristic principle which underlies canonical approaches to quantum gravity—just as there isn't for any other research program in quantum gravity.

Where things gets interesting—and controversial—is when we turn to the second point and so look at what different research programs take to be central innovations of past physics on which to build future physics. What is the approach taken in LQG? Most advocates of LQG maintain that GR's central insight, which is to be preserved into future physics, is the idea of 'background independence'. In fact, it is often claimed that its background independence constitutes one of the major advantages of LQG over string theory. Even accepting that background independence is, in fact, an advantage in a theory of quantum gravity, whether LQG indeed has an edge over string theory here turns on one's concept of 'background independence' as well as on subtle points regarding the interpretation of string theory. We will return to the issue of whether string theory is in any problematic way background *dependent* in §10.4. Our conclusion there will be that this is not the case on a suitably loose but physically relevant reading of background independence.

The notion of background independence at work in programs of QG based on GR is somewhat narrower, as shall be examined in §5.2. In §5.3, we discuss the reformulation of GR in Hamiltonian form, a necessary condition for canonical quantization to be applied. At last, we introduce LQG in §5.4. The last substantive section of the chapter, §5.5, presents the covariant approach to LQG, which has become dominant in recent years. §5.6 concludes.

Let us finish with a remark concerning the technical level of the material in this chapter. We generally presuppose the basics of GR and some differential geometry, particularly in §5.2–5.4. These three sections are rather technical, as is §5.5. They can be skipped by readers with little appetite for the technical nitty-gritty. Those readers can skip right to the next chapter, although §5.3.3 may still be of interest.

5.2 Background independence and general covariance

This section gives a detailed analysis of the related notions of background independence and general covariance and thus of what many take to be the chief lesson of GR. Readers who are not interested in the minutiae of these ideas but instead want to move to the development of LQG proper can skip this section.

Background independence is widely accepted—even among string theorists— as an important criterion in theory choice, in particular as we seek a fundamental theory of gravity. Its relevance, however, we think, derives both from its 'backward'

114   OUT OF NOWHERE

as well as its 'forward' role in the construction and interpretation of theories. Its backward function, to be discussed in this section, is to encode the equivalence principle in the general framework of GR. In this sense, it captures important aspects of GR's central idea of the geometrization of spacetime. Its forward function is to deliver a guiding principle for what we take to be the orthodox interpretation of LQG, to be discussed in §6.1.1 but also in the next chapter when we discuss the emergence of spacetime in LQG. By playing an important role in the physical interpretation of LQG's fundamental structures, background independence in its forward-looking role contributes to the understanding of the (putative) physical salience of LQG.

The general idea of background independence is best explained through its emergence from the considerations regarding the equivalence principle.[2] Disregarding its many (inequivalent) formulations and their relations, the equivalence principle can be understood as the heuristic principle which guided Einstein to GR and according to which effects due to inertia and to gravity are manifestations of the same structure, the 'inertio-gravitational field'. This unitary structure can be split up into inertial and gravitational components differently, depending e.g., on the kinematical state of an observer. In the literature, this idea is sometimes interpreted to imply the existence of one of these components only, at the expense of the other: either it is said that there really just is the gravitational field; or it is claimed that GR shows how gravity is not a force, but instead is subsumed under geometrical aspects of spacetime. Beyond a terminological debate, is seems clear that according to GR, there is just one fundamental physical field here, which is responsible for both what are interpreted as inertial and gravitational phenomena.

Whatever its name, there is an important sense in which this field is dynamical, i.e., it interacts with other physical fields. If, however, the field which is responsible for inertial, spatiotemporal effects is dynamical, then there is no longer any fixed inertial background as we find it in Newtonian physics, quantum physics, and special relativity. At root, 'fixed' here means that the background field is identical across all the models of the theory (as has been clarified by Pooley 2017), which can be given the intuitive gloss in terms of interactions.

For instance, conventional quantum field theory (QFT) requires a fixed inertial spacetime background. Standardly, Minkowski spacetime offers such a background, but in principle many fixed spacetime metrics, including curved ones, may be used. The background metric enters in most equations of QFT; it underlies the formulation of the commutation relations, the operator product expansions, the propagators, etc. Whatever the specific choice, *some* background metric must be assumed in conventional QFT. Conventional QFT applied to gravity cannot,

---

[2] An approach similar to the one proposed here can be found in Rovelli (2004, ch. 2) and Belot (2011), even though our notion of background independence is more circumscribed than those found in these references (and much more so than that in Smolin 2006).

therefore, evade excising the inertial component from the inertio-gravitational field and so assumes some split of the metric field,

$$g_{ab} = b_{ab} + h_{ab},\qquad(5.1)$$

where b_{ab} is a non-dynamical background field and h_{ab} are those perturbations on the inertial background field which one considers the dynamical gravitational field of interest.[3] The background b_{ab} encodes spatiotemporal relations and thus causal structure. On this smooth classical background b_{ab} lives the field h_{ab}, which—if it is a quantum field—must commute at spacelike separations as determined by the background geometry encoded in b_{ab}.

Quantized linear gravity with its division of the gravitational field into a fixed background structure and a dynamical quantum field as in equation (5.1) delivers useful approximations for sufficiently weak gravitational fields and is thus widely believed to be valid for low-energy applications (see Wallace 2022b). However, such a split stands in contradiction with the idea of the equivalence principle as formulated above. A theory in line with such an equivalence principle must be *background independent*, in the sense that it does not assume a split of the inertio-gravitational field into a fixed inertial and a gravitational component. To the extent to which a fundamental theory ought to honor the equivalence principle, it should thus not involve such a split. Background independence, often invoked as a selling pitch by advocates of LQG, can thus be seen to essentially amount to a postulate demanding a generalized equivalence between inertial and gravitational masses.

It should be noted that one can make an argument—common in string theory and discussed below in §10.4—that if the split is arbitrary, i.e., if no particular split is physically privileged, there is a sense in which background independence is maintained, as in those cases the split is somewhat formal, not physical. Of course, advocates of LQG may well not accept that this weakened form of background independence fully answers to their motivations. Moreover, it should be noted that theories which still formally require a physically idle split have arguably yet to find their most perspicuous formulation.

How can background independence be formally implemented in the mathematical language of a spacetime theory? Many authors equate, without due explication, background independence with *general covariance*, which in turn is often equated with *invariance under active diffeomorphisms*.[4]

Let us introduce the relevant notions somewhat more carefully.[5] Let us call a theory *generally covariant* just in case its equations retain the same form under

---

[3] We follow the usual convention of using Latin letters in indices when we use the abstract index notation to designate coordinate-free geometric objects and Greek letters when we refer to a coordinate-dependent entity.

[4] We will come back to our views on the relationships between these notions below. Note that dissent to this common view is voiced, e.g., by Pooley (2010).

[5] Our treatment is based on Wüthrich (2006, ch. 3).

116 OUT OF NOWHERE

general transformations, where 'general transformations' will be characterized below. Retaining 'the same form under general transformations' can be understood in at least two different senses: passively and actively. Passively, the equations keep the same form in any coordinate system, i.e., they are covariant under arbitrary coordinate transformations. Active transformations, on the other hand, do not primarily concern the expression or re-expression of the same physical content in some different coordinate systems, but instead they are maps between manifolds which move around the fields themselves. As such they are best considered in a coordinate-independent way.[6]

The *principle of general covariance* demands that a physical theory be generally covariant under active diffeomorphisms, where *diffeomorphisms* are one-to-one and onto C^{∞}-maps from a manifold to itself with a C^{∞}-inverse. Such a diffeomorphism ϕ then 'pulls back' or 'pushes forward' scalar functions and vector and tensor fields on the manifold.[7] The considerations presented in footnote 6 can be generalized. For any two tensor fields defined on a manifold and related by an active diffeomorphism, we can find a corresponding passive diffeomorphism encoding a coordinate transformation such that the functions which represent one of the tensor fields' components in one coordinate system are identical with the ones that represent the other tensor field's components in the other coordinate system.[8] If a theory is thus generally covariant under active transformations, then its equations of motion are the same in both coordinate systems. This implies, in turn, that for a dynamical tensor field which satisfies the equations of motion of a generally covariant theory, its relative by an active diffeomorphism must be a solution of the same equations too. Let us call a theory *invariant under active spacetime*

---

[6] In his explanation to distinguish between active and passive diffeomorphisms, Rovelli (2004, §2.2.4) considers Earth's surface and two points on it—the city of Paris and the village of Quintin in Brittany—as well as a temperature field defined everywhere on Earth's surface. He associates a passive diffeomorphism with a re-expression of the temperature field at a fixed point as a consequence of a mere re-coordinatization of Earth's surface (choosing the origin of the coordinate system in, e.g., Paris or Greenwich). Covariance under passively interpreted transformations thus only means that nature does not care which coordinate system we impose upon her. Actively interpreted transformations, on the other hand, relate two a priori distinct fields with one another. For the sake of illustration, Rovelli considers a constant breeze which has uniformly transported yesterday's temperature field to today's. Yesterday's temperature field and today's temperature field are a priori distinct fields. Therefore, postulating covariance under actively interpreted transformations amounts to more than just claiming that nature does not mind about her coordinate dress. The two interpretations are often confused because for any two fields related by an active diffeomorphism, one can always find a passive transformation such that in the new coordinate system one field is expressed by the same functions as the other was in the old coordinate system. Imagine that the wind has moved the air from Quintin to Paris within 24 hours, and thus rotated the temperature field by a given angle α around Earth's axis. If we think of this constant breeze which brings yesterday's temperature field into today's as an active diffeomorphism relating the two, then today's temperature field at Paris can be given by the same expression as yesterday's field at Quintin by rotating the coordinate system in which we write the temperature field by the same angle α.

[7] See Malament (2012, §1.1–1.5) for a detailed development of these notions.

[8] It is worth pointing out that active diffeomorphisms are maps defined on the manifold $\mathcal{M}$, while passive diffeomorphisms are defined on $\mathbb{R}^n$.

diffeomorphisms just in case active spacetime diffeomorphisms map a solution of the dynamical equations to another solution of the same equations.

Thus, the demand for general covariance together with the uncontroversial codification of arbitrary coordinate transformations qua passive diffeomorphisms imply invariance under active diffeomorphisms. Conversely, if a theory is invariant under active diffeomorphisms and arbitrary coordinate transformations are captured by passive diffeomorphisms, then the theory is generally covariant. General covariance and invariance under active diffeomorphisms will therefore, somewhat imprecisely, often be used interchangeably here—and elsewhere in the literature—because we assume that arbitrary coordinate transformations are passive diffeomorphisms.

Let us now take these concepts and make the idea of general covariance more precise. To that end, we conceive of GR as a set of models, where each model consists in a triple $\langle \mathcal{M}, g_{ab}, T_{ab} \rangle$ of a manifold $\mathcal{M}$, and a metric field g_{ab} and a stress-energy tensor T_{ab} defined on $\mathcal{M}$ which satisfy the Einstein field equation (EFE) $G_{ab}[g_{ab}] = 8\pi T_{ab}$. A symmetry operation is a map from models to models such that (i) the set of models which characterizes GR is preserved, i.e., if a model is in that set, then its image under that map is also in the set, and (ii) it arises from a corresponding spacetime diffeomorphism $\phi \in \text{Diff}(\mathcal{M})$, where $\text{Diff}(\mathcal{M})$ is the group of diffeomorphisms on $\mathcal{M}$. In this sense, a symmetry of GR maps solutions of the field equation to solutions of the field equation such that a model $\langle \mathcal{M}, g_{ab}, T_{ab} \rangle$ is mapped to a corresponding model $\langle \mathcal{M}, \phi^* g_{ab}, \phi^* T_{ab} \rangle$ for which the diffeomorphism has 'pulled back' the fields on $\mathcal{M}$.[9] If a symmetry transformation arising from $\phi \in \text{Diff}(\mathcal{M})$ acts on the set of models in the described manner, then every element of $\text{Diff}(\mathcal{M})$ corresponds to a unique such symmetry. We can call a theory *formally generally covariant* just in case the set of symmetries of the theory consists in those mappings which correspond to active spacetime diffeomorphisms. Accordingly, demanding formal general covariance simply amounts to postulating that for each diffeomorphism $\phi \in \text{Diff}(\mathcal{M})$ there exists a unique symmetry such that if $\langle \mathcal{M}, g_{ab}, T_{ab} \rangle$ is a model of the dynamical equations, then so is $\langle \mathcal{M}, \phi^* g_{ab}, \phi^* T_{ab} \rangle$.

As the Einstein-Hilbert action in GR is invariant under transformations $\phi \in \text{Diff}(\mathcal{M})$, if $\langle \mathcal{M}, g_{ab}, T_{ab} \rangle$ solves the EFE, so does $\langle \mathcal{M}, \phi^* g_{ab}, \phi^* T_{ab} \rangle$. GR is thus a formally generally covariant theory.

However, the principle we are after is a *physical* principle, encoding a substantive condition on the physics, not just a formal requirement on the formulation of a theory. It is concerned with the interpretation of the formalism, and it demands that active spacetime diffeomorphisms be interpreted not merely as symmetries,

---

[9] These diffeomorphisms preserve the Einstein-Hilbert action of GR and therefore carry a solution of the Einstein field equation into another solution of the same equation, which derives from the Einstein-Hilbert action by varying the metric field g_{ab}.

118   OUT OF NOWHERE

but as gauge symmetries of the theory. A gauge symmetry of a theory is a symmetry such that models of the theory related by the gauge transformation correspond to the same physical situation. Identifying a symmetry as a gauge symmetry thus requires an interpretation of the formalism.

The *substantive principle of general covariance* then demands that the group $\text{Diff}(\mathcal{M})$ of active spacetime diffeomorphisms is the gauge group of GR. This is a strictly stronger demand than requiring mere formal general covariance, as the latter is consistent with a symmetry relating two distinct physical situations.[10]

There are two main motivations for postulating substantive general covariance (Wüthrich 2006, §3.2). First, the 'hole argument' suggests that in order to avoid an unpalatable failure of determinism, two diffeomorphically related models of GR ought to be identified as representing the same physical situation, thus turning the elements of $\text{Diff}(\mathcal{M})$ into gauge symmetries of GR.

The second, and related, motivation to see diffeomorphism invariance as a gauge symmetry arises from a general analysis of theories with so-called variational symmetries. Variational symmetries are just those transformations which leave the action of a theory (quasi-)invariant. These transformations form a group. Theories whose dynamical equations can be derived from an action principle and are therefore cast as Euler-Lagrange equations fall under the purview of Noether's theorems (Noether 1918). For these theories, gauge symmetries are just those transformations which absorb the underdetermination arising in the context of the second Noether theorem. Since the group $\text{Diff}(\mathcal{M})$ is a variational symmetry of the Einstein-Hilbert action, GR falls under Noether's second theorem. The underdetermination is taken care of if $\text{Diff}(\mathcal{M})$ is seen as the gauge group of GR.[11]

Both motivations behind demanding (substantive) general covariance—the hole argument and the more general considerations originating from Noether's second theorem—thus offer reasons for endorsing the demand based on a desire to avoid an indeterministic dynamical evolution. Although avoiding this 'faux' indeterminism undoubtedly delivers a sensible justification for general covariance, one should be aware of the fact that GR is arguably not a deterministic theory, as has become increasingly clear over the past years.[12]

What is the relationship between background independence and general covariance? Does a background-independent theory automatically satisfy general

---

[10] The term 'substantive general covariance' is borrowed from Earman (2006a). Weinstein (1999) has argued that GR should not be considered a gauge theory because the group $\text{Diff}(\mathcal{M})$ is not a gauge group in the sense of particle physics, i.e., it is not the automorphism group of a principal fiber bundle. Although of course what is meant by the 'physics' that remains invariant under gauge transformations is very different in GR and in the standard model, we still find it useful to consider $\text{Diff}(\mathcal{M})$ as GR's gauge group.

[11] For a fuller explanation, see Wüthrich (2006, §3.2). Pooley (2010, 203) points out that in order for this motivation to qualify as an *argument*, it would need to be shown that being a diffeomorphism is necessary to be a variational symmetry, not sufficient. So it remains open how seriously this motivation ought to be taken.

[12] See Earman (2004, sec. 6; 2006c, sec. 6), Doboszewski (2020), and Smeenk and Wüthrich (2021).

covariance and vice versa? The standard rhetoric, particularly in canonical quantum gravity, certainly suggests that the two notions are equivalent. However, this identification is somewhat frivolous and, strictly speaking, false.

In order to see the equivalence fail, it suffices to realize that background independence by itself is not sufficient for (substantive) general covariance. Clearly, there could be background-independent theories of gravity with (dynamical) symmetries different from those of GR, which would not even be formally generally covariant. If anything then, background independence is the weaker criterion—one which may be necessary for general covariance.

So let us consider whether background independence is necessary for general covariance. To this end, let us assume a split of the metric field as given in (5.1) into a fixed inertial structure b and a field h on this fixed background. Correspondingly, we replace g_{ab} in the Einstein-Hilbert action by $b_{ab} + h_{ab}$ and vary the action not with respect to g_{ab} in order to obtain a field theory of g_{ab}, but instead vary h_{ab} to obtain a field theory of h_{ab} on a fixed inertial background structure b_{ab}. The resulting Euler-Lagrange equations of motion will in general differ if the action is varied with respect to different fields. The exact difference, of course, depends on what is chosen as the fixed background in the second case. However, for any particular split, the symmetry group of the theory will not consist in the full spacetime diffeomorphisms but be restricted to those transformations leaving the background intact. The symmetry group is thus broken down to the symmetry group of the background spacetime, provided that the dynamical fields also obey the symmetry. If, for instance, the background metric is η_{ab}, then the remaining symmetries of the theory will be those of Minkowski spacetime, i.e., the Poincaré group. The larger group $\text{Diff}(\mathcal{M})$ will no longer be a symmetry of the theory. Breaking background independence will thus generally lead to a violation of general covariance.[13]

So it appears as if general covariance implies background independence. That this implication indeed holds can be seen by considering a challenge to it in the form of a toy theory described by Sorkin (2002) and discussed in Earman (2006a). Sorkin considers a classical scalar field Φ with mass m propagating on a Minkowski spacetime.[14] The dynamics of this field is captured by the Klein-Gordon equation, which reads in a formally generally covariant variant

$$\eta^{ab}\nabla_a\nabla_b\Phi - m^2\Phi = 0, \qquad (5.2)$$

---

[13] Introducing a fixed background also entails that the group of variational symmetries of the action is at most a finite-dimensional Lie group. Consequently, such theories are beyond the scope of Noether's second theorem and so have also lost an important motivation to demand general covariance. For a fuller explanation including specific examples, see again Wüthrich (2006, §3.2).

[14] More correctly, he considers a *massless* scalar field. Earman (2006a) generalizes the toy theory such as to include scalar fields with non-vanishing mass m. Here we discuss Earman's generalization.

120   OUT OF NOWHERE

where ∇_a is the covariant derivative operator determined by the Minkowski metric η_{ab}. The variation of the action

$$S[\Phi, \eta] = \frac{1}{2} \int d^4x \sqrt{-\eta} \left(\eta^{ab} \nabla_a \Phi \nabla_b \Phi + m^2 \Phi \right) \tag{5.3}$$

with respect to Φ while treating η_{ab} as fixed (and so there is no variation with respect to it) yields equation (5.2). The variational symmetry group of this action, and thus its gauge group, is the Poincaré group $P(1, 3)$. In order to render the toy theory of this action not only formally generally covariant, but also substantively so, we need to enlarge its gauge group such that it becomes $\text{Diff}(\mathcal{M})$. This enlargement will necessitate the following reformulation. First, we need to replace η_{ab} by a general pseudo-Riemannian metric field g_{ab}. Second, the covariant derivative operator ∇_a must now be determined by g_{ab}. Finally, equation (5.2) must be adjoined by a condition constraining g_{ab} to be flat,

$$R_{abcd} = 0. \tag{5.4}$$

Any set of solutions of $g^{ab} \nabla_a \nabla_b \Phi - m^2 \Phi = 0$ coupled with equation (5.4) is also a solution of (5.2), and vice versa. In order to remain faithful to the principled way of determining the gauge group of the new theory, the variational symmetries of its action must be studied. It turns out that if an auxiliary tensor field λ^{abcd} with the same symmetries as R_{abcd} is introduced and the action is rewritten such that λ^{abcd} plays the role of a Lagrange multiplier,

$$S[\Phi, g_{ab}, \lambda^{abcd}] = \frac{1}{2} \int d^4x \sqrt{-g} \left(g^{ab} \nabla_a \Phi \nabla_b \Phi + m^2 \Phi + \lambda^{abcd} R_{abcd} \right), \tag{5.5}$$

the gauge group of the toy theory becomes $\text{Diff}(\mathcal{M})$. Thus, the toy theory satisfies a substantive form of general covariance. The variation of the auxiliary field λ^{abcd} immediately gives the required flatness condition (5.4), while variation with respect to Φ gives the Klein-Gordon equation, as before. Furthermore, varying g_{ab}, which is now a dynamical field, leads to field equations expressing that the stress-energy tensor for Φ acts as a 'source' for the auxiliary field.

We concur with Earman (2006a) that the two actions (5.3) and (5.5), which have different variational symmetries and contain different sets of fields, encode not two different formulations of the same theory, but two distinct theories. In the substantively generally covariant theory given by (5.5), every solution contains a metric field and a scalar field such that in each case, the metric field is the Minkowski metric η_{ab} and the scalar field obeys the Klein-Gordon equation as determined by η_{ab}. Therefore, it could be argued, the theory (5.5) should count as background-*dependent*. If that were the case, the above claimed implication of

THE ROAD TO LOOP QUANTUM GRAVITY 121

background independence from (substantive) general covariance would no longer obtain.

However, Sorkin's toy theory (5.5) should arguably not be deemed background-dependent for the following reason. Background independence, as a reminder, requires that a theory does not assume a split of the inertio-gravitational field into a fixed inertial field, the 'background', and a gravitational field propagating on this background. The demand that the background remains 'fixed' means that the theory's action $S[\phi_1, \phi_2, ...]$ is not varied with respect to the background field b_{ab}, which is not a dynamical variable of the theory, but only with respect to the gravitational field 'proper' h_{ab}, and perhaps other dynamical fields. Translated to the case at hand, the Minkowski spacetime is presumably the fixed background field b_{ab}, while the scalar field Φ and the tensor field λ^{abcd} are additional dynamical fields. Interpreted like this, there is no gravity proper in Sorkin's toy example.[15] Thus, we see that the dynamical equations of Sorkin's toy theory are not obtained from splitting the dynamical metric field g_{ab} into a fixed component b_{ab} and a dynamical component h_{ab}, re-expressing action (5.5), and varying it with respect to h_{ab} and the other dynamical fields. It is just that the auxiliary field constrains the metric field to be the Minkowski metric. But the metric field g_{ab} is perfectly dynamical and interacts with the other dynamical fields. So there is not much motivation for counting Sorkin's example theory as background-dependent.

If this assessment of Sorkin's toy example is right, then we can maintain that general covariance entails background independence, but not vice versa, as we will for the rest of this chapter and the next one. We understand that this stance depends on subtle interpretations of the physics and of the precise content of the principles of background independence and general covariance. But our interpretation is certainly in line with the one at least implicit in canonical quantum gravity.

A central concern in canonical quantum gravity, including in LQG, is to preserve substantive general covariance and thus background independence in this more demanding sense as the central lesson of GR. Although a violation of these principles may result in a viable theory, such as quantized linear gravity, such a theory would never be a candidate for a fundamental theory from this perspective. This perspective even finds support in string theory. In fact, we will argue in §10.4 that there is no unequivocal sense in which string theory, as standardly interpreted, is background dependent. In order to arrive at this conclusion, however, we will need to insist that background independence, and in particular the principle of general covariance, cannot be naively read off a theory's surface formulation, but requires a sound physical interpretation of the mathematics.

---

[15] This interpretation is incautious insofar as background independence demands that no split of the metric is made into background and gravity, which is what is implicitly done when one says that 'there is no gravity in Sorkin's toy example'.

122 OUT OF NOWHERE

5.3 Hamiltonian general relativity

Much of the literature in physics (and, perhaps to a lesser extent, in philosophy of physics) proceeds as if the name 'GR' has an unambiguous referent, i.e., that it is somehow clear what exactly 'GR' is. However, this is not at all the case: the truth is more complicated—and much more interesting. All hands agree that the usual suspects such as Minkowski spacetime, Friedmann-Lemaître-Robertson-Walker spacetime, Schwarzschild spacetime, de Sitter spacetime, and a few others are models of GR (with their appropriate energy-matter content). Many (though not all) would agree that basic mathematical conditions such as the Hausdorff condition must be satisfied by all relativistic spacetimes. But the heart of GR, the EFE, only gives us a local condition. How permissive should we be when we move from the comparatively clear local requirements on the theory to the wide open field regarding the global structure of spacetime? What causal or topological conditions should be imposed?

Although we will not delve into these deep questions, we cannot avoid them altogether as different answers to these questions will lead to distinct theories which may all have some claim to be 'GR'. If we construct a quantum theory of gravity starting out from GR, the question of the exact nature of our starting point thus becomes urgent. Some decisions need to be made, some precisification of what version of GR we use as our vantage point is required. As we saw in the last two chapters, causal set theory arguably starts out from the assumption that relativistic spacetimes are future- and past-distinguishing. In standard textbooks on GR such as Wald (1984), one often identifies the collection of smooth, 4-dimensional, connected Hausdorff Lorentzian manifolds $\langle \mathcal{M}, g_{ab} \rangle$ with relativistic spacetimes and so triples $\langle \mathcal{M}, g_{ab}, T_{ab} \rangle$ consisting of spacetimes with their appropriate energy-matter content T_{ab} satisfying Einstein's field equation with models of GR.[16,17]

5.3.1 Reformulating general relativity

Standard prescriptions to find a quantum theory from a classical one start either from a Lagrangian (e.g., for the path integral method) or a Hamiltonian (e.g., for canonical quantization) formulation of the classical theory. As mentioned in the introduction to this chapter, LQG seeks to apply canonical quantization to GR and

[16] See Manchak (2021a) for a compelling contemporary perspective on how to think about what GR is.

[17] This section and the next are based on Wüthrich (2017).

THE ROAD TO LOOP QUANTUM GRAVITY   123

therefore casts GR as a Hamiltonian theory.[18] This choice implies certain decisions about which version or sector of GR we are considering; in particular, it restricts its attention to globally hyperbolic spacetime (and so imposes a rather strict causal condition on relativistic spacetimes). As it turns out, this amounts to imposing that the topology of $\mathcal{M}$ is $\mathbb{R} \times \Sigma$, where Σ denotes any Cauchy surface. Although this sector of GR is clearly physically very important (for instance in the formulation of an initial value problem in GR), it should also be noted that its selection severely restricts the scope of the theory, and does so in an a priori fashion—'by hand'.[19]

Apart from giving us the best justification for thinking of Einstein's field equation as a *dynamical* equation, governing the evolution of a spatial geometry over time, casting GR as a Hamiltonian system has various other advantages mostly to do with sharpening the concept of 'gauge' in the context of GR, as John Earman (2003) claimed: it makes precise the otherwise vague distinction between 'local' and 'global' transformations, it explains how the fiber bundle formalism arises when it does, it allows us to relate GR to Yang-Mills gauge theories, it delivers a formalization of the gauge concept, and it directly leads to foundational questions such as the nature of observables and the status of determinism in GR. The main advantage of the Hamiltonian formulation, however, accrues when we attempt to quantize classical GR.

However, casting GR as a Hamiltonian system is forcing it "into the Procrustean bed of the Hamiltonian formalism" in the words of Tim Maudlin (2002, 9). The reformulation incurs a cost, directly connected to its main benefit of making GR look like an ordinary dynamical theory. Hamiltonian GR reinterprets the 4-dimensional spacetimes of standard GR as 3-dimensional 'spaces' evolving in a 'time' according to the dynamics as given by Hamilton's equation. Pulling space and time asunder, however, contravenes what is often considered to be the deepest insight of (special and general) relativity: 4-dimensional spacetime is fundamental and any separation of this unity into space and time may at best be contingent or merely relative to a frame of reference. This blatant violation of 4-dimensionalism, however, is not as deep as it appears: the imposition of constraints in some sense undoes the split. A closer look at what exactly goes on in the Hamiltonian reformulation is clearly philosophically worth our while.[20]

---

[18] See Wald (1984, appendix E) for a systematic introduction to Lagrangian and Hamiltonian formulations of GR. This textbook of 1984 only deals with the ADM version of Hamiltonian GR and does not treat Ashtekar's version, pioneered in 1986.

[19] At least this is the case for Hamiltonian GR and for LQG as it was originally conceived. Arguably, by taking LQG to be a merely 'regional' theory describing processes in finite regions of spacetime without regard of what happens 'elsewhere', we can evade any commitment to a definite global topology. We will return to this issue in §6.2.

[20] In connection with what follows, chapter 1 of Henneaux and Teitelboim (1992) is recommended reading. For a less formal and hence more accessible treatment of the problem of time, cf. Huggett et al. (2013, §2) and references therein.

124　OUT OF NOWHERE

A cautionary note of rather great importance: for the rest of this chapter and the next, we will restrict ourselves to vacuum spacetimes, i.e., models of GR in which $T_{ab} = 0$. We do this initially to ease our presentation, but note that historically LQG has been based on vacuum spacetimes. Obviously, how to incorporate matter into the resulting theory then poses an important challenge to the research program.[21] Although GR's vacuum sector is enormously important in physicists' practice— many of the most important spacetimes are vacuum spacetimes—a fundamental theory will have to integrate matter. As the main motivation for seeking a quantum theory of gravity in the first place was GR's inability to incorporate the quantum nature of matter, it would be ironic to end up with a quantum theory of gravity which can ultimately only deal with vacuum worlds. Let us restrict ourselves to the vacuum case, although we will say a few more words about it.

The Hamiltonian formulation of GR is best understood by starting with its Lagrangian formulation, as the constraints are derived from the Lagrangian side. Starting out from the Einstein-Hilbert action $S[g_{ab}] = 1/16\pi \int_M d^4x \sqrt{-g} R$ for gravity without matter, where g is the determinant of the metric tensor g_{ab} and R the Ricci scalar, one obtains a Lagrangian version of GR with the dynamical Euler-Lagrange equations in terms of a Lagrangian function $L(q, \dot{q})$ of (generalized) coordinates q and the (generalized) velocities $\dot{q}$. The vacuum field equations of GR are obtained by varying $S[g_{ab}]$ with respect to g_{ab}, i.e., they are the equations of motion of Lagrangian GR, the Euler-Lagrange equations. They are second-order differential equations.

The solutions to the Euler-Lagrange equations will be uniquely determined by $q, \dot{q}$ just in case the so-called 'Hessian' matrix $\partial^2 L(q, \dot{q})/\partial \dot{q}^{n'} \partial \dot{q}^n$ of $L(q, \dot{q})$, where n labels the degrees of freedom, is invertible. This is the case if and only if its determinant does not vanish. In case it does vanishes, which means the Hessian is 'singular', the accelerations $\ddot{q}$ will not be uniquely determined by the positions and the velocities and the solutions to the Euler-Lagrange equations will contain arbitrary functions of time (Henneaux and Teitelboim 1992, §1.1.1). Thus, the impossibility of inverting $\partial^2 L(q, \dot{q})/\partial \dot{q}^{n'} \partial \dot{q}^n$ is an indication of gauge freedom.

5.3.2 Hamiltonian systems with constraints

How does such gauge freedom arise in Hamiltonian systems? Skipping over many of the details on how constraints arise in some Hamiltonian systems (which can be found in Henneaux and Teitelboim 1992, ch. 1), let us nevertheless mark some of the main points as they are central to recognizing how the so-called 'problem of

---

[21] A challenge that has been addressed, see e.g., Rovelli (2004, ch. 7), Thiemann (2007, ch. 12), Rovelli and Vidotto (2015, ch. 9) and references therein. We will briefly return to this point in §5.4.2.

time' arises in canonical quantum gravity. For the reader familiar with the theory of Hamiltonian systems, the following may serve as a reminder.

A Hamiltonian system is characterized by Hamilton's equations of motion, $\dot{q} = \partial H/\partial p$ and $\dot{p} = -\partial H/\partial q$, which are of first order, and where $H = H(q, p)$ is the Hamiltonian function of coordinates q and momenta p. The Euler-Lagrange equations of the Lagrangian formulation can be transformed into Hamilton's equations by the introduction of canonical momenta,

$$p_n = \frac{\partial L}{\partial \dot{q}^n}, \tag{5.6}$$

where $n = 1, ..., N$ labels the degrees of freedom, N being the total number of degrees of freedom and the Hamiltonian function is defined as

$$H(q, p) = \dot{q}^n p_n - L(q, \dot{q}). \tag{5.7}$$

If the Hessian is singular, these momenta are not all independent. In those situations, there is gauge freedom. The dependencies are captured by M so-called *primary constraints*

$$\phi_m(q, p) = 0, \quad m = 1, ..., M. \tag{5.8}$$

The relations (5.8) define a smooth submanifold of the phase space Γ—the space of solutions of the equations of motion. If the equations (5.8) are linearly independent, then the submanifold will be of dimension $2N - M$ and equations (5.6) define a mapping from a $2N$-dimensional phase space to the $(2N - M)$-dimensional manifold defined by (5.8). In order to render the transformation bijective and thus invertible, the introduction of M extra variables u^m is required, as the dimensions differ by M. These variables turn out to be the Lagrange multipliers which enforce the primary constraints (5.8).

Their introduction leads to the Hamiltonian equations of motion for arbitrary phase space functions $F(q, p)$ of the canonical variables

$$\dot{F} = \{F, H\} + u^m \{F, \phi_m\}, \tag{5.9}$$

where $\{,\}$ is the usual Poisson bracket

$$\{F, G\} := \frac{\partial F}{\partial q^i} \frac{\partial G}{\partial p_i} - \frac{\partial F}{\partial p_i} \frac{\partial G}{\partial q^i}.$$

Consistency requires that the primary constraints ϕ_m be preserved over time, which leads to secondary constraints, defining a submanifold of the submanifold defined by the primary constraints, the so-called *constraint surface* C.[22]

---

[22] See Henneaux and Teitelboim (1992, §1.1.5).

126 OUT OF NOWHERE

Any two functions F and G in phase space that coincide on the constraint surface are said to be *weakly equal*, symbolically $F \approx G$. In case they agree throughout the entire phase space, their equality is considered *strong*, expressed as $F = G$. This leads us to a classification of constraints into first-class and second-class constraints, defined as follows:

Definition 17 (First-class constraints). *A function $F(q, p)$ is termed* first class *if and only if its Poisson bracket with every constraint vanishes weakly,*

$$\{F, \phi_j\} \approx 0, \quad j = 1, ..., J. \tag{5.10}$$

If that first-class function is a constraint itself, then we call it a first-class constraint. *A function in phase space is called* second class *just in case it is not first class.*

The property of being first class is preserved under the Poisson bracket, i.e., the Poisson bracket of two first-class functions is first class again.

The arbitrary functions u^m encode the gauge freedom which arises for systems with a singular Hessian. It can be shown that the first-class primary constraints generate gauge transformations as they do not change the physical state, at least for theories considered to be deterministic. Although it turned out to be generally false, Dirac conjectured that all first-class constraints generate gauge transformations. As the counterexamples against Dirac's conjecture are rather exotic, we can presently assume that all first-class constraints do generate gauge transformations.[23] In fact, unlike second-class constraints, first-class constraints generate transformations within C, a so-called 'gauge orbit': a *gauge orbit* is a submanifold of C which contains all those points in C which form an equivalence class under a gauge transformation. These gauge orbits form curves in C, with the first-class constraints as tangents to these curves.

The equivalence classes of points in $C \subset \Gamma$ under gauge transformations constitute the so-called *reduced* or *physical phase space* $\Gamma_{phys} \subset C$. In other words, the physical phase space is obtained by identifying all points on the same gauge orbits. In a Hamiltonian system with constraints, the constraint equations are equations which the canonical variables must satisfy in addition to the dynamical equations of the system in order to capture the physical content of the theory. Solving the constraints means to use these additional equations in order to eliminate variables corresponding to 'unphysical' degrees of freedom. Once all constraints are solved, the remaining variables uniquely specify physical states and thus capture the physical content of the theory. They parametrize the *physical* phase space.

---

[23] For a counterexample, see Henneaux and Teitelboim (1992, §1.2.2). Pitts (2014) argues against this orthodoxy on different grounds; orthodoxy is defended by Pooley and Wallace (2022).

The total number of canonical variables ($= 2N$) minus twice the number of first-class constraints equals the number of *independent* canonical variables. Equally, the number of *physical* degrees of freedom is the same as half the number of independent canonical variables, or the same as half the number of canonical variables minus the number of first-class constraints.[24]

5.3.3 Gauge freedom and the problems of time and change

In order for GR to answer to the usual standard in physics of delivering dynamical theories describing how physical systems evolve over time, the 4-dimensional structure of spacetime must be broken up into 'space' that evolves in 'time'. This is certainly required in a well-posed initial value formulation of a theory or in a Hamiltonian theory, where Hamilton's equations explicitly solve for time derivatives. This severance of space and time threatens the general covariance so central for GR, a point to which we will return.[25]

In order to determine whether a meaningful dynamical formulation of GR exists, a gauge condition must be imposed, eliminating unphysical degrees of freedom. When GR is recast as a constrained Hamiltonian system, we have six independent components of the three-metric q_{ab} and six independent components of the corresponding conjugate momentum π^{ab} for a total of 12 dynamical variables. Using the method to calculate the number of physical degrees of freedom given at the end of the previous subsection, we thus arrive at two physical degrees of freedom per point in space for the gravitational field, as there are four first-class constraint equations.[26]

Einstein's field equation can be understood (with an appropriate gauge fix) as a kind of hyperbolic second-order differential equation for which there are existence and uniqueness theorems. It may thus appear as if GR is a reasonably deterministic theory. However, singularities or 'holes' may arise to frustrate these theorems.[27] In order to sidestep problems of this kind, one can demand that a spacetime $\langle \mathcal{M}, g_{ab} \rangle$ be *globally hyperbolic*, i.e., it contains a *Cauchy surface* Σ—a closed achronal set Σ whose 'domain of dependence' is the entire manifold $\mathcal{M}$.[28] Roughly, imposing this condition means that there exists a global 'instant of time' Σ such that the full physical state on Σ suffices to fully determine the full physical state throughout the

---

[24] Strictly speaking this is true only in the absence of second-class constraints, but those are eliminable without loss of generality (Henneaux and Teitelboim, 1992, §1.4.3).

[25] How general covariance gets implemented in Hamiltonian GR and the subtleties that arise in doing so are discussed in Wüthrich (2006, §4.4). What follows explicates the gist of this implementation.

[26] Cf. Wald (1984, 266) for a slightly different way of calculating the physical degrees of freedom of the gravitational field (with the same result).

[27] Cf. e.g., Doboszewski (2020) and Smeenk and Wüthrich (2021).

[28] For a fuller definition of these causal notions see Wald (1984, 201).

128   OUT OF NOWHERE

entire spacetime. Unsurprisingly, in order to cast GR as a Hamiltonian system, we must thus demand that the spacetime be globally hyperbolic.

We remind the reader that globally hyperbolic spacetimes are topologically '3 + 1', i.e., the topology of $\mathcal{M}$ is $\Sigma \times \mathbb{R}$, where Σ is a 3-dimensional submanifold of $\mathcal{M}$.

How troubling is the restriction to global hyperbolicity? Many physicists do not find it problematic, and perhaps even necessary in light of the need for a physical theory to have well-behaved dynamics. Philosophers tend to be more critical, as imposing global hyperbolicity as a condition of physical reasonableness amounts to asserting a rather strong form of the (unproven!) cosmic censorship hypothesis. These philosophers include for instance Earman (1995), Curiel (2001), Smeenk and Wüthrich (2011), and Manchak (2011).[29] A respectable reason to resist the demand for global hyperbolicity is epistemic and is furnished by Manchak (2011), who proves that any spacetime which is not "causally bizarre" is observationally indistinguishable from another, non-isometric spacetime with all the same local properties but which is not globally hyperbolic.[30][31] From this, Manchak concludes that "[i]t seems that, although our universe may be ... globally hyperbolic ... we can never know that it is." (414) In the light of this result, some caution seems appropriate. However, it is clear that if LQG were to succeed, this would certainly deliver a post facto justification for the demand. Moreover, the issue of global hyperbolicity may become otiose if we think of LQG not as a global theory, but instead as describing only what is appropriately 'regional', as we will discuss in §6.2. Let us proceed then under this presupposition.

Having required global hyperbolicity, we may reasonably assume that we will obtain a well-behaved dynamics. Unfortunately, this assumption seems frustrated by the infamous 'problem of time'. Let us explain what the problem is, even though we will later (in §6.1.4) see that both aspects of the issue depend upon interpretative choices and so can be avoided.

As it turns out, in Hamiltonian GR the Hamiltonian H is itself a constraint, which directly leads to Hamiltonian GR's most stubborn interpretational challenge, which plagues canonical quantizations of GR. This 'problem of time' comes essentially in *two* strands: the apparent disappearance of time as a physical magnitude, and the 'freezing' of dynamics. The first aspect of the problem of time, the vanishing of time as a fundamental magnitude, already arises at the classical level, even though there is a sense in which it only comes into full bloom at the quantum rung. We will return to this issue in §6.1.4, where we will see that there are plausible ways of resolving it.

---

[29] Although there are notable exceptions, such as Maudlin (2007b, 175 and 188ff).

[30] A spacetime is *causally bizarre* if it contains an event $p \in \mathcal{M}$ such that the chronological past of p is $\mathcal{M}$—all of it. This implies that all of spacetime may be observed from one point. Furthermore, causally bizarre spacetimes necessarily contain closed timelike curves.

[31] The converse does not hold.

The second version of the problem, the 'problem of change', as the freezing of the dynamics is more aptly called, fully arises at the classical level. It can be understood as an argument whose conclusion is that there is no change in any genuine physical quantity. In order to arrive at this conclusion, however, two substantive premises are needed.

First, it is assumed that only gauge-invariant quantities can capture the true physical content of a theory.[32] If two distinct mathematical models are related by maps which are interpreted as 'gauge' transformations, then they describe the same physical situation. In fact, this is what we mean by 'gauge'. However, which transformations ought to be counted as gauge is not a priori given, but may instead be substantive to controversy. Either way, it is natural to assume that the physical content of a theory is exhausted by the gauge-invariant quantities of the theory. In the context of Hamiltonian systems, this idea is captured by so-called *Dirac observables*, functions in phase space which are constant along gauge orbits on the constraint surface. Thus, if the premise is true, then the physical content of a constrained Hamiltonian theory is exhausted by its Dirac observables.

Consequently, we need to identify the first-class constraints of Hamiltonian GR. Without going into the mathematical details,[33] this brings us back to the principle of general covariance, which can count as a second substantive premise of the argument. As a reminder, this principle requires invariance under active spacetime diffeomorphisms. These form a group, $Diff(\mathcal{M})$ of maps between 4-dimensional manifolds. In the Hamiltonian formalism, $Diff(\mathcal{M})$ breaks down into a group of 3-dimensional 'spatial' diffeomorphisms and a group of 1-dimensional 'temporal' diffeomorphisms.[34] For example, in the ADM version of Hamiltonian GR (to be discussed in §5.4.2), the spacetime diffeomorphisms are generated by normal and tangential components of the Hamiltonian flow. Since the constraints generating the diffeomorphisms must vanish (weakly), these components of the Hamiltonian vanish (weakly). Furthermore, as in any Hamiltonian theory, it is the Hamiltonian which generates the dynamical evolution via the Hamilton equations. As the Hamiltonian is a first-class constraint and so is constrained to vanish, we immediately arrive at the conclusion that the dynamics gets 'frozen'.

The Hamiltonian generates the gauge orbits in the constraint surface. The Dirac observables are constant along these orbits, as they are invariant under gauge transformations. In other words, all genuinely physical magnitudes must be constants of the motion, i.e., they must remain constant over time. Change, as it turns

---

[32] We take this to be the orthodox view, even though there is dissent, e.g., by Rovelli (2014).

[33] These can be found, for the ADM and Ashtekar-Barbero versions of Hamiltonian GR, in Wüthrich (2006), §4.2.1 and §4.2.2, respectively.

[34] Strictly speaking, this results in a different theory as the symmetry group of Hamiltonian GR differs from the standard one (Wüthrich, 2006, §4.2). It should also be noted that the group $Diff(\mathcal{M})$ depends on the manifold $\mathcal{M}$, and so for different manifolds, we have different groups.

130 OUT OF NOWHERE

out, is merely representational redundancy, not physical fact. Thus, the argument concludes, there is no change.[35]

It should be noted that the argument does not depend on the specific form of the Hamiltonian, but on the fact that it is a constraint. Thus, it is general covariance which carries the lion's share in the argument. It should also be noted that how to understand the dynamics properly is already challenging in standard GR, but in Hamiltonian GR these difficulties stare you in the face.

Of course one may respond to this argument (as does Maudlin 2002) that it shows, by reductio, that the Hamiltonian formulation of GR has no serious standing as a physical theory. In fact, this response can be supported by the fact that Hamiltonian GR is not equivalent to standard GR, however theoretical equivalence is formalized. At best, Hamiltonian GR only represents the sector of GR of globally hyperbolic spacetimes, and so only part of GR. Moreover, as a careful technical analysis reveals, the symmetry groups are not exactly the same.

However, brushing aside Hamiltonian GR would also come at a considerable cost: Hamiltonian GR serves as the basis of canonical quantum gravity, one of the main approaches to quantum gravity. By virtue of this fact alone, it deserves to be taken seriously, and this is what we will do here. Furthermore, the extension of the Hamiltonian formalism to the general relativistic context is, arguably, empirically equivalent to the standard formulation of GR if one no longer insists on conceptualizing dynamics with respect to coordinate time, but instead with respect to some appropriate internal variable, as we will discuss in §6.1.4.

5.4 Loop quantum gravity

Let us look more closely to LQG and its 'canonical' core and origin. After some preliminary remarks in §5.4.1 on the canonical quantization of the Hamiltonian formulation of GR, we turn to the basics of LQG in §5.4.2 and the important spin network states in §5.4.3.

5.4.1 The canonical quantization of Hamiltonian GR

Canonical quantization requires that the classical theory to be canonically quantized must be cast as a Hamiltonian system. It transposes the canonical variables of the classical theory into quantum operators acting on a Hilbert space, turning the classical Poisson bracket algebra into the canonical commutation relations obtaining between the elementary operators in the quantum theory. If the Hamiltonian

---

[35] For a discussion of philosophical reactions to this situation, cf. Huggett et al. (2013, §2.3). There is of course a vast literature on the subject—a useful recent entry point is given by Thébault (2022).

THE ROAD TO LOOP QUANTUM GRAVITY   131

system possesses constraints, the constraint functions are converted to compound operators, built from the basic ones, turning the constraint equations into wave equations. These constraint equations state that the operators annihilate the states on which they are acting, in correspondence to the classical constraint equations. Only those states satisfying these constraints qualify as *physical* states. They form the *physical Hilbert space* $\mathcal{H}$.

LQG is a canonical quantization of Hamiltonian GR.[36] In Hamiltonian GR, the Hamiltonian itself becomes a constraint, as explained in §5.3.3. In the quantum theory, this takes the following form:

$$\hat{H}|\psi\rangle = 0, \tag{5.11}$$

which must therefore be satisfied by any state in $\mathcal{H}$. Equation (5.11), also called the 'Wheeler-DeWitt equation', gives a very direct intuition of both the problem of time and that of change discussed above. Concerning the problem of time strictly so called, comparing (5.11) to the ordinary Schrödinger equation,

$$\hat{H}|\psi\rangle = i\hbar\frac{\partial}{\partial t}|\psi\rangle, \tag{5.12}$$

we notice the absence of the time parameter t in (5.11). This absence is not merely due to its being 'hidden' in $\hat{H}$ (or in $|\psi\rangle$) and reappearing when the concrete form of $\hat{H}$ is spelled out. In any known concrete form of $\hat{H}$, physical time seems to be absent. Quite literally, time drops out of the equation in Hamiltonian quantum gravity.

This form of the problem of time is closely related to its second guise, i.e., what in §5.3.3 above we called the 'problem of change'. If one thinks, as is standard in Hamiltonian mechanics, that the Hamiltonian drives the dynamics and that therefore equation (5.11) plays the role of the dynamical equations of quantum Hamiltonian GR, in close analogy to (5.12) for ordinary quantum mechanics, then we observe that the time derivative of the quantum state vanishes.

Let us elaborate. Just as in the classical case, constraint operators generate the gauge symmetries of the theory. In this sense, if we require gauge-invariant observables, then functions $\hat{F}$ of operators represent Dirac observables as defined above just in case they commute with all constraint operators $\hat{C}_i$

$$[\hat{F}, \hat{C}_i]|\psi\rangle = 0,$$

for all $i = 1, ..., m$, where m is the number of constraints, and for all $|\psi\rangle$ in $\mathcal{H}$. Since the Hamiltonian is a constraint, Dirac observables must commute with it

---

[36] Textbooks on LQG include Rovelli (2004), Gambini and Pullin (2011), and Rovelli and Vidotto (2015). For its mathematical foundations, see Gambini and Pullin (1996) and Thiemann (2007).

132   OUT OF NOWHERE

and so be constants of the motion, i.e., not change over time. If, in line with the argument above, Dirac observables exhaust the physical content of the theory, no genuine physical quantity changes over (coordinate) time, 'freezing' the dynamics. The conclusion then appears to be that there is no change in a physical system described by quantum Hamiltonian GR. Change, if there is any, only arises as an artifact of representation, and not in the actual physical system being described. Or so it seems.

At the classical level, one could dodge this conclusion by rejecting the Hamiltonian formulation of GR altogether. In the spirit of Maudlin's resistance to Hamiltonian GR discussed above (§5.3.1), such a rejection would be well motivated by noting that the problem of time illustrates how the project of casting GR like an ordinary dynamical theory has gone awry. A dynamical theory without dynamics!

However, there are plausible ways of conceptualizing dynamics in Hamiltonian GR, as we will see in §6.1.4. Furthermore, at the quantum level, the move of rejecting the Hamiltonian formulation is not available: canonical quantum gravity requires a Hamiltonian formulation of GR to get going. Without it, the project is dead. Perhaps the problem should have been expected, as GR teaches us that time is not external to the physical system studied, but is part of the system we are trying to quantize. It is hardly surprising, then, that there remains no time external to the system which 'ticks' its dynamical evolution. If there is time, we must find it within (again, see §6.1.4). In fact, the disappearance of time we witness here is a harbinger of things to come: the dissolution of time and space in the fundamental ontology of LQG, to which we will return in chapter 6.

5.4.2 LQG: the basics

Before we introduce the basics of LQG, let us note several key limitations of the approach as it stands today.

First, to stress the point again, LQG is a quantization not of what would ordinarily be considered GR, but only of its globally hyperbolic sector. This sector enjoys obvious physical importance, but as argued above, it seems problematic to neglect non-globally hyperbolic spacetimes when we move to quantum gravity. It should be emphasized that so neglecting them does not automatically invalidate the non-globally hyperbolic spacetimes in GR as 'unphysical': it could well be the case that from a fundamental theory which was obtained via a recipe starting out from only globally hyperbolic spacetimes, some non-globally hyperbolic spacetimes of GR nevertheless turn out to be the best approximations to the fundamental physics in some cases.[37]

---

[37] This case is considered in Wüthrich (2021).

We postpone a debate on the merits and demerits of the first major limitation of LQG. In the meanwhile, a second seems clearly unwelcome: at least historically, only vacuum spacetimes were considered. As a result of this, it might seem unclear whether the resulting quantum theory adequately takes into account the obviously non-zero matter and energy content of the universe. However, there are two reasons for hoping that the situation may not be quite as bad as one might fear. First, vacuum models of GR and vacua more generally are often central to our understanding of the theory at stake. In GR, many of the most relevant models are vacuum solutions of the Einstein equation. Thus, understanding the vacuum situation may well shed light on quantum gravity. Second, just as fundamental models constructed from the globally hyperbolic sector of GR may nonetheless give rise to non-globally hyperbolic spacetimes, something analogous could happen with fundamental models built from the vacuum sector. In this sense, even though the quantum theory was originally drafted from the vacuum sector of the classical theory, it may come to be interpreted to contain matter. Thus, matter may be absent at the fundamental level and nevertheless emerge from the properties of the fundamental structure—perhaps from its topological or combinatorial properties. (However, it may of course turn out that the emerging matter is highly non-local or violates energy conditions.) In addition, we may argue that the distinction between spacetime (or the gravitational field) and matter fields is superficial and somewhat conventional. In fact, we may take it to be one of the lessons of GR that the gravitational field resembles matter fields much more than we might have thought, and then offer methods of extending the treatment the gravitational field receives in canonical quantum gravity to include matter fields, as does Rovelli (2004, §7.2).[38]

A quantum theory of a physical system can be naively characterized by three elements: the states the system can be in, its (state dependent) properties, and the dynamics under which it evolves. The first are captured in a Hilbert space, whose elements are the physical states of the system. The system's properties are expressed by its 'observables', which are mathematically elements of an operator algebra. These observables should make contact with empirical tests of the theory. Finally, the dynamics is usually given by a special operator, the Hamiltonian, and a Schrödinger- (or Heisenberg-)type equation. However, in quantum gravity, we should not expect to capture the theory's dynamics by such an equation, as there is no time external to the system at stake (and no internal time with all its properties), as in non-relativistic or special-relativistic contexts. The dynamics will have to be captured differently: as we shall see in §5.5.2, this will involve transition amplitudes between 'initial' and 'final' states calculated in analogy to Feynman graphs.

Starting from a Hamiltonian formulation, one would normally select a pair of canonically conjugate variables that coordinatize the classical phase space.

---

[38] See also Thiemann (2007, ch. 12), Rovelli and Vidotto (2015, ch. 9) and references therein.

134   OUT OF NOWHERE

Although different choices lead to *prima facie* different quantum theories, it seems as if the choice should not have physical significance and that therefore, the resulting theories would in principle be equivalent.[39] A historically important choice was the 'ADM variables' introduced by Arnowitt et al. (1962). In ADM, the 'configuration' variables are given by the induced three-metric $q_{ab}(x)$ on the 3-dimensional hypersurface, the 'lapse' function which gives a measure of the proper time elapsed along the normal to the hypersurface, and the 'shift' function determining the spatial shift within hypersurfaces. The conjugate momenta of the lapse and shift functions vanish, and that of the three-metric is constructed from the extrinsic curvature of the hypersurface (and the three-metric). The ADM variables lead to two constraints: the scalar or Hamiltonian constraint, as well as the vector or diffeomorphism constraints.[40] Unfortunately, the ADM approach leads to insurmountable technical difficulties as an approach to *quantum* gravity and has largely been given up as a result.

The ADM variables are sometimes also called 'metric' variables, for the simple reason that they express the spacetime geometry in terms of the 4-dimensional metric g_{ab}, from which the three-metric and the lapse and shift functions are defined. In contrast, LQG starts out from an (equivalent) expression of the spacetime geometry in terms of connections. Based on earlier work by Sen (1982), Ashtekar (1986) discovered his elegant 'new variables' for canonical quantum gravity: a connection A_a^i and its conjugate, a densitized triad 'electric field' E_i^a. From these variables, one can construct a 'holonomy' and its conjugate 'flux' variables.[41]

Let us offer a bit more detail. Although some of it may appear to be interpretationally loaded, the following is intended to be formal. The first step is to enlarge the phase space of GR by introducing a locally inertial frame in the form of a triad field $e_a^i(x)$ via $q_{ab}(x) = e_a^i(x)e_b^j(x)\delta_{ij}$, with $i, j = 1, 2, 3$, indicating a sense in which the triad field is the square of the three-metric. The three vectors of the triad field can be physically interpreted as the (spatial) axes of a local orthonormal inertial frame. The variables of this now extended phase space will carry both tensorial (or spacetime) indices as well as 'internal' ones, denoted by letters from the middle of the alphabet. Together with a conjugate momentum essentially consisting of the extrinsic curvature π_{ab}, these triads form the basic variables. The Einstein equation is then re-interpreted as a statement about a (spacetime) connection rather than

---

[39] In that the resulting quantum theories would be unitarily equivalent. However, counterexamples to this expectation are known (Ruetsche 2011, particularly ch. 3).

[40] The scalar constraint is really just the normal component of the full Hamiltonian, which generates a motion of the hypersurface both in the normal and the tangential direction. For a fuller explanation, see Wüthrich (2006, §4.2.1).

[41] Soon after Ashtekar published his work on the variables, Jacobson and Smolin (1988) found an infinite number of loop-like, exact solutions of the Wheeler-DeWitt equation in Ashtekar's formulation. These results provide the basis for the "loop representation of quantum general relativity" (Rovelli and Smolin 1990) and thus mark the birth of what became LQG.

about a metric. The Ashtekar variables are then obtained from the variables of the thus extended ADM formalism by

$$A_a^i = \Gamma_a^i + \beta \pi_{ab} e^{bi}, \tag{5.13}$$

where Γ_a^i is the so-called 'spin connection' of the triad field e_a^i, and β is the Immirzi parameter (which we assume to be real), the only free parameter of LQG; and by

$$E_i^a = \frac{\sqrt{|q|} e_i^a}{\beta}, \tag{5.14}$$

where $|q|$ is the determinant of q_{ab}. The only non-vanishing 'equal-time' Poisson brackets for each spacelike hypersurface are

$$\{A_a^i(x), E_j^b(y)\} \propto \delta_a^b \delta_j^i \delta^3(x, y). \tag{5.15}$$

It turns out that these canonical variables—the connection A_a^i and its conjugate momentum, the triads E_i^a—are the only pair of connection variables which do not lead to second-class constraints (Gambini and Pullin 1996). In sum, we replace the 12 variables q_{ab}, π^{ab} of the ADM phase space by 18 variables A_a^i, E_i^a parametrizing the extended ADM phase space. This extension is bought at the price of introducing another type of constraint—the so-called 'Gauss' constraints, which 'remove' the six additional degrees of freedom and so 'undo' the extension from the original ADM phase space. Through the extension, the theory becomes a gauge theory with a (compact) gauge group—the group of triad rotations—making the powerful techniques for the canonical quantization of gauge theories available.

Classically, the algebra of the Poisson brackets among the canonical variables encapsulates the geometrical structure of the phase space. In order to move to the quantum theory, one fixes a Hilbert space of quantum states $|\psi\rangle$ and turns the basic variables into quantum operators $\hat{A}_a^i, \hat{E}_i^a$ defined on this Hilbert space, with their algebra of commutation relations modeled on the classical structure. The quantum constraint equations are formulated in analogy to their classical counterparts, with the canonical variables of the constraint functions turned into quantum operators acting on states in the Hilbert space. Classically, the constraint functions are set to zero; in the quantum theory, they annihilate the states, i.e., $\hat{C}|\psi\rangle = 0$. As we saw in §5.4.1, only states annihilated by all the constraints are 'physical'. In other words, the physical Hilbert space is the kernel of the constraint operators in the original Hilbert space.

This delivers the physical Hilbert space, and the observables are operators constructed from the basic variables. What about the theory's dynamics? On the canonical approach, the dynamics is captured by the Hamiltonian. However, the Hamiltonian also turns out to be a constraint (5.11), which means that it sets

136 OUT OF NOWHERE

the physical states to zero, and there is no dynamical evolution over time, as adumbrated at the classical level in the problems of time and change discussed in §5.3.3.

In LQG, we find a total of three (families of) constraints, as mentioned above: the same ones as for the ADM variables, plus the Gauss constraints. These latter indicate a rotational gauge freedom of the triads. They generate an infinitesimal $SU(2)$ transformation in the internal indices. The Gauss constraint equations can be solved. Second, we have the 'diffeomorphism' constraints, which generate the spatial diffeomorphisms on the spacelike hypersurfaces. Although more challenging, these constraints have been solved as well. Jointly, the two steps lead us to the 'kinematical Hilbert space' $\mathcal{H}_K$, which contains only those states which are annihilated by the Gauss and diffeomorphism constraints. We will study the structure of $\mathcal{H}_K$ in a moment.

As a last step, one would like to solve the Hamiltonian constraint in order to arrive at the physical Hilbert space. For a long time, it was not even clear what the Hamiltonian concretely looked like. Although Thiemann (1996) managed to formulate an explicit Hamiltonian for the vacuum case, the Hamiltonian constraint has so far defied solution.[42] As a research program in canonical quantum gravity, LQG thus remains an incomplete theory. Thus, all results (including our interpretation and the lessons we draw below) must remain preliminary.

Physicists have pursued two workarounds to circumvent this problem. First, they have sought to simplify the classical theory by moving from the full phase space to the so-called 'mini-superspace' of only highly symmetric spacetimes, and then seeking to work through the canonical quantization of this simplified theory. This simplification is appealing because the Hamiltonian constraint equation can then be solved. Since the spacetimes in mini-superspace are homogeneous and isotropic, it is natural to think of the reduced theory as the cosmological sector of the full theory. Consequently, this approach has been dubbed 'loop quantum cosmology'.[43]

The other workaround follows the canonical quantization scheme up to the construction of the kinematic Hilbert space, but instead of trying to solve the final constraint, it reconceives the dynamics in a 'covariant' way.[44] One way of studying the kernel of the Hamiltonian in $\mathcal{H}_K$ would be to find a projection operator $\hat{P}$ that projects the states of $\mathcal{H}_K$ to the space of solutions of (5.11). Such a projection operator can in principle be constructed from 'transition amplitudes'

---

[42] See Rovelli (2004, §7.1) for an outline of how the concrete construction of the Hamiltonian works in principle.

[43] For an authoritative introduction, see Bojowald (2011a). For a philosophical presentation, see Huggett and Wüthrich (2018, §3).

[44] See Rovelli and Vidotto (2015) for an excellent textbook length presentation, but see also Rovelli (2004, ch. 9) and Wüthrich (2006, §5.3.2). We will return to the covariant perspective in §5.5 and in §6.2.

$W(s_f, s_i)$ between 'initial' states $|s_i\rangle$ and 'final' states $|s_f\rangle$, which are both elements of $\mathcal{H}_K$ (Rovelli 2011, §3). It turns out that the elements of the projection operator define an inner product of the physical Hilbert space $\mathcal{H}$, $W(s_f, s_i) = \langle s_f|s_i\rangle$. It should be noted that the transition amplitudes, qua elements of the projection operator, remain well-defined even in the absence of time. However, these transition amplitudes are very hard to compute. The covariant approach uses the Lagrangian spinfoam formalism, which permits in principle the perturbative calculation of transition amplitudes. In sum, the covariant, Feynman-inspired sum-over-graphs approach called spinfoams is motivated, to a large extent, by the sheer mathematical difficulties in solving the Hamiltonian constraint of the canonical theory. Unfortunately, its relation to canonical quantization is unclear and it is therefore still an open question how, if at all, the covariant spinfoam models relate to canonical gravity (Wüthrich 2006, §5.3.3). Here, as elsewhere, the proof is in the pudding, and the much more important—indeed decisive—question is not what the relation between the canonical and covariant heuristics turns out to be, but rather by which (mix of) methods do we end up with a consistent, viable, and ultimately empirically confirmed quantum theory of gravity with the right classical limit.

On both the canonical and the covariant approaches, the kinematical Hilbert space $\mathcal{H}_K$ plays an important role and captures much of the established state of the art in LQG. So let us have a closer look at it.

5.4.3 Spin network states: the technical background

As we discussed above, states in the kinematic Hilbert space $\mathcal{H}_K$ solve the first two constraint equations, the Gauss and diffeomorphism constraints. A natural way to start the construction of a Hilbert space is to consider wave functions $\Psi(A_a^i)$ of the configuration variable A_a^i, i.e., to directly seek a connection representation. However, when physicists explored this strategy, it turned out that this representation is plagued by a number of technical problems. For instance, the Hamiltonian operator $\hat{H}$ was not well defined, as standard regularization techniques from QFT do not work in background-independent theories. Furthermore, it was impossible to define an inner product of suitably gauge-invariant wave functions in the connection representation.[45] These difficulties led to the development of a novel representation, the 'loop representation', by Rovelli and Smolin (1988, 1990), which forms the basis of LQG.

Connections govern the parallel transport of geometric objects such as tangent vectors along spatial curves in the manifold. Although the resulting parallel-transported object will in general depend on the curve, vector addition and scalar multiplication are preserved. Let us consider a 3-dimensional hypersurface Σ in

---

[45] For details, see e.g., Gambini and Pullin (2011, §7.4).

138   OUT OF NOWHERE

$\mathcal{M}$ with a fixed topology. For an oriented smooth path γ from $p \in \mathcal{M}$ to $q \in \mathcal{M}$ and a vector (or fiber) bundle structure E over $\mathcal{M}$ with a connection A, the result of parallel-transporting a vector $u \in E_p$ from p to q along γ is $U[A, \gamma]u$. The linear map $U[A, \gamma] : E_p \to E_q$ is called a *holonomy* along the path γ. The holonomy of a given path is a functional on the space of smooth, 3-dimensional, real connections defined on Σ. We consider loops, i.e., closed curves such that $p = q$.

It turns out that all the gauge-invariant information packed into a given vector potential is implicitly contained in the trace of the holonomy along all possible loops on a manifold for that vector potential.[46] Thus, traces of holonomies (which are called, as in Yang-Mills theory, 'Wilson loops') form a natural basis of any gauge-invariant function of a connection. They allow us to transform the connection representation into the loop representation, where one works with a space of functions of loops.

Moving to the quantum theory, let us define the states in this loop representation now and thereby start the construction of the Hilbert space. For an ordered collection Γ of oriented smooth paths γ_l (with $l = 1, ..., L$) and a smooth function $f(U_1, ..., U_L)$ of L holonomies, a pair $\langle \Gamma, f \rangle$ defines a 'cylindrical functional' of the connection A:

$$\Psi_{\Gamma, f}[A] = f(U(A, \gamma_1), ..., U(A, \gamma_L)). \qquad (5.16)$$

This leads to the linear space S of all such functionals for any Γ and f. Any reversal of the orientation of a path γ_l amounts to replacing the corresponding argument of f with its inverse. Any change of the ordering in Γ is the same as changing the order of the arguments of f. Since we can define a scalar product on S, the completion of S under the norm defined by that scalar product yields a Hilbert space (Rovelli 2004, §6.2). This is the kinematic Hilbert space $\mathcal{H}_K$. Its inner product was chosen to be invariant under diffeomorphisms and local gauge transformations and must be such that real-valued classical observables become self-adjoint operators. These conditions are necessary for the theory to be consistent and to have the desired classical limit.

At this point, we are skipping over many technical details. What matters for present purposes is that there are elements of $\mathcal{H}_K$, called 'spin network states', which provide a basis in $\mathcal{H}_K$.[47] These spin network states can be thought of as the quantum states of the gravitational field. Since the initial hypersurface Σ in which we 'placed' the graph structures is a 3-dimensional spacelike submanifold, it is also natural to interpret the spin network states as quantum states of 3-dimensional physical space. If this interpretation is correct, physical space generically arises from quantum superpositions of spin network states.

---

[46] This is 'Giles' theorem' (Giles 1981); cf. also Gambini and Pullin (2011, ch. 5).

[47] The technical details can be found in Rovelli (2004) and Thiemann (2007). See also Rovelli (2011, §2.3) for the technical aspects of spin network states and the interpretation we are about to offer in the next subsection.

THE ROAD TO LOOP QUANTUM GRAVITY   139

Spin network states are constructed from spin networks, which in turn are based on the collection Γ of oriented paths γ_l. The construction proceeds as follows. We assume that the paths γ_l overlap at most at their endpoints, which thus become 'nodes' of a connected graph Γ in Σ, i.e., a collection of points n of Σ ('nodes') joined by curves l in Σ ('links'). The *valence* of a node is the sum of its outgoing and incoming links. For a given graph Γ with a given ordering and an orientation for each link, an irreducible representation j_l can be associated with each link l. Similarly, i_n is an assignment of a so-called 'intertwiner' to a node n, connecting the representations of the links adjacent to that node. A *spin network* S (embedded in Σ) is a triplet (Γ, j_l, i_n), where j_l and i_n are said to 'color' the links and nodes, respectively. It is then technically straightforward to define a spin network state $|S\rangle$ in S which corresponds to a spin network $S = (\Gamma, j_l, i_n)$. See figure 6.1 for a graphical representation, to be discussed in §6.1.1.

How should this formalism of spin networks be interpreted? How should we think of them physically? We will return to interpretive questions in the next chapter and treat the physical interpretation of spin networks in particular in §6.1.1.

5.5 The covariant perspective

Simplifying somewhat, we can state that the research programme of LQG, at least as conceived originally, faces one major technical roadblock and one interpretative challenge. Both of them arise with the Hamiltonian constraint, $\hat{H}|\Psi\rangle = 0$. Technically, the obstacles in articulating the concrete form of the Hamiltonian, let alone solving the constraint equation have so far been too formidable to have admitted exact resolution. Philosophically, it directly leads to the closely related problems of time and of change, to which we will return in the next chapter.

As noted in §5.4.2, physicists have pursued two major alternative approaches in order to circumvent both issues at once. First, they have simplified the system to be subjected to quantization; second, they have modified the quantization strategy. Either the original procedure is maintained while the system to be input into the procedure gets modified, or else the procedure is modified but the system maintains full generality. Here, our focus will be on the second approach, called 'covariant loop quantum gravity', which shifts gear after the Gaussian and diffeomorphism constraints are resolved and replaces the original canonical strategy of solving the Hamiltonian constraint with a possibly equivalent, 'covariant' approach to the dynamics of the theory.

The first approach is 'loop quantum cosmology'. Here, a very high degree of symmetry is introduced to reduce the complexity of the relativistic spacetime models to be quantized. These symmetry-reduced models (in 'mini-superspace') result from imposing full isotropy and homogeneity and represent highly idealized

140   OUT OF NOWHERE

cosmological models (hence loop quantum *cosmology*) of a perfectly symmetrical universe with just one degree of freedom—its size as captured by the scale factor. One then introduces a scalar field to act as a fiducial time with respect to which the scale factor then varies. It is the simplest possible model of an expanding or collapsing cosmos. This symmetry-reduction renders the Hamiltonian constraint amenable to explicit articulation and resolution. The resulting quantum models suggest a cosmos which first collapses onto itself only to then re-expand after the big-bang phase (which is then more aptly called the 'big transition' phase). An alternative interpretation of the resulting model reads it as proposing the birth of connected twin universes (Huggett and Wüthrich 2018).

The hope in loop quantum cosmology is that we glean valuable lessons for the full theory from simple toy models like this. The concern, quite naturally, is that we don't; the highly idealized toy model may well be rather unrepresentative of the general case, and the high degree of symmetry may mask important features of quantum gravity. Recently, researchers in loop quantum cosmology have started to address this worry by studying slightly more complicated models with some degree of anisotropy. In this book, we will not pursue an analysis of loop quantum cosmology. The interested reader is directed to the relevant sources.[48]

5.5.1 Canonical-covariant methodology

The other major road is a mix and match of quantization methods pursued by *covariant LQG*. As also noted above, the central idea here is to use the kinematical Hilbert space of the canonical programme as described in §5.4, but replace the Hamiltonian constraint equation with transition amplitudes characteristic of a covariant approach. Before we go into the details of this leading idea and the consequences for LQG, let us address a worry one might have: why should we expect to arrive at any useful result by combining different quantization methods in a rather ad-hoc manner driven by what appears to be nothing but an opportunistic avoidance of mathematical challenges? The worry is that haphazardly throwing together distinct methods in what appears an unprincipled way does not seem to be a promising strategy in the face of formidable challenges.

This worry, however, does not stick. For starters, quantization methods give us merely heuristic recipes, and one can easily argue that in heuristics 'anything goes'. Whether we find our new theory by conscientiously following a method that had previously succeeded, through a reverie featuring dancing atoms snakelike, or from monkeys randomly hitting keys on typewriters, the only thing that counts

---

[48] See Ashtekar and Singh (2011), Bojowald (2011a), and Bojowald (2011b, ch. 6) for a presentation of loop quantum cosmology, Bojowald (2008) and Bojowald (2010) for popular presentations, and Wüthrich (2006, chs. 7-8), Huggett and Wüthrich (2018), and Wüthrich (2022) for philosophical assessments.

is the resulting theory and its intrinsic scientific value. In this spirit, we may mix and match any methods for finding a quantum theory of gravity whatsoever, as long as it is empirically adequate and satisfies certain scientific criteria. In fact, quantization methods are much more principled than random somnambulism or anything of the kind. Furthermore, we have some inductive evidence that they work—at least in some circumstances.

So what are the relevant scientific criteria by which to judge our theories in quantum gravity? First, we will ultimately want them to be empirically adequate. As important as this is, since empirical tests of quantum gravity may be some time,[49] this criterion may not be so useful for now. Second, we would want them to be as simple as possible (but no simpler) while at the same time as explanatory as feasible. We will not here address the criteria of simplicity and explanatory strength. Furthermore, the resulting theories should be reasonably well-defined quantum theories. This criterion remains somewhat vague, and there is a sense in which the standards here cannot be too high, given that even our currently best quantum theories tend to suffer both from mathematical insufficiencies and interpretational troubles.

Perhaps the single most important scientific criterion in quantum gravity to date is that the candidate theory has the correct classical limit, i.e., that is reduces to something sufficiently like GR in the limit in which the quantum nature of the fundamental structures matters much less. In this limit, the assumption that both spacetime and matter are classical will work to a very good approximation, as testified by the empirical success of GR. Satisfying this criterion can thus also be seen as a particular aspect of the requirement that a theory be empirically adequate; in particular, it boils down to a demand that the theory be empirically adequate in the domain in which the theory it replaced was empirically adequate (but not necessarily beyond this domain).

Let us add two remarks. First, satisfying the requirement of the classical limit ipso facto means to circumvent the problem of empirical incoherence. If a theory meets the requirement of the classical limit, it will have managed to relate the structures it postulates, under the right circumstances, to relativistic spacetimes, at least to the extent to which they are responsible for the empirical success of GR. Assuming that we know how to extract phenomena and predictions from those relativistic spacetimes, the problem of empirical incoherence dissolves for a theory meeting this requirement.

Second, we should expect that in general there exist many quantum theories for a given classical limit, even empirically inequivalent ones. Thus, the relationship between quantum theories and corresponding classical theories is many-to-one.

---

[49] Though perhaps not as long as often thought—see Huggett et al. (2023). It should also be noted that current quantum theories of gravity do make predictions, although these tend to be quantitatively rough or altogether qualitative. Furthermore, we do not necessarily need full theories to arrive at usable predictions, permitting us to test theories before they are complete

142 OUT OF NOWHERE

Consequently, the quantum theory of gravity is underdetermined by its classical limit and we should thus expect that additional criteria of selecting our theory will eventually be needed.

Alas, it is not as if we are faced with an embarrassment of riches in quantum gravity. While there are many distinct research programmes, all of these have met with immense challenges. In this sense, it would be a significant success to formulate just a single concrete, complete, and consistent quantum theory of gravity. We can save worrying about the problem of non-uniqueness once we have two such theories. Presumably, at this point, we would want to employ empirical tests of tiebreakers between different candidate theories. But we are not there yet.

In sum, although we do not have any guarantee of success by mixing and matching different quantization methods, shifting to the covariant approach halfway through the canonical procedure is legitimate. The charge of methodological impurity against it does not stick: if the strategy can be brought to success in that it delivers a reasonably well-defined quantum theory with the correct classical limit, then the end will justify the means. Indeed, if we jump ship to the covariant approach, the problem of time may disappear altogether as the dynamics is no longer given by the Wheeler-DeWitt equation.

5.5.2 Covariant dynamics

In the canonical approach, we can think of the dynamics as encoded in the Hamiltonian operator, which has turned out to also be a constraint operator, as a projection operator from the kinematic Hilbert space to the physical Hilbert space which projects spin network states in the kinematic Hilbert space onto states solving the Hamiltonian constraint (or Wheeler-DeWitt) equation. But in analogy to ordinary, background-dependent QFT, we can capture the quantum dynamics in terms of transition amplitudes between 'initial' and 'final' states. This is precisely the guiding idea behind covariant LQG: code the dynamics as transition amplitudes between spin network states. As we will note again in §9.1, if they satisfy the Wightman axioms, the transition amplitudes specify a QFT via the Wightman reconstruction theorem (Streater and Wightman 1964, ch. 3).

As discussed in Rovelli (2004, ch. 9), covariant LQG expresses these transition amplitudes $W(s_f, s_i)$ between initial spin networks $|s_i\rangle$ and final spin networks $|s_f\rangle$ in a Lagrangian manner as sums over paths which is known as the 'spinfoam formalism', because the 'paths' being summed over are 'spinfoams'. The spinfoam idea follows Feynman in that it interprets the transition amplitudes as sums over paths of spin networks, based on the intuitive idea that a spinfoam is a "worldsurface swept out by a spin network" (Rovelli 2004, 320). The resulting spinfoams are combinatorial structures which do not live on some background spacetime, but instead can be thought of as what gives rise to spacetime analogously to how spin networks gave rise to space. The paths represent possible combinations of actions of the Hamiltonian $\hat{H}$ on the nodes of the spin network $|s_i\rangle$ such that it matches up with $|s_f\rangle$ and thus constitutes possible spinfoams.[50]

The Hamiltonian can be concretely understood as acting on the nodes of a spin network state such that edges split or merge. Its action is combinatorial in that it either collapses three nodes into one or it multiplies an existing node into three (see figure 5.1; the numbers a, b depend on the labels of the spin network acted upon). In this way, it affects the structure of the spin network and allows it to grow or to shrink.

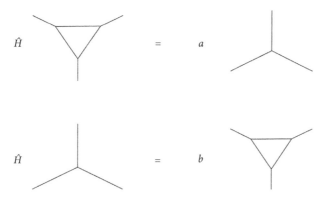

Figure 5.1 The basic action of the Hamiltonian on nodes of a spin network.

A spinfoam σ can be understood as a labeled Feynman graph of spin networks encoding the interactions occurring at the nodes, bounded by $|s_i\rangle$ and $|s_f\rangle$. In comparison with a Feynman graph which consists of vertices and edges connecting the vertices, a spinfoam has an additional structure: it collects vertices, edges, and *faces*. Faces are the world histories of links in the spin networks, and they join at edges, the world histories of the nodes. Edges, in turn, meet at vertices, which represent the interactions among the nodes, i.e., the actions of the Hamiltonian as represented in figure 5.1. The edges and the faces of a spinfoam are labeled with the corresponding quantum numbers for volumes and areas, respectively.

In this perturbative approach, the perturbative expansion of the transition amplitude $W(s_f, s_i)$ will be as a sum over all possible spinfoams bounded by the given spin networks $|s_i\rangle$ and $|s_f\rangle$. More precisely, the expansion sums over the weighted amplitudes associated with each spinfoam. The idea is schematically illustrated in figure 5.2. $\mathcal{A}(\sigma)$ represents the amplitude corresponding to a particular spinfoam σ. These spinfoams are endowed with quantum numbers on the vertices, edges, and faces. These numbers have been suppressed in the figure.

---

[50] Apart from Rovelli (2004, ch. 9), we recommend the more recent Rovelli and Vidotto (2015).

144   OUT OF NOWHERE

In the example in the figure, both $|s_i\rangle$ and $|s_f\rangle$ are a simple spin network with two trivalent nodes.

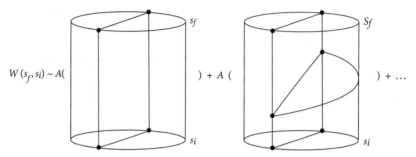

Figure 5.2 Schematic example of how to calculate a transition amplitude $W(s_f, s_i)$ perturbatively.

Let us sketch in the roughest of outlines how the calculation of these spinfoam amplitudes $\mathcal{A}(\sigma)$ works. First, one decomposes the entire spinfoam σ into its vertices v, each of which has a vertex amplitude $A_v(\sigma)$, which is in turn determined by (the matrix element of) the Hamiltonian $\hat{H}$ between the incoming and the outgoing spin networks. Similarly, one constructs the amplitudes of the faces $A_f(\sigma)$ and of the edges $A_e(\sigma)$. The amplitude $\mathcal{A}(\sigma)$ of a spinfoam σ is then obtained by the product of the amplitudes of all its individual vertices, edges, and faces:

$$\mathcal{A}(\sigma) = \mu(\sigma) \prod_f A_f(\sigma) \prod_e A_e(\sigma) \prod_v A_v(\sigma) \tag{5.17}$$

where $\mu(\sigma)$ introduces the weight given to the amplitude of spinfoam σ. In fact, the amplitudes A_f, A_e, A_v do not depend on the entire spinfoam σ, but only on the quantum numbers of the adjacent faces and edges.[51] The goal of a spinfoam model is to explicate equation (5.17) such that the transition amplitudes becomes calculable.[52]

This concludes the brief technical presentation of the covariant approach to LQG. We will return to its physical interpretation in the next chapter.

---

[51] For a more detailed discussion of spinfoam models and how they relate to LQG, see Rovelli (2004, ch. 9); for a succinct account of full covariant LQG, see Rovelli and Vidotto (2015). As far as we know, Reisenberger and Rovelli (1997) were the first ones to derive the formal framework that encapsulates the dynamics in terms of sums over histories from canonical LQG. They started out by expanding the exponential projection operator as a sum and gave each term of the sum a geometrical interpretation as a spinfoam. A rigorous introduction to spinfoams is given by Baez (1998).
[52] We will not go into a discussion of different spinfoam models here; for more information, see Wüthrich (2006, §5.3.2) on which this section is based.

5.6 Conclusion

LQG is a major approach to quantum gravity, which powerfully embraces central lessons of GR and transliterates them into the context of quantum physics. At the core of these lessons is GR's background independence in the form of substantive general covariance. We have attempted to present the core of the argument leading to LQG (and to some degree to canonical quantum gravity more widely) and trace it through the Hamiltonian form of GR to arrive at spin networks and, as the canonical quantization procedure hit major roadblocks, the spinfoams resulting from the covariant perspective on the theory. A recurring theme of our presentation was that the somewhat heterogeneous heuristic strategy was legitimate, even though of course it does not guarantee success. Thus, the heuristics of LQG are plausible, but uncertain.

Some attention should be paid to the problems of time and of change. Unlike at the classical level, where arguably the strictures of the argument can be evaded, at least to some extent, by avoiding Hamiltonian formulations of GR, this is evidently not possible for quantizations based on them, as the problem is built right into the framework. Perhaps we ought to have expected such an outcome—after all, GR teaches us that time is not external to the physical systems of interest but itself forms part of spacetime in dynamical interactions with the material content of the universe, which constitute the usual physical systems physics describes. In other words, time is part of the physical system we are trying to quantize. This thread will be continued in the next chapter.

In the next chapter, we will develop the physical interpretation of both the canonical and the covariant aspects of the theory and argue that there is a relevant and rather substantive way in which the structures postulated by the resulting theory are not spatiotemporal. In order to see how spacetime emerges from these structures, and thus how the problem of empirical incoherence (and of the classical limit) is addressed, we will rely, once again and unsurprisingly, on the services of spacetime functionalism.

Chapter 6
The disappearance and emergence of spacetime in loop quantum gravity

As we have seen in the previous chapter, loop quantum gravity (LQG) builds on what it takes to be the main lesson of general relativity (GR)—the demand for general covariance as captured by the substantive principle of general covariance—recasts GR in the Hamiltonian formalism, and proceeds with a canonical quantization of Hamiltonian GR to arrive at a quantum theory of gravity.

In this chapter, we will find that LQG postulates fundamental structures which are non-spatiotemporal, or at least not fully and straightforwardly spatiotemporal (§6.1). The problems of time and of change challenge the fundamental existence of time: it appears as if there is no direct correlate of physical time at the fundamental level.

Furthermore, we will argue that spin networks—the current best guess of the fundamental structures postulated by LQG—diverge quite significantly from what we would think of as space. The argument will be two-pronged. First, the quantum system will generically be in a state of superposition of distinct spin network states and so will generically have indeterminate geometry. Second, the phenomenon of 'disordered locality' implies that there is no direct connection between the adjacency relations at the fundamental level and contiguity at the emergent level.

We will argue that this shows that LQG trades in fundamental structures which are not essentially spatiotemporal. In order to avoid the problem of empirical incoherence, or, ultimately, to connect the scientific with the manifest image of the world, it must be established that relativistic spacetimes are excellent approximations to the LQG structures, at least in those regimes in which GR is empirically highly accurate. In other words, it must be shown that spacetime emerges in LQG. To outline how this can be accomplished is the main task of this chapter, to be mainly addressed in §6.3, where we introduce what we will call the 'Butterfield-Isham scheme of emergence', and in §6.4, where we cast the emergence of spacetime in LQG in functionalist terms.

Before we analyze how spacetime may or may not emerge in LQG in §6.3 and §6.4, §6.2 presents the covariant approach to LQG and how it conceives of the dynamical aspects of the theory. This presentation will include a particular take on relativistic dynamics which has been forcefully advocated as being part and parcel of covariant LQG and may get around the problems of time and of change.

Out of Nowhere. Nick Huggett and Christian Wüthrich, Oxford University Press.
© Nick Huggett and Christian Wüthrich (2025). DOI: 10.1093/oso/9780198758501.003.0006

Returning to a topic already addressed in the context of causal set theory (§4.3), in philosophy of time, one is often presented with three main options regarding the fundamental metaphysics of time: presentism, eternalism, and the growing block view. In a nutshell, presentism takes all and only present, but not past or future, entities to exist. We say 'entities', rather than 'objects', 'events', or 'times', hoping to convey a liberal attitude not committed to any particular ontological categories. On this view, the present is normally understood as dynamically updating in that some of its entities cease to be and so are no longer part of the present, while others come to be and so are being added to it. The growing block view conceives of the sum total of existence as of a growing block to which new layers of entities are continually being added by a dynamical process of becoming. The present is just the top layer of the sum total of existence. This metaphysics thus admits past and present, but not future, entities as existing. In contrast to these two views, eternalism does not fundamentally include a dynamic aspect of becoming or passage and so does not require a fundamental distinction between past, present, and future. Instead, it accepts all entities irrespective of their temporal status as existing. In the context of GR for example, eternalism is the view that anything that exists somewhere in this 4-dimensional spacetime exists.

Relativistic physics is usually thought to have decided this debate in favor of eternalism. Recently, the literature on what the implications of relativity or other theories in fundamental physics are supposed to be has exploded, with many substantive and some unexpected contributions. We will not follow all these developments, fascinating though they may be, and instead focus on how this debate may be affected by results in LQG. We will do this in §6.5.

§6.6 concludes.

6.1 The non-spatiotemporality of canonical LQG

In the previous chapter, we introduced the technical apparatus and the main conceptual framework necessary to articulate LQG as it stands today. It is time to offer a philosophically informed interpretation of the theory. Our primary focus in this section will be to develop a sense of just how non-spatiotemporal the fundamental structures of LQG are. For this, we will restrict ourselves to canonical LQG for now, moving to the covariant perspective in the next section.

Having presented LQG, including its classical guiding ideas, in the previous chapter, we will develop in §6.1.1 the physical interpretation of the spin network states. As such, it is still quite close to their technical introduction in §5.4.3. Readers more interested in the philosophical implications of LQG may wish to skip subsection §6.1.1. Its main point is that spin network states are interpreted as discrete, combinatorial, background independent geometric structures which

148   OUT OF NOWHERE

in some sense represent 'space', despite their divergence from what we would normally identify as spatial.

The remainder of this section considers to what extent and in what form space and time have disappeared from the fundamental ontology of LQG. Before we do this, let us emphasize that we are not interested in a purely terminological dispute about whether the fundamental structure according to LQG is to be called 'spacetime' or not. In this sense, this section is not about (meta-)semantics. If you insist on calling fundamental structures 'spacetime', perhaps because they vaguely and remotely remind you of spatial or temporal structures, or because it is these structures that have to give rise to something like relativistic spacetime, and certainly to the spatiotemporality of the manifest world, then so be it.

Instead, this section contains our attempt to describe in some detail in what sense the world according to LQG is less spatiotemporal than it is according either to GR or to the world as it manifests itself to us. We would like to note, however, that the structures found in LQG are significantly different from such relativistic spacetimes, and are not so obviously and directly related to the spatiotemporality of the manifest world. These facts suffice to justify speaking of the 'disappearance' of space and time, even though this disappearance is of course not total.

The case for the disappearance of space and time in LQG rests on three arguments, covered in subsections §6.1.2, §6.1.3, and §6.1.4, respectively. First, since LQG is a quantum theory, it will introduce quantum indeterminacy. Since it is spacetime itself which is being quantized, we should expect this quantum indeterminacy to pertain to spacetime. Consequently, we should not expect our world, or any part thereof, to be necessarily in a state of determinate geometry and in this sense of determinate spatiotemporality. To be sure, quantum indeterminacy of the spacetime geometry is not something special about LQG; indeed, any quantum treatment of gravity beyond semi-classical quantum gravity is presumably going to exhibit this form of quantum indeterminacy. This issue will be developed in §6.1.2.

Second, in §6.1.1, we note that the links in spin networks represent a relation of 'adjacency' between (what physicists call) 'fundamental atoms of space'. There is also a rich structure of spacetime events being 'near' one another in relativistic spacetimes. However, there is no guarantee that these two notions neatly map onto one another, and instead we find a 'disordered locality'. We will discuss disordered locality in §6.1.3.

Third, the problems of time and of change introduced in the last chapter in §5.3 can be taken to indicate that time (and change) are not fundamental in canonical quantum gravity, but instead must somehow emerge from a fundamentally non-temporal structure. This will be the topic in §6.1.4.

In fact, it seems as if there is a fourth way in which spacetime disappears in LQG: we should expect there to be models of LQG which are not approximated by any relativistic spacetime, i.e., models without a classical limit, but in a more

THE DISAPPEARANCE AND EMERGENCE OF SPACETIME IN LOOP   149

radical way than merely because of a quantum superposition. Although of course we live in a world which could only be described by a fundamental model that is very well approximated by relativistic spacetime (for those regimes for which GR is a valid approximation), it is a generic feature of quantum theories of gravity that they admit models in which spacetime does not emerge. This occurs for example when the quantum system is in a non-geometric phase.[1] As we discussed in §1.2, group field theory, a generalization of LQG, admits such non-geometric phases and thus states which are not even in the running for collapsing into a determinate geometry. Insofar as LQG may be based in a more general group-field theoretic account, we may also face, in the context of LQG, a more radical departure from spatiotemporality than the three ways to be addressed below. We will not dwell on this point, but instead turn our attention to the three arguments above.

In §6.1.5, we will consider (and dismiss) three objections against our interpretation of LQG as seriously non-spatiotemporal. In §6.1.6, we will summarize and conclude this section.

6.1.1 Spin network states: interpretation

As we have seen in the previous chapter (in §5.4.3), the spin network states form a basis in the kinematical Hilbert space of canonical LQG. As such, they represent quantum states of the gravitational field. More precisely—and to repeat—, they can be considered to be quantum states of 3-dimensional physical space.[2] These spin network states can be represented by combinatorial graphs consisting of nodes and links between them, together with additional properties of the nodes and links represented by numbers i_n and j_l, respectively, as illustrated in figure 6.1. We will get to the physical interpretation of these properties shortly.

A few facts about spin network states. First, only overall global orientations of the links of a spin network matter physically. Second, spin network states can be decomposed into finite linear combinations of basal multiloop states. Thus, there is a sense in which these multiloop states build up the spin network states. Third, the spin network states are gauge invariant, i.e., invariant under the SU(2)-transformations associated with the Gauss constraints.

In order to obtain states invariant under the physically crucial 3-dimensional—i.e., 'spatial'—diffeomorphisms, and thus the kinematical Hilbert space $\mathcal{H}_K$, one moves to equivalence classes of unoriented graphs Γ under spatial diffeomorphisms, so-called *knots* as they are studied in knot theory. One can then construct

---

[1] And so we are at level 2 in Oriti's hierarchy of non-spatiotemporality (Oriti 2021).

[2] It should be noted, however, that on the canonical program they can also be considered quantum states of *4-dimensional spacetime*, since physical Hilbert space consists just of those states in the kinematical Hilbert space (among them, spin network states), which satisfy the Hamiltonian constraint. On the covariant version of LQG, only spinfoams represent quantum states of spacetime.

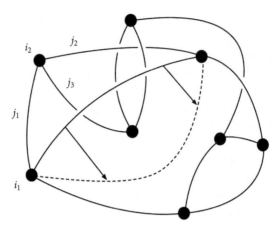

Figure 6.1 A spin network state is characterized by an abstract graph with 'spin'-representations on the nodes and the links between them: deforming the spin network as indicated does not change the state.

states called *s-knot states* based on these knots, and distinguish states within a knot by their distinct labelings or 'colorings' of the links and nodes of Γ.

Spin network states are represented by labeled graphs embedded in a background space Σ, while *s*-knot states are represented by equivalence classes of labeled graphs under spatial diffeomorphisms. We refer to these equivalence classes as *abstract* labeled graphs. There is some ambiguity in the literature, as 'spin network states' are sometimes meant to refer to what are strictly speaking *s*-knot states. As 'spin network state' is more evocative than '*s*-knot state', we will henceforth also use the former to refer to the latter. What matters is that spin network states in this sense are completely characterized by three types of information: the abstract graph Γ, the irreducible $SU(2)$-(hence 'spin') representations j_l on the links and the $SU(2)$-representations on the nodes, the intertwiners, denoted by i_n.

The abstractness of the graphs is central to the physical interpretation of the spin network states as quantum states of physical space, rather than of some physical system *in* space. This abstractness is illustrated in figure 6.1: although a spin network is represented to be somewhere on the paper or on the screen, the arrows indicate how different embeddings are representationally equivalent. Although it is impossible for us to directly depict an abstract spin network, the particular spatial embedding shown in the figure is a circumstantial and representationally idle artifact. We will return to this aspect in §6.1.5.

The spin network states $|\Gamma, j_l, i_n\rangle$ are eigenstates of two important operators defined on $\mathcal{H}_K$: the 'area' and 'volume' operators. The area operator $\hat{A}$ is an $SU(2)$-gauge-invariant self-adjoint operator with a real, discrete spectrum. Similarly, the volume operator $\hat{V}$ is also a self-adjoint non-negative operator with a real, discrete spectrum (Rovelli 2004, §6.6). Both operators have been constructed in close

THE DISAPPEARANCE AND EMERGENCE OF SPACETIME IN LOOP   151

analogy to classical quantities and thus permit a straightforward interpretation. The classical quantities give a measure of the area of a 2-surface and the volume of a 3-region, respectively. Correspondingly, the area and volume operators are interpreted to be observables of surface area and volume (at least in certain states; see 6.4)—hence their names.

We should emphasize that it is precisely this geometric interpretation of the operators which endows the emergence of spacetime with physical salience beyond the merely formal relationship we will discover in this section and in §6.3. In this sense, it is crucial for the physical salience that this geometric interpretation be postulated of the mathematical formalism. We should note that this geometric postulate is extremely natural, as we take the discussion in the rest of this section and in §6.3 to show. Nevertheless, it must be navigated with care, as will be argued in §6.1.2 and particularly in §6.1.3. And it *is* an additional postulate.

But area and volume of what? Crucially, the volume operator obtains contributions only from the nodes of a spin network state, and not from the graph's links. If applied to a spin network, or to a part thereof, the result is a sum of the contributions from the nodes of the spin network. Rovelli (2004, §6.7) interprets this as each node representing a "quantum of volume". It thus appears as if a node is a mereological simple with a fundamental volume property determined by its intertwiner. Each of the nodes can thus be thought to represent a 'chunk' of space with a volume given by the eigenvalue V_{i_n} of the volume operator applied to that node. The atomic chunks of quantized volume are connected, but also separated, by the links relating them. These links represent adjacency relations between the elementary volumes they connect—linked nodes are 'spatially' contiguous, furnishing the fundamental spatial relations. The area operator only receives contributions from the links, but not from the nodes. Thus, it is natural to interpret the area operator applied to a single link as giving an elementary measure of its surface area, i.e., the surface area separating two adjacent elementary volumes. This area is determined by the spin representation j_l of link l through

$$\mathbf{A}_l = 8\pi\sqrt{j_l(j_l + 1)}, \tag{6.1}$$

(in natural units), which means that the smallest non-zero eigenvalue will be (for $j_l = 1/2$) of the order of 10^{-66} cm^2 in SMI units. Just as the volume operator, the area operator has a discrete spectrum (Ashtekar and Lewandowski 1997, 1998, 1999; Rovelli and Smolin 1995a, 1995b). In sum, the intertwiners on the nodes are the quantum numbers of the volume, while the spin representations on the links are the quantum numbers of the area. The graph Γ determines the structure of adjacency relations between the elementary chunks of space: each node represents an 'atom' of space separated from its neighbors by surfaces of contiguity. The granular, 'polymer' geometry of space is a direct result of the discreteness of the spectra of the area and volume operators. In this sense, it appears as if the smooth

152 OUT OF NOWHERE

space of the classical theory has been replaced by a discrete quantum structure representing the granular nature of space at the Planck scale. At the very least, the continuous space(time) we find in GR must emerge from these discrete structures.

Let us close this section with a few remarks. First, the geometric interpretation of the spin network states just offered remains, although eminently natural, an *interpretation*, which rests, in turn, on the geometric interpretation of the geometric operators. If for some reason the geometric interpretation of the area and volume operators fails, we cannot interpret spin networks in this straightforward geometric (though discrete) way.

Second, one reason to remain cautious about the geometric interpretation is that the area and volume operators are not Dirac observables. They are partial observables in the sense of Rovelli (2002). Therefore, the geometric interpretation remains conditional, among other things, upon confirmation of the full physical theory, and in this sense of the theory's 'dynamics', which includes either resolution of the Hamiltonian constraint on the canonical version of the theory, or the alternative concrete realization of the transition amplitudes on the covariant one, which we have introduced in §5.5 in the previous chapter and to which we will return in §6.2 below.

Finally, it is worth noting that it is a *result* (rather than an *assumption*) of LQG that areas and volumes operators can only take discrete values. This result is, of course, based on the computation of their spectra, which were found to be discrete. Extrapolating this finding, one could therefore claim that LQG predicts that physical space is not a continuum as it was in relativistic spacetimes, but a discrete structure instead. In this sense, LQG seems to predict the discreteness of space— importantly—conditional on the geometric interpretation offered in this section. According to LQG, measurements of the Planck-scale geometry of space will thus yield values corresponding to the possible eigenvalues of the operator at stake.

It is important to note that the claim of discreteness is *not* based on the fact that spin network states are represented by discrete, combinatorial graphs. These graphs should be best thought of as calculational devices, in analogy to Feynman diagrams, rather than as a direct representation of reality of how physical space really 'is' (again, arguably, in analogy to Feynman diagrams). They are technical tools to analyze what are ultimately interpreted as discrete geometric structures. This interpretation, importantly, and to repeat, is based on the fact that geometric observables captured by the area and volume operators have discrete spectra.

6.1.2 Indeterminate space

As we are looking for a quantum theory of gravity, we expect the physical system that is being described to be in a state which can be represented by an element

THE DISAPPEARANCE AND EMERGENCE OF SPACETIME IN LOOP   153

(more precisely, a ray) in Hilbert space.[3] If we limit our focus to the kinematic part of LQG, then we find that it is spin network states which form a basis of such a Hilbert space. Thus, just as the electromagnetic field consists of a superposition of n-photon states, we could thus assume that physical space is described by a quantum superposition of spin network states, and we would expect quantum phenomena like interference and entanglement.

However, there is a sense in which a 'spacetime' superposition seems worse than ordinary quantum superpositions. If an electron is in a superposition of spin up and down in a particular direction, then although it is not in a determinate spin state in that direction, it has a determinate spin in some *direction*. Generally speaking, we should expect that at least expectation values of state-dependent properties are meaningful for ordinary quantum superpositions. It is not clear that this remains the case for superposition states of spacetime geometry: in such a state, we would not have some spacetime region of which this would be the quantum state—the very notion of 'region' is inadmissible in such a case. This general feature is the first point that needs to be navigated with care.

Let us focus on superpositions in kinematical LQG. Suppose we start out from a fixed but very large Hilbert space, which is sufficiently large to contain elements able to represent the state of a large system with many gravitational degrees of freedom.[4] The spin network basis $|\Gamma, j_l, v_n\rangle$ contains spin networks of a fixed graph Γ, but with different colorings on the nodes and the links. Clearly, the total state could be a superposition $\alpha|\Gamma, j'_l, v'_n\rangle + \beta|\Gamma, j''_l, v''_n\rangle$. As we see from equation (6.1), the eigenvalue of the area operator $\hat{A}$ depends on the spin representations j_l, and as the superposition state is not an eigenstate of $\hat{A}$, we conclude that the corresponding surface area is indeterminate if $j'_l \neq j''_l$. Similarly, distances and volumes are generically indeterminate as well. Although the structure of adjacency relations is given by the fixed graph, and so remains determinate throughout, the geometric properties of the structure exhibit quantum indeterminacy. Not only is 'space' fundamentally discrete, but it also generically has an indeterminate geometry.

So far, so expected for a full quantum theory of gravity. Potentially, however, there is a deeper sense in which spacetime may disappear that is unique to LQG: could the structure of adjacency relations itself suffer from quantum indeterminacy? In other words, could the system be in a state of superposition of different graphs, each with its own, distinct structure of adjacency relations among nodes (as we suggested in our Huggett and Wüthrich 2013a)? In other words, could there be a superposition $|\psi\rangle = \alpha|\Gamma', j'_l, i'_n\rangle + \beta|\Gamma'', j''_l, i''_n\rangle$ with distinct graphs Γ' and Γ''? If this were the case, then the fundamental adjacency of the atomic elements would itself be indeterminate, and not just the geometry.

---

[3] If it is in a pure state; otherwise, we need density matrices to represent states.
[4] Some might be tempted to seek a 'universal' wave function describing the full state of the universe, but as this would invariably lead us into an interpretational thicket we wish to circumvent, we try to avoid any suggestion of this. We will return to this topic in §6.2.

154 OUT OF NOWHERE

The problem here is how we should understand this superposition $|\psi\rangle$.[5] By virtue of their distinct graphs, $|\Gamma', j'_l, i'_n\rangle$ and $|\Gamma'', j''_l, i''_n\rangle$ are elements of distinct Hilbert spaces $\mathcal{H}_{\Gamma'}$ and $\mathcal{H}_{\Gamma''}$, respectively, and so the expression $\alpha|\Gamma', j'_l, i'_n\rangle + \beta|\Gamma'', j''_l, i''_n\rangle$ is ungrammatical. The most straightforward way to identify a Hilbert space in which $|\psi\rangle$ could live is to consider the unique graph Γ which is formed by the two disconnected components Γ' and Γ''. This graph determines the Hilbert space $\mathcal{H}_\Gamma$ with a basis $|\Gamma, j'_l, j''_l, i'_n, i''_n\rangle$. In this Hilbert space, $|\psi\rangle$ can now be expressed as $|\psi\rangle = \alpha|\Gamma, j'_l, 0, i'_n, 0\rangle + \beta|\Gamma, 0, j''_l, 0, i''_n\rangle$. In $|\psi\rangle$, no links or nodes receives a non-zero coloring in both terms. Thus, a superposition of distinct adjacency structures can be given a grammatical expression, and turns out to be qualitatively different from a 'mere' superposition of spin network states of the same graph.

There seems to be a second way in which we could articulate the idea of an indeterminate adjacency structure. A state like $|\psi\rangle$ could live in much bigger Hilbert spaces, of which $\mathcal{H}_\Gamma$ is a subspace. This is the case for any Hilbert space $\mathcal{H}_\Delta$ if the (connected) graph Δ contains Γ' and Γ'' as non-overlapping subgraphs. $\mathcal{H}_\Delta$ has a basis $|\Delta, j_l, i_n\rangle$, and we can ask again whether there could be superpositions $|\psi'\rangle = \alpha|\Delta, j_l, i_n\rangle + \beta|\Delta', j'_l, i'_n\rangle$, where Δ' is a graph distinct from Δ. Of course there could be, but in order to give a syntactically correct expression of $|\psi'\rangle$, we would have to move to the larger Hilbert space $\mathcal{H}_E$, where E is the unique graph consisting of the two disconnected graphs Δ and Δ'. And so on and so forth. It seems as if at each stage there must be a given graph which holds fixed the fundamental adjacency structure, and that, therefore, the world must have a fixed, determinate adjacency structure at the fundamental level. This view seems unwarranted, since there is no end to the nesting of any Hilbert space into ever larger ones of which the Hilbert space is a subspace. Given that there is no end point to this nesting, there is no basis for thinking that the world has a well-grounded and determinate (albeit contingent) adjacency structure. Moreover, at no stage can it capture the idea of an indeterminate adjacency structure; in order to do that, we revert to the first option described in the previous paragraph.

We thus see that according to LQG, physical space generically has an indeterminate geometry and may well have an indeterminate structure of adjacency relations. The theory has the resources to express both these ideas unambiguously. Fundamentally speaking, space is thus discrete, of indeterminate geometry, and possibly even of indeterminate adjacency structure. Whether or not one continues to call it 'space', there is no doubt that we are faced here with a structure radically different from what we encounter in anything we might wish to call 'space' in relativistic spacetimes.

---

[5] We wish to thank Carlo Rovelli for helpful correspondence on this point.

6.1.3 Disordered locality

In the previous subsection, we have seen that locality is threatened in LQG in the sense that the fundamental form of contiguity, encoded in the structure of adjacency relations, may suffer from quantum indeterminacy. The phenomenon of so-called 'disordered locality' (Markopoulou and Smolin 2007; Huggett and Wüthrich 2013a) undermines locality in a second way: the notion of locality native to LQG can come apart from any natural way to conceive of locality in the emerging spacetime. Any concept of locality native to LQG should presumably be based on adjacency relations obtaining between nodes of spin network states, two such adjacent nodes may give rise to events which are far apart as judged in the emergent spacetime, as illustrated in figure 6.2. In other words, we have two concepts of locality—one fundamental, one at the level of emergent spacetime—and they may not line up with one another.

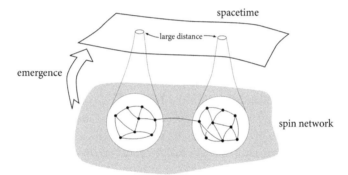

Figure 6.2 A spin network (shown as shaded) with two regions (shown enlarged) connected by an adjacency relation and how they are embedded in an emerging spacetime.

Disordered locality means that whatever notion of locality or localisability is empirically relevant at the emergent level cannot directly be that of the fundamental level. Instead, 'macroscopic' locality at the emergent level will have to arise from the fundamental structures just as the apparent continuity and the determinate spatial geometry.

Let us describe disordered locality in more detail. For this, we need to get clear on what 'locality' means both at the fundamental and the emergent levels. Locality at the fundamental level is captured by these relations of adjacency between the nodes of the spin network, now assuming them to be determinate. We furthermore assume that LQG has a well-defined classical limit which gives us a recipe for relating fundamental spin networks to emergent relativistic spacetimes—in case such a relation obtains. We should generally expect there to be spin networks without classical limit, i.e., fundamental ways the world could be according to LQG

156   OUT OF NOWHERE

in which spacetime does not emerge (this was the fourth way in which space-time disappears in the introduction to this section). Clearly, if the emergence of spacetime fails in this sense, locality is maximally disordered in that there are determinate locality relations in the fundamental spin network, but there is no emergent spacetime, and a fortiori no sense of emergent locality. We will return to the issue of the emergence of spacetime in LQG and the connected question of the classical limit in §6.3. For now, we assume that there exists such a limit and a relativistic spacetime emerges from our fundamental spin network. The point in the present subsection is that even in this case, the emergent locality structure may be 'disordered'.

What is the emergent locality structure? 'Locality' is notoriously hard to cap-ture in GR. A relativistic spacetime will of course have a topological structure, and so have a neighbohood structure. Furthermore, its metrical relations obtain-ing between spacetime events determine spatiotemporal 'distance' between them, at least locally. Of course, there is no physically privileged way in which this spatiotemporal distance decomposes into spatial and temporal distances (or dura-tions). At least, that is the case in general; if we focus on specific examples, such as the important class of FLRW spacetimes, then we do find that they may admit a privileged decomposition, such as a foliation into 'spaces' of constant spatial curvature, totally ordered by 'cosmological time' such as for FLRW spacetimes. Given this preferred foliation, e.g., in the FLRW case, (spatial) 'locality' can be expressed in terms of what is 'nearby' according to the induced 3-metric of the folia. What counts as 'nearby' can then be precisified to suit the empirically rele-vant context, but we may have to resort to different concepts in different families of spacetimes.

If a relativistic spacetime emerges from a spin network, and with it metric rela-tions, then the emergence relation fixes the relationship between the nodes of the spin network and the spacetime events. This relation can be understood as a map-ping from nodes to events. If we now look to the images of pairs of adjacent nodes, we will in general discover that not all such pairs are mapped into Planck-sized neighborhoods of one another in the emergent spacetime: the images of adjacent nodes may not be 'nearby' as judged by the emergent metric relations in the space-time. And possibly vice versa: nearby events may be the images of nodes rather removed from one another in terms of adjacency. This is disordered locality.

One can think of disordered locality to arise in those cases in which the macro-scopic distance between events is not monotonic in the number of adjacency links required to get from one to the other of their fundamental counterparts. This seems to open the possibility of superluminal signaling: if we are fortunate enough to sit near one end of such a 'shorter' fundamental connection between events, we may be able to send a signal along that connection such that it arrives earlier than a light signal in spacetime which was simultaneously released would arrive at the corresponding spacetime event.

Also given these potentially grave consequences, one may wonder why we should think that there exist such pairs of events whose locality is disordered. And we may base our wondering on the fact that a continuum spacetime contains much more 'information' than a discrete structure. It seems as if we can embed any spin network whatever into a continuum spacetime which preserves the adjacency relations at the fundamental level. Whatever the (average) grid spacing of a possibly highly irregular discrete structure may be, there will be infinitely many events 'between' the images of any two nodes of the embedding into a continuum spacetime. In this sense the continuum spacetime structure has much more scope to accommodate any wild, disordered, non-local adjacency between nodes: it can basically 'track' any such non-local adjacency at the fundamental level, and much more on top. Clearly, in order to arrive at the conclusion of disordered locality, we need an additional condition on emergence: not every spacetime will do.

The condition we are looking for is the requirement, similar to that demanded in causal set theory (condition 3 in definition 16 in §4.1.3), that the emergent spacetime not contain any sub-Planckian funny business.[6] Since quantum gravity appears to preclude physics below the Planck scale anyway, we should not permit its introduction at the emergent level of classical spacetime. This condition can be expressed in terms of some averaging or coarse-graining. Imposing it effectively precludes the continuum spacetime from using its small-scale structure to track any Planck-scale non-locality and so enables the phenomenon of disordered locality. With this condition in place, we recognize a second way in which the geometric interpretation of the area and volume operators must be accepted with care: given the phenomenon of disordered locality, these operators are 'area' and 'volume' operators really only in fortuitous circumstances where disordered locality is sufficiently suppressed; and of course even then they deserve their geometric interpretation only in virtue of their indirect relation with emergent geometry. We will discuss this indirect relation in §6.3.

Although in principle measurable, since we observe no effects, these non-localities should be expected to be suppressed in any large-scale approximation relevant to the emergence of spacetime. Adjacencies giving rise to disordered locality must be sufficiently weak for them not to matter at the level of the approximative emergent spacetime. If they are sufficiently weak, then they will be crowded out by those adjacencies sufficiently strong and 'local' in the emergent spacetime.

One might object that fundamental spin networks exhibiting disordered locality vis-à-vis their corresponding emergent relativistic spacetime should not count as physically significant in any way, precisely because they seem to show pathological features in their locality structures. But if what was asserted in the previous paragraph is correct and these non-localities are indeed strongly suppressed at the

---

[6] But unlike in causal set theory, where the condition is given explicit expression and consideration, we are not aware of corresponding explicitness in LQG.

158 OUT OF NOWHERE

macroscopic level, then we see no basis on which to rule out these spin networks as 'unphysical'.

While we thus have reason to expect disordered locality to be sufficiently weak so as not to frustrate the emergence of an effective relativistic spacetime, the phenomenon nevertheless shows how the locality structures at the fundamental and emergent levels can come apart. This does not mean that the fundamental structures are not spatial in *any* sense; but it does establish that they are not directly spatial as macroscopic, emergent space is. This is also the case for reasonably 'geometric' spin network states. Consequently, geometry is emergent.

6.1.4 No change, no time?

In §5.3.3 in the previous chapter, we distinguished between the *problem of time* and the *problem of change*. The problem of time arose in Hamiltonian GR because time seemed to disappear as a background variable with respect to which physical systems would evolve. This problem was exacerbated when we canonically quantized Hamiltonian GR, as we did in §5.4.1; there, a comparison between the Wheeler-DeWitt equation and the Schrödinger equation made the disappearance of time apparent. The problem of change, on the other hand, surfaced because what qualified as genuine physical quantities turned out to be constants, i.e., quantities that did not 'change'. Any change, it appeared, was an artifact of the mathematical representation of the physics.

Prima facie, we seem to have a rather direct case against the existence of time, at least at the fundamental level. Perhaps this is not all that surprising, given that standard GR already put to rest any Newtonian prejudice that physics describes the evolution of spatial systems over some external, independent time. Hamiltonian mechanics is a powerful mathematical apparatus designed to deliver on this Newtonian promise. Given the failure of this Newtonian project in GR, it is not surprising that the Hamiltonian formalism suggests strong, and possibly misleading conclusions regarding the status of time. In other words, just because there is no apparent external time, we cannot infer that there simply is no time. Rather, as one would expect in GR, we have to identify 'time' in the internal structures of the physical system we are studying itself. In GR, there exists an 'instrumentarium' of doing that, from situations in which there is a physically privileged notion of time that can be extracted as in the case of cosmic time, to globally hyperbolic spacetime where we can identify highly non-unique 'moments in time' in the form of Cauchy surfaces, to situations where we can only hope to capture a highly local sense of time which cannot be projected to any global notion of time.

Of course we should expect to be able to similarly identify features of spin networks which give rise to the temporal aspects of relativistic spacetime and, ultimately, of our experience of the world. Essentially, we have to establish that LQG,

THE DISAPPEARANCE AND EMERGENCE OF SPACETIME IN LOOP   159

like any other theory of quantum gravity, reduces to classical GR in an appropriate limit. In fact, it is necessary to be able to do so in order to circumvent the threat of empirical incoherence. Given the problem of time and given that we should also expect any para-temporal features of spin networks to be subject to quantum indeterminacy (similarly to their para-spatial features as discussed in §6.1.2), this will be a highly non-trivial task, with the features of spin networks giving rise to temporality far from evident. In light of this, it seems altogether appropriate to speak of the emergence of time.

What about the second, related, problem of change? In principle, our reaction stays the same as for the problem of time: it shows how far removed the structures of quantum gravity are from anything temporal, but nevertheless the dynamic aspects of our world must ultimately be recovered if LQG is to remain viable as an approach to quantum gravity. As a direct consequence of the problem of change, features of fundamental physics which are to account for 'change' cannot merely rely on Dirac observables to capture the physical content of the theory. Rather, a more indirect and complicated way of teasing out those aspects of spin networks which are responsible for the emergence of change is required. The common approach is to conceive of change of quantities with respect to one another, rather than in some absolute sense (Rovelli and Vidotto 2024, §5): we do not seek the temporal evolution of quantities over some absolute time, i.e., we do not conceive of these quantities as functions of some independent time variable T, but instead we accept that T is just another variable ontologically on a par with all other physical variables, and consequently understand motion or change implicitly by relating these variables with one another. In this latter approach, the variables so related denote so-called 'partial observables' (Rovelli 2002), which capture the physical content of the theory in the relations in which they stand to one another.

The point that matters for present purposes is the following. Particularly to the extent to which they are deeper problems than they are at the level of classical GR, the problems of time and of change illustrate just how far removed from a notion of physical time the fundamental structures of canonical quantum gravity are. In this way, given that LQG is a canonical quantization program, they add to the sense in which space and time disappear in LQG. As noted before, this renders establishing the emergence of spacetime all the more urgent. To sketch how this might work in LQG will be the central concern of §6.3 and §6.4.

6.1.5 Objections

Before we outline the rest of the chapter, let us briefly consider recent philosophical criticism of the idea that spacetime is not fundamental in LQG. First, Le Bihan and Linnemann (2019) argue that the "explanatory gap" between fundamental, non-spatiotemporal structures in quantum gravity and spacetime in GR is not as wide as it is sometimes made to be. On the one hand, spacetime is not fully present already in GR, and so only needs to be recovered to the extent to which GR is committed to it. To what extent this is the case will depend on one's interpretation of GR, and will not concern us here (although it is evidently true that spacetime only needs to be recovered in the sense in which GR is committed to it). On the other hand, they analyze several approaches to quantum gravity and diagnose the presence of a fundamental distinction between 'quasi-spatial' and 'quasi-temporal' structures in these approaches. We concur that there exist such traces, both in causal set theory and in LQG. However, the mere existence of such (possibly very minimal) traces of an asymmetry between what in some sense may be considered 'spatial' and 'temporal' structures falls way short of there being spacetime at the fundamental level. As argued in the last chapter and as summarized above, there are strong reasons to doubt the presence of anything like spacetime in LQG. Of course, the fundamental structures will have to be sufficient for spacetime, or something close enough, to emerge in the appropriate limit. As will be argued in §6.3.2 and §6.4, they will ultimately (and necessarily) have to sustain the relevant functional roles of spacetime. In fact, the viability of the geometric interpretation of spin networks is central for that story to work, as we will argue in §6.4. In this sense, it is not at all surprising that we can identify some traces of space and time (and their difference) in quantum gravity. Similar remarks apply for string theory.

Second, Norton (2020) argues for two theses. First, he claims that whether or not spacetime really disappears from the fundamental ontology of LQG depends on our interpretation of the theory and that there exist legitimate interpretations on which it does not disappear. Second, he maintains that on many of these interpretations, including what he takes to be the "received" one, spacetime does not emerge from spin networks. This is either because spacetime was not absent at the fundamental level in the first place, or because the spin networks, if not understood as embedded in a background manifold, are not "substantival" (23). As we hope to have made clear in the previous subsections, we do not think that the first disjunct is realized in LQG: spacetime *is* absent at the fundamental level. As for the other horn, we have argued in §6.1.1 that spin networks ought not to be understood as embedded in a background manifold. Thus, the second disjunct does not get any traction. So let us focus on the first thesis here.

Norton introduces several new interpretations of LQG, which he dubs versions of "substantivalism". Never mind the details, but these interpretations have in common that they all reify in some sense the background manifold on which the spin networks are defined. In this way, they resemble versions of 'manifold substantivalism' in the context of GR. These interpretations are united in their refusal to discard the manifold (or the spacelike hypersurface Σ in the construction of the spin network states in §5.4.3 and §6.1.1) when we moved from particular graphs embedded in their background space Σ to what are strictly speaking 's-knot states'

THE DISAPPEARANCE AND EMERGENCE OF SPACETIME IN LOOP   161

rather than spin network states, i.e., equivalence classes of labeled graphs under spatial diffeomorphisms.[7] It is clear that these equivalence classes themselves do not depend on the background manifold and the embedding any more—that is precisely the point of taking these equivalence classes—and so there remains no reason to take the manifold 'ontologically seriously' as do these substantivalist interpretations. The background manifold Σ was just a constructive aid, which we discard like the metaphorical ladder once we climbed it. Thus, we see little reason to follow Norton in concluding that there are "well-motivated" interpretations of LQG in which spacetime (in the form of this background manifold) does not disappear from its fundamental ontology.

Finally, Rovelli (2020) asserts that space and time are present in LQG and thus appear to be at odds with our conclusion above that spacetime disappears in LQG. The basis for his assertion is his claim that space and time are stratified concepts with a multiplicity of attributes, each of which could be associated with what one means by 'space' and 'time'. By untangling the different aspects of these concepts, he seeks to clarify the sense in which space and time are present or absent in LQG.

First, space. Rovelli distinguished three (sub-)concepts of space. The relational concept of 'being located relative to', he claims, is fully present in LQG in the form of the adjacency relations in a spin network. However, both the problems of indeterminate space and disordered locality revolt against this conclusion, even though of course there is a sense in which adjacency relations can be interpreted as 'spatial'. The second concept, of absolute, substantival Newtonian space is not present in the foundation of LQG, even though it appears as an approximation in some circumstances. We agree with this, and see no way in which this challenges our conclusions. The third concept, relativistic spacetime, consists of two parts, the relational part covered above, and the relativistic counterpart of Newtonian space, namely the dynamical entity Rovelli identifies as the gravitational field. Since LQG is a quantum theory of *gravity*, and thus of the gravitational field, he finds this sense of space to be fully present in LQG. Since relativistic spacetime (or the gravitational field, if you prefer) is radically transformed into rather different structures, we find an ample sense in which space in this sense disappears in LQG.

Second, time. Here, Rovelli distinguishes five (sub-)concepts of time. The first is the relational one, where his point and our reply proceed analogously to the case of space, mutatis mutandis. We will return to the expectation that there are 'temporal' analogs to quantum indeterminacy and disordered locality in §6.5. The second and third concepts of time, the Newtonian and relativistic ones, again proceed in close analogy to their spatial analogs. We agree with Rovelli's take on the problem of time and change, as we discuss in §6.1.4. The fourth concept concerns time's irreversibility, which Rovelli admits is only to be found at a higher statistical level.

---

[7] We hope to have made it clear that even though we continue calling the states 'spin network states', we are actually referring to *s*-knot states which do not 'live' on background spaces.

162　OUT OF NOWHERE

Rovelli also accepts that the final concept, of experienced passage or 'becoming', is not present in LQG.[8]

In sum, our conclusion above according to which spacetime is absent from the ontology of LQG, still stands, even in light of these objections.

6.1.6　The disappearance of space and time

Concluding this section, we reiterate that there are significant ways in which space and time are absent from the ontology of LQG, at least in a direct interpretation of the fundamental structures as they stand today. Having said that, the spin network states of these structures are 'spatial' in that they (1) are built on the basis of relations of 'adjacency', which is the fundamental physical relation of LQG par excellence, and (2) are eigenstates of operators with a natural geometric interpretation as area and volume operators. We should flag that spin network states are spatial courtesy of this geometric interpretation only. Since it is a very natural interpretation, we will adopt it and later recognize its centrality for spacetime functionalism and thus for the emergence to be physically salient, rather than just formal. We will return to this issue in §6.4, where we will restrict this geometric interpretation to a class of spin networks states—the 'weave states'.

The abstract graphs used to represent spin networks, however, should not taken too literally as spatial as they appear to be; rather, they serve as a calculational device similar to Feynman graphs in quantum field theory. The physical content of the theory is to be teased out via an interpretation of the operators.

Given this apparent spatiality of the fundamental structures of LQG, one might be encouraged to join John Earman, who insists that

> although classical general relativistic spacetime has been demoted from a fundamental to an emergent entity, spacetime per se has not been banished as a fundamental entity. After all, what LQG offers is a quantization of classical general relativistic spacetime, and it seems not unfair to say that what it describes is quantum spacetime. This entity retains a fundamental status in LQG since there is no attempt to reduce it to something more fundamental. (2006b, 21)

We agree with Earman that the fundamental structures maintain some claim to be 'spacetime', based on that they are quantizations of classical spacetime structure. However, since this argument is ultimately based merely on the heuristics of theory formulation, we do not take it too seriously. Much stronger is the other direction of the relation between quantum and classical structures: one could argue that what

---

[8] We will return to becoming in the context of LQG in §6.5.

LQG describes is quantum *spacetime*, precisely because it deals with those structures from which relativistic spacetime emerges. Thus, as we will argue in §6.4, these fundamental structures can be taken to fulfill the physically relevant spacetime roles, and it is by virtue of *that fact* that they could legitimately be considered spacetime.

Either direction relies on the relation between the fundamental structures of LQG and relativistic spacetimes. What about the intrinsic nature of the LQG structures themselves? As mentioned at the outset of this section, they clearly retain some spatiality. Nevertheless, it is a vast exaggeration to claim that we have space and time fundamentally present in LQG. The main evidence for this skepticism has been presented above in §6.1.2, §6.1.3, and §6.1.4. It was threefold:[9]

1. Although spin network states form a basis of the Hilbert space, generically the quantum states will be superpositions of spin network states and so not eigenstates of the geometric operators. Thus, generic quantum states will have indeterminate geometry, and perhaps even indeterminate adjacency relations, undermining the claim to spatiality.

2. Both spin network states and relativistic spacetimes possess a structure of 'vicinity' or 'contiguity' relations, which is captured by the adjacency relations in LQG and by topological and metric relations in GR. However, in general, these structures do not neatly map onto one another, leading to the phenomenon of 'disordered locality'. This phenomenon makes clear just how much work needs to be done to understand the non-trivial ways in which the fundamental structures can give rise to space as we know it.

3. The combined problems of time and of change suggest that time and anything temporal—any change—is absent in canonical quantum gravity. Although there are good reasons to think that this is too quick, it clearly establishes, once again, that the way in which time, change, and any familiar dynamics emerge from fundamental LQG are far from direct and require careful technical and interpretive work to peel out the emergence of time.

Our point is the following: although the fundamental structures of quantum gravity maintain aspects of spacetime to some degree, they are so far removed from relativistic spacetimes that this loss of spatiotemporality threatens the empirical coherence of quantum gravity. It is therefore incumbent on any approach to quantum gravity to establish how something sufficiently like relativistic spacetime emerges from what the approach postulates as fundamental structures. It will be the task of sections §6.3 and §6.4 to give some account of how that is supposed to work in the case of LQG.

---

[9] And we didn't even mention that the geometry of spin networks is discrete!

6.2 Interpretation of covariant LQG

Let us move to the physical interpretation of the spinfoam models of the covariant perspective on the problem of quantizing GR. Feynman graphs are normally thought to represent ways in which an initial state may interact such as to evolve into the subsequently measured final state. If we add up the weighted contributions of all graphs to the transition amplitude between the initial and the final state, we can predict the probability with which the final state will occur *given the initial state*.

It would seem as if an analogous interpretation cannot be imposed on spinfoams: the spin network states that mark off the spinfoam are, after all, supposed to correspond to (instantaneous) three-spaces. Furthermore, under what we label the *global interpretation*, they encompass all of space, universe-wide. To think of these structures in such a global manner is a point of view naturally inherited from GR. However, at least once we move to quantum gravity we recognize that preparing or even just measuring the entire quantum state of the universe is not only fiscally irresponsible, but physically impossible. Of course, one could insist on the global interpretation and retort that neither our incapacity of manipulating nor of observing the global state invalidate the global analysis. A justification for this stance could be inductive: given that the orthodox interpretation has worked seamlessly in other, non-cosmological contexts where it faced no real competitor, its extension to the present case is only natural, indeed rational.

However, there are two systematic differences between the earlier case of this inductive basis and the target case of interest. First, the 'globality' of the target case seems to preclude a distinction between a system and an external observer. Apart from the fact that it would seem to be necessary to resolve the problem of coupling matter to spin networks in order to understand the interaction between the material observer and the observed spatial geometry, if the system encompasses literally everything, then it is not clear how we could have a separate observer, somehow outside of the spatial slice at stake. At the very least, the defender of globality would owe us an account of how to conceive of the relationship between system and observer.[10]

Second, and arguably just as fatally, the single-shot nature of the situation precludes a reasonably controlled repetition of measurements involving the same initial state. This means that we cannot hope to collect the data necessary to make ordinary statistical inferences so central to science. Both of these reasons suggest that we drop the pretense of globality.

---

[10] For instance, one could argue that we would need to identify some material system whose dynamics is sensitive to, but which (at least for practical purposes) does not affect the gravitational dynamics. We thank an anonymous reader for pointing this out.

THE DISAPPEARANCE AND EMERGENCE OF SPACETIME IN LOOP   165

Let us distinguish a cosmological from an astrophysical scope of a theory.[11] The cosmological scope appears to be global, or 'universal', and may thus run into the problems just described. In order to circumvent them, and presumably also to resolve the measurement problem every quantum theory faces, a separation between observer and the observed must somehow be introduced. Although the observer is of course also *in* the all-encompassing space they observe, a natural restriction on the side of the observed system will enable the required separation: what we observe cosmologically are really just a rather small number of coarse-grained degrees of freedom relevant to specify the global cosmological structure of the universe, and not literally all degrees of freedom of the system. Thus, it seems as if there is plenty of scope to resolve the tension also on a cosmological, and in some sense global, reading of (covariant) LQG, provided we do not confuse a global, cosmological perspective with a 'total' theory.[12]

However, what we wish to emphasize is the possibility of an 'astrophysical' scope on which LQG describes the fundamental physics just of a compact finite region of spacetime, without regard of anything outside that region, let alone of what happens globally in space.[13] On this perspective, we delimit the scope of the theory to the 'experimental situation' in which we examine a 'medium-sized' region of spacetime, perhaps of particular astrophysical interest. This could be the region inside a detector or a similar measuring device, or an observed star or similar object. We assume that the observer suffers from no interactions with the region under the 'microscope' that would invalidate this perspective. Let us call this perspective 'regional' and oppose it to the global perspective outlined above.

In fact, the covariant approach to LQG has predominantly included the regional perspective, often implicitly and sometimes explicitly. Taking the regional perspective seriously also means that the region at stake is bounded not just in the timelike direction with 'initial' and 'final' states, but also 'sideways' in the spacelike direction. In fact, as is argued in Rovelli and Vidotto (2015, §2.4.2), the regional perspective suggests that we focus on a compact region R. The quantum state on the region's boundary $\Sigma = \partial R$ is an element of the boundary Hilbert space, and the transition amplitude between 'in' and 'out' states is then a linear functional on this Hilbert space, depending only on the state on the boundary Σ, which consists in both the 'in' and 'out' states. In fact, it is this state, which also fully determines the geometry of Σ. In full consistency, the regional perspective is then taken in the context of (classical) ADM variables (Rovelli and Vidotto 2015, §3.4.1), where the Hamilton function (their equation 3.64, and indeed 3.97) is a functional of the metric on a 3-sphere. As the Hamilton function codes all solutions to the EFE with the given (finite!) 3-sphere as boundary, the regional physics can be completely

---

[11] For more on this distinction, see also Wüthrich (2021, §19.4).
[12] See Halliwell and Hawking (1985) for an positive example of what we have in mind.
[13] We wish to thank Carlo Rovelli for discussions on this point.

modeled without any global quantities. The regional perspective is maintained throughout the discussion of classical discretization in chapter 4 of Rovelli and Vidotto (2015), where the lattices considered in lattice gauge theory are all finite and serve as models for finite spinfoams. The regional perspective is epitomized in the picture of a lens-shaped, finite spacetime region enclosed by two spacelike boundaries, joined at their common boundaries (see figure 6.3). If the spacetime region is 4-dimensional, then the spacelike boundaries are 3-dimensional, and their intersection is 2-dimensional. As Rovelli and Vidotto (2015) make clear, the idea is to use finite regions like this in astrophysical applications such as in black hole physics in order to describe quantum gravity processes.[14]

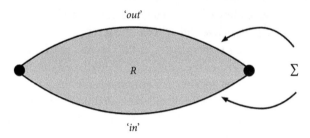

Figure 6.3 The finite region R with spacelike boundary 'in' and 'out' states jointly forming Σ.

As should be evident, the covariant program is thus not just a set of technical tools to progress toward a quantum theory of gravity (although it is that too), but also a wholesale *perspective* on the field in which the ambition moves from articulating a fundamental theory of global, all-encompassing models, as it is frequently encountered in relativistic cosmology, to solving concrete problems where quantum gravity must be applied. One localizes the process where quantum gravity cannot be neglected in a finite region of 4-dimensional spacetime, enveloping the physical goings-on where the classical, general-relativistic approximation fails and seeks to define the transition amplitudes that deliver the probability distribution for the boundary states. These can in principle be computed in the style of Feynman, using the spinfoam formalism.

Moving to a regional perspective has several implications at the more philosophical level. First, it seems as if we regain something closer to the prototypical measurement situation which gives rise to the traditional measurement problem in quantum physics. If this is the case, then on this perspective any quantum theory of gravity will have to address the measurement problem. However, we should

---

[14] Already when CW visited Rovelli's group in Luminy for an extended research stay in 2004, we were drawing compact spacetime regions on the blackboard. The idea of moving away from global hypersurfaces to arbitrary finite regions is already explained in Rovelli (2004, §5.3.3). See also figure 10.3 in Rovelli and Vidotto (2015, 214).

like to emphasize that the measurement problem will arguably arise in some guise or other for any quantum theory of gravity. It is entirely possible, however, that it will take a somewhat different form; after all, the standard measurement problem, as we find it clearly explicated for instance in Maudlin (1995), arises because distinct determinate measurement outcomes cannot result from a linear evolution of the quantum state, if that state gives a complete description of the state of the system. How we should think of the measurement problem in a context in which the dynamics is less explicit than Schrödinger evolution remains the question for another day. Given that quantum states will generally be superpositions in a geometrical basis, at least in LQG, the question of how an apparently classical world with a determinate spacetime geometry emerges will still be pressing. We will address it in the next two sections—without, disappointingly, solving the measurement problem.

Second, the central assumption of the regional perspective idealizes the true situation, where a chunk of 'space' can hardly be prepared and measured and still remain completely isolated from the rest of the universe. Since spin networks are eigenstates of the area and volume operators, the delimiting spin network states $|s_i\rangle$ and $|s_f\rangle$ can be interpreted as describing the results of geometrical measurements of these states in medium-sized regions. The area and volume operators do not commute with the Hamiltonian and are therefore not Dirac observables. They constitute, however, partial observables in Rovelli's sense (Rovelli 2002). The orthodox physical interpretation of spinfoams therefore maintains that the initial and final spin network states represent spatial regions and that the transition amplitudes predict the probability with which a given initial state evolves to the measured final one.

Third, back in §5.3.3 we complained about the restriction to globally hyperbolic spacetimes in canonical quantum gravity. This restriction seemed deplorable in the light of the sincere possibility that actual physical phenomena might turn out to best be described by non-globally hyperbolic spacetimes. However, on the regional perspective, this restriction is much less problematic. Although there is no guarantee that the much milder restriction to regionally globally hyperbolic spacetimes will not miss any physically relevant situation, considerations regarding the global topology need not concern us. The regional perspective avoids any commitment to a definite global topology and so deflects much of this criticism.

Fourth, the regional perspective is a very natural approach to astrophysics. For astrophysical phenomena such as black holes, the regional perspective offers not only a viable approach, but one that is much closer to actual practice in physics.[15] However, the study of cosmology seems to require a global perspective; at the very

---

[15] Which perspective best fits the description of black holes depends on how we define a black hole; the standard GR definition of a black hole as the complement of the causal past of future null infinity is global, yet actual calculations in astrophysics are done on the basis of regional definitions. For more on different concepts of black holes, see Curiel (2019).

least, it is unclear how to do cosmology in a regional setting. In (quantum) cosmology, the goal is to isolate some cosmic degrees of freedom from the rest of the universe, i.e., from other degrees of freedom, and to study their properties and evolution. The regional perspective seems limited in that we should expect there to be entanglement between cosmic degrees of freedom and the environment, which means that the regionally studied degrees of freedom will in general be mixed states. As the covariant approach is so closely tied to the regional perspective, it seems as if it will be unable to account for situations in which such entanglement cannot be neglected.[16] Of course, this may just mean that we have to be careful when we partition the system at stake from its environment, and have to be prepared to adjust this separation, most likely by changing our coarse-graining to a finer grain. Thus, it would be a mistake for the regional perspective to hold fixed, once and for all, our identification of the system of interest.

Finally, let us return to the question of whether the covariant perspective offers a fundamentally spatiotemporal picture. It seems as if it is ineliminably spatiotemporal, given that the delimiting spin network states are clearly identified as 'initial' and 'final' states of a spatially extended geometry and the inside is supposed to represent the spacetime region at stake. But such an interpretation would be rash. Just as it would be problematic to interpret the 'paths' in a Feynman graph as concrete spatiotemporal paths of material bodies,[17] we cannot hastily identify the spinfoam with a concrete region of spacetime. Furthermore, two of the three reasons against interpreting the fundamental structures of LQG spatiotemporally we adduced in §6.1.6 still apply here: space is still indeterminate and locality still disordered.

In contrast, the dual problems of time and change seem to have evaporated on the covariant perspective; after all, part of the point of the move to the covariant perspective was exactly to sidestep the technical and conceptual difficulties of the canonical program to do with the Hamiltonian. If we jump ship from the canonical to the covariant program, the problem of time seems to just evaporate; after all, there is no Wheeler-DeWitt equation anymore to worry about. Of course, we still will not have an external time, which determines which state is 'initial' and which 'final'. And the point needs to be treaded carefully: if the covariant theory is really equivalent to the canonical one, as it is supposed to be, then if we have a problem of time or of change on the canonical side, we will surely have the corresponding problem on the covariant side. However, as we argued in §6.1.4, these problems only illustrate that the fundamental structures are not straightforwardly (spatio)temporal. Emphatically, there is no implication that no dynamics can emerge—despite the absence of an absolute, external time.

---

[16] At least on the standard conception of cosmology—if one equates the big bang to a white hole, or to a black hole-to-white hole transition, as do Bianchi et al. (2018), then arguably the regional perspective does not face this difficulty.

[17] See for instance Redhead (1990).

6.3 The emergence of spacetime in LQG

6.3.1 Emergence, reduction, and the classical limit

The fundamental structures postulated by LQG are generically non-spatiotemporal. If this is correct (and we wish to avoid empirical incoherence), then spacetime must be shown to emerge from these non-spatiotemporal structures. Let us return to our discussion of emergence in chapters 1 and 2 for this purpose. As is standard in the relevant literature both in physics, but also in recent philosophy of physics, we take 'emergence' as a name or placeholder for a relation to be explored, rather than as a substantive claim regarding the relationship between the fundamental and non-fundamental levels. In particular, as we have emphasized on many occasions and in line with Butterfield (2011a, 2011b), *emergence is not intended to flag a claim that the relation is non-reductive.* Although it is common to use the term in this way particularly in the philosophy of mind, we follow the common usage in physics and philosophy of physics and explicitly take it to be consistent with there being a reductive relation between the descriptions of the non-spatiotemporal (fundamental) and higher spatiotemporal levels. In fact, we will argue that the relation can best be captured by functional reduction, applying our template introduced in chapter 2. If this is right, then the relation is clearly reductive. In the present context, we should expect that the 'emergence' in play is a collective designation for broadly reductive relationships: reduction, as an inter-theoretic relation, is the *working hypothesis* of the quest in quantum gravity to regain spacetime.

Let us try to positively characterize emergence. Butterfield (2011a) defines it to mean "properties of a system which are *novel* and *robust* relative to some appropriate comparison class" (921, his emphasis). For Butterfield, properties are *novel* if, roughly, they show striking features not found in the comparison class, or if they are not definable from that class, and properties are *robust* if, roughly, they are "the same for various choices of, or assumptions about, the comparison class" (ibid.).[18]

Crowther (2016, 2018) has developed this account of emergence inspired by effective field theory and high-energy physics, which largely fits the case of quantum gravity as well (as she has noted, of course). Emergence is a bipartite concept, which includes aspects of dependence as well as of independence between levels or theories. Crowther analyzes both aspects, putting more weight on the ways in which the theories are independent from one other, while noting that dependence can be understood as an asymmetrical relation. For her, the independence is cashed out by the two individually necessary and jointly sufficient conditions

---

[18] Butterfield makes both terms more precise for the physical situations he is interested in in a companion paper (2011b). He does this in terms of limits. As the somewhat more general framework of his earlier work with Isham is more suitable for LQG, we will rely on that in §6.3.2.

of *novelty* and *robustness*.[19] A higher-level theory is *novel* if it is "suitably distinct" (Crowther 2018, 82) from the more fundamental theory, describing some physics which cannot cleanly be expressed in terms of the more fundamental theory using different degrees of freedom. The robustness of the emergent physics, on the other hand, "means that it is largely impervious to changes in the details of the more fundamental physics" (ibid.).

However, it is not clear why robustness should really be necessary for emergence: if the physics at the emergent level appropriately depended on the fundamental physics and would be sufficiently novel and different from it, then it would seem perfectly adequate to speak of 'emergent physics'. Having already assumed dependence (in the form of a reductive relation), robustness does not seem to add anything we need. Without quarreling about words, we will thus not consider to what degree the spacetime physics at the emergent level is "impervious" to quantum gravity and speak of 'emergence' whether or not this condition is satisfied. We expect that a certain degree of robustness will in fact be present in the emergence of spacetime from quantum theories of gravity; we do not recognize that it *must* be present for the relation to qualify as emergence.

It is evident, however, that an appropriate sense of novelty must be present to justify speaking of 'emergence'. In the present case of spacetime, we take it that this is established by our interpretation of the ontology of LQG above in §6.1 (and of causal set theory and string theory in the corresponding chapters) as being to some relevant degree non-spatiotemporal. If that's the case, then the ontologies of quantum gravity differ from relativistic spacetimes sufficiently for the condition of novelty to be met easily. This leaves us with the dependence between the two levels or theories.

We noted in chapter 2 that in order to evade the problem of empirical incoherence, it must be shown how the fundamental structures of quantum gravity give rise to something which is sufficiently like relativistic spacetimes. Expressed at the level of theories, it must be established that GR is a good approximation to the underlying theory in at least those regimes in which GR has proved to be a successful theory. As was already noted in Wüthrich (2017, 321), the task is discharged by 'taking the classical limit' of the fundamental theory, i.e., by showing that the classical theory (or something sufficiently like it) results from an appropriate procedure showing how the native effects of the fundamental theory are hidden behind the phenomena so well accounted for by the classical theory—GR in our case. If, to speak with Reichenbach (1938), quantization procedures belong to the "context of

---

[19] The terminology shifts from Crowther (2016) to Crowther (2018): in the former, robustness is called 'autonomy', while in the latter 'autonomy' is the conjunction of novelty and robustness. In order to eliminate the ambiguity, we avoid speaking of 'autonomy' altogether.

discovery", then taking the classical limit constitutes a central element of the "context of justification". If we cannot show that the quantum theory of gravity at stake reduces to GR in those domains where GR is a well established, successful theory, and thereby explain why GR is as successful as it is in those domains in which it is successful, then the project has failed to meet a basic methodological principle of physics. This basic principle demands for a theory T_1 to be acceptable as replacing another theory T_2 as more fundamental, more explanatory, more universal, T_1 must account (a) for the successes of T_2, as well as (b) for at least some of its failures.

'Taking the classical limit' means finding a mapping between models of the fundamental, 'reducing', theory and models of the higher-level, 'reduced', theory. The models thus being related can be individual models, sets of in some sense 'generic' models, or the near totality of models of the two theories. As should be clear from our discussion in chapter 2, it will not suffice to procure a merely mathematical expression of such a mapping; instead, we will also need to supply a demonstration of its 'physical salience'.

Clearly, the map from the set of quantum states to the set of classical spacetimes should not be expected to be bijective, but rather many-to-one, as there will generally be multiple quantum states with the same classical limit. Moreover, there will be some quantum states with no corresponding classical states. Finally, as Butterfield and Isham (2001, 80) note, interpretational subtleties may mean that the quantization of a classical theory does not return the original classical structure as emerging from the quantum theory. Thus, rather than arriving at GR, we should expect to find a close cousin of GR.

The classical limits of candidate theories of quantum gravity remain incompletely understood. This is not different for LQG. The difficulties tend to come in two different kinds. First, there are technical obstacles. Second, there are conceptual and interpretational challenges. This is where philosophers can hope to make contributions, e.g., by exploring the conceptual landscape, mapping possibilities, and, more concretely, by bringing the philosophical literature on emergence, reduction, and functionalism to bear on our present concern of the emergence of spacetime. As noted in the last chapter, at least on its original canonical version, LQG remains an incomplete theory lacking a 'dynamics'. Consequently, much of what is being said in the remainder of this section and the next rests on the kinematical part of the theory, and in particular on its spin network states with their geometric interpretation. As we are about to introduce and discuss the scheme of emergence of time proposed by Butterfield and Isham (1999, 2001), for whom the problem of time in canonical quantum gravity constitutes a major motivation, the restriction to kinematics has some of the advantages of theft over honest toil. However, the focus on spin network states can be justified by the central role they also play in the covariant perspective. We expect the main lessons to carry over to covariant LQG.

172   OUT OF NOWHERE

6.3.2 The Butterfield-Isham scheme of emergence

Butterfield and Isham (1999, 2001) consider various notions of emergence, which may be of potential help in the context of canonical quantum gravity.[20,21] As it turns out, they all cast emergence as a broadly reductive relation, in line with our approach throughout this monograph, but in dissonance with sections of the philosophical literature. The literature on reductive relations between theories is rich and diverse.[22] From this, Butterfield and Isham (1999) conclude that there may not be a single concept of reduction to fit all cases, even if the analysis is confined to physics. They distinguish three major ways in which theories can stand in a reductive relation: definitional extension, supervenience, and emergence.

To see one theory as a definitional extension of another, one presupposes a syntactic concept of scientific theories: theories are deductively closed sets of propositions. In this approach, GR is a *definitional extension* of LQG just in case it is possible to add definitions of all non-logical symbols of GR to LQG such that every theorem of GR can be proven in LQG-plus-GR symbols.

Understanding a theory as a definitional extension of another is attractive because the program promises all the rigor and clarity of logic. If a theory T_2 is a definitional extension of another theory T_1, then this arguably goes some way to explaining why T_2 was as successful as it was, eventually breaks down, and is replaced by a superior theory.

However, the tool of definitional extension cannot handle cases such as the relation between GR and LQG, as they are not nearly as neat as would be necessary for the application of the concept of definitional extension. Even the case of the relationship between Newtonian mechanics and special relativity, perhaps with its c-to-infinity limit, one of the simplest cases in all of physics, gets significantly more complicated (though arguably still manageable) if we relate Newtonian (or Galilean) spacetime to Minkowksi spacetime. In order to determine whether GR is a definitional extension of LQG, the mathematical relation between the two theories (including the classical limit) ought to be understood much better. Unless this is the case, the concept of definitional extension cannot be gainfully applied here.

It is clear that an elucidation of the mathematical relation between GR and LQG will involve approximations, in the sense that at least some statements of GR are approximately true in LQG. As a first step, one would presumably try to formulate an intermediate theory between GR and LQG such that it would be possible to prove all requisite theorems of this intermediate theory in LQG augmented by

---

[20] Butterfield and Isham (1999) consider quantum geometrodynamics and Euclidean quantum gravity, in Butterfield and Isham (2001) the focus is on the old particle-physics approach, superstring theory, and canonical quantum gravity.

[21] This section is based on Wüthrich (2017).

[22] We will not even attempt to field this literature here; cf. Spector (1978) for an analysis of many proposals for reduction as an inter-theoretic relation, with a particular eye to the physical sciences.

some definitions. In a second step, one would then proceed to show that GR, or at least its relevant parts, can be recovered as an approximation from this intermediate theory. Following Butterfield and Isham (1999), we can define this process of approximation as follows:

Definition 18 (Approximating procedure). *An approximating procedure is the process of either neglecting some physical magnitudes, and justifying such neglect, or selecting a proper subset of states in the state space of the approximating theory, and justifying such selection, or both, in order to arrive at a theory whose values of physical quantities remain sufficiently close to those of appropriately related quantities in the theory to be approximated.*

It should be emphasized that the clause "appropriately related quantities in the theory to be approximated" bears a lot of weight and occludes substantive work to show that the pertinent relations are 'appropriate'.[23] Clearly, if relating two theories includes an approximating procedure, then we have gone beyond mere definitional extension. We will return to approximating procedures below when we include them in our toolbox when we bring in emergence.

Let us move to the second form of reduction considered by Butterfield and Isham (1999): supervenience. GR supervenes on a quantum theory of gravity such as LQG just in case all its predicates supervene on the predicates of LQG for a fixed set $\mathfrak{A}$ of objects on which the predicates of both GR and LQG are defined. Given such a set $\mathfrak{A}$ of objects, the set of GR predicates *supervenes* on the set of LQG predicates just in case any two objects in $\mathfrak{A}$ which differ in what is predicated of them in GR must also differ in what is predicted of them in LQG. This straightforward characterization of how GR could be said to supervene on LQG runs into the obvious problem that by requiring a fixed set of objects, the characterization only finds application for a relation between two theories sharing a joint ontology of objects on which the ideologies of the theories are then defined. Given that LQG's ontology consists in, perhaps, loops or spin networks or spinfoams or whatever some physical interpretation of the fundamental Hilbert space and the mathematical machinery of the theory suggests, and that GR's ontology includes, arguably, spacetime structures and perhaps matter fields, the above characterization of supervenience seems rather hopeless for the present case.

The requirement that the same ontology $\mathfrak{A}$ must underlie both theories can be relaxed.[24] A natural relaxation would be to partition the set $\mathfrak{A}$ into subsets $\mathfrak{A}_1$ and $\mathfrak{A}_2$ such that each subset contained the ontology of one of the theories, and then demand that $\mathfrak{A}$ be closed under compositional operations such as the mereological sum or set formation. The two sets of predicates of the two theories, let us call

---

[23] We thank Erik Curiel for discussions of this point.
[24] We thank Jeremy Butterfield for having suggested this relaxation to us.

174   OUT OF NOWHERE

them $\mathfrak{P}_1$ and $\mathfrak{P}_2$ respectively, would then be primarily defined on their respective base sets of objects. In order for the notion of supervenience to get some traction, it would then be necessary that the predicates of LQG, which are initially just defined on the fundamental ontology of LQG, would somehow induce predications on the non-fundamental ontology of GR such that we could judge whether GR's predication on its ontology supervenes on LQG's. However, the ontology of GR does not merely consist in composites of the ontology of LQG and so it would remain utterly unclear how the predication induced by LQG on GR's ontology would work. We see no interesting way to make this work.

Thus, neither is GR a definitional extension of LQG nor does it supervene on LQG. However, the third, broadly reductive but more liberal, relation of 'emergence' proposed by Butterfield and Isham is promising for our case:

Definition 19 (Emergence). *For Butterfield and Isham, a theory T_1 emerges from another theory T_2 iff there exists either a limiting or an approximating procedure to relate the two theories (or a combination of the two).*

Apart from the definition of 'approximating procedure' given in definition 18 above, we need one for 'limiting procedure':

Definition 20 (Limiting procedure). *A limiting procedure is taking the mathematical limit of some physically relevant parameters, in general in a particular order, of the underlying theory in order to arrive at the emergent theory.*

We should certainly not expect that the emergence of GR from LQG be captured by a simple limiting procedure. According to Rovelli (2004, §6.7.1), this can be gleaned from the remarkable failure of the original interpretation of the newly discovered loop representation of GR around 1988. Originally, physicists believed that the classical, smooth geometry of spacetime could be recovered from the lattice of loops by taking the limit of a vanishing lattice constant, in direct analogy to letting the lattice constant of a lattice field theory go to zero and thus connect it to a conventional QFT. In other words, they did not expect quantum space to actually be discrete.

However, this changed after the derivation of the spectra of the area and volume operators in 1995 when researchers tried to construct 'weave states' (discussed in more detail below) which approximate a classical geometry: with quantum states defined as the limit one obtains when the spatial loop density grows without bounds (i.e., when the loop size goes to zero), the approximation turned out not to become more accurate as the limit was taken. Therefore, this limit was evidently not physically relevant. Instead it was found that in this limit, the eigenvalues of the area and volume operators increased, suggesting that the surface area and the volume of the physical system under consideration grows in this limit. What did not happen was anything that could have been interpreted as the physical density of the loops to increase when the lattice constant was decreased: it turned out that the physical density of loops remained unaffected by the size of the lattice constant—it was simply determined by a dimensional constant of the theory, Planck's constant.

Rovelli interprets this result to mean that "more loops give more size, not a better approximation to a given [classical] geometry" (ibid.). The loops, it seems, have an intrinsic physical size. Manipulating the lattice constant does thus not change the structure from quantum states to smooth manifolds. It does not affect anything in the physics, except that larger volumes of quantum space are considered. As noted above, the natural conclusion was that there is a minimal physical scale of a fundamentally discrete structure. Given that some essential features of the classical geometry such as its smoothness cannot be reduced to or identified with properties of the quantum states of LQG, GR does not arise from LQG by a simple limiting procedure. As we will see, limiting procedures will play some role in recovering GR from LQG, but are insufficient by themselves. GR is not just a mathematical limit of LQG.

There is another reason that a simple limit does not suffice to recover a classical theory from a quantum theory: a limiting procedure cannot eliminate superposition states, which are of course generic in a quantum theory. As argued by Landsman (2006), the classical world only emerges from the quantum theory if some quantum states and some observables of the quantum theory are neglected, *and* some limiting procedure in the sense of definition 20 is executed. According to his view, to be discussed below, relating the classical with the quantum world thus takes both the limiting as well as the approximating procedures.

Consider approximations. What is the 'approximandum', i.e., the classical theory to be approximated? It is obviously GR, or at least part of it. In the original formulation of LQG, no matter was assumed to be present and the relativistic spacetime was assumed to be globally hyperbolic. One might thus think that the approximandum was GR's sector of globally hyperbolic vacuum spacetimes. However, for reasons given in §5.4.2, such a conclusion might be rash, and it may well be better not to insist on globally hyperbolic vacuum spacetimes.

As Landsman (2006) lists, there are three major ways in which classical physics is typically thought to relate to quantum physics: (i) by a limiting procedure which involves the limit $\hbar \to 0$ for a finite system, (ii) by a limiting procedure which uses the limit $N \to \infty$ of a large system of N degrees of freedom while $\hbar$ is held constant, and (iii) by decoherence or by a consistent histories approach. In fact, as Landsman shows, the 'factual' limit $N \to \infty$ (which is in some sense physically realizable in our world) is, mathematically speaking just a special case of the 'counterfactual' limit $\hbar \to 0$ (which is not physically realizable in our world). In this sense, (i) and (ii) are not as distinct as one might have thought.

176   OUT OF NOWHERE

Landsman argues that while none of these three ways is individually sufficient to understand how classicality emerges in quantum physics, they jointly suggest that classicality results from ignoring certain states and certain observables from the quantum theory. We have already argued above that taking limits such as (i) or (ii) is insufficient for regaining classicality as no such limit can resolve a quantum superposition. This is where some have thought that 'decoherence' (iii) comes into play. In general, 'decoherence' designates a process in which a quantum system's state has its interference suppressed by the system's interaction with the 'environment', such that a pure quantum state, by virtue of its interaction with the system's 'environment', evolves, over a rather short time span, from a superposition state to an 'almost' mixed state with classical probability distributions but 'almost' no quantum interference left. Roughly, decoherence leaves the quantum system in an approximate mixture of eigenstates of some macroscopically relevant operator.[25]

In §6.2, we distinguished a cosmological from an astrophysical scope of a theory, as well as between a global and regional perspective taken on it. If a QG theory's scope is deemed astrophysical and a regional perspective is taken, then it seems natural enough to separate the system from its environment. However, if the scope is considered cosmological and the perspective taken is global, then it seems difficult to make the separation. In this situation, the relevant 'system' must consist in some cosmologically privileged internal degrees of freedom of the cosmos. In other words, we 'coarse grain' to wash out almost all degrees of freedom, singling out a few as the 'system' against the rest, which act as the 'environment'. In this case, the system's environment is not spatially external to it.

This brings us to a second way in which the notion of decoherence at play in quantum gravity must be generalized. Not only can decoherence no longer be an interaction between spatially cleanly separated subsystems, but also its usual conception as of a dynamical process in time must be revisited to fit a context in which the fundamental description of the world does not include the dynamical evolution of systems against some external time. The solution here will involve understanding how dynamical processes such as decoherence can co-emerge with spacetime when both seem to presuppose, or at least enable, the other.[26]

Taken together, a theory of decoherence needs to identify the relevant (astrophysical or cosmological) degrees of freedom, discern them from those attributed to the environment, and explain how the interaction between the two kinds of degrees of freedom works to suppress the quantum interference, even in the absence of a normal kind of 'dynamics'. In other words, decoherence, or something like it, will then deliver a mechanism, which 'drives' the system at stake to the right kind of semi-classical quantum states, and so justify the selection of a

---

[25] For a review of decoherence, see Bacciagaluppi (2020).
[26] For an example of what we have in mind, see Kiefer (2004, particularly chs. 8 and 10, and indeed §10.1.2).

subset of states (and a subset of physical magnitudes) made in the approximating procedure at play in the emergence of spacetime in the case of LQG.

Using the terminology proposed by Butterfield and Isham, the emergence of spacetime in LQG will follow a procedure with two main steps. First, fully at the level of the quantum theory, but with a view to macroscopic classicality, a physical mechanism will drive the generic quantum states of the fundamental structures into the semi-classical states. which bear a close relationship to classical states. The first step is described by an approximating procedure. Second, a limiting procedure relates these semi-classical states to states in the phase space of the classical theory by showing how magnitudes of the classical states arise as a limiting case.

Let us end the section by emphasizing two points. First, the two-step procedure above is but a crude general template of a largely formal nature. In the next section, we will argue that spacetime functionalism will give us a more detailed understanding of what exactly it is that needs to be recovered from the fundamental quantum structures, thus providing information of how the vague procedure can be filled in and in particular how it can be endowed with physical salience.

Finally, apart from the technical issues to be resolved and the physical salience to be established, there also remain deep interpretive questions to be addressed. In particular, neither the technical two-step procedure nor spacetime functionalism in themselves should be expected to fully solve the quantum measurement problem, which will resurface in quantum gravity.

6.4 Spacetime functionalism

In §2.4, we characterized spacetime functionalism as consisting in two essential steps (not to be conflated with the two steps of the Butterfield-Isham scheme above): in (SF1), the spacetime entities, properties, or states required to recover all relevant empirical content (but not necessarily more than that) are functionalized by specifying their identifying roles; in (SF2), an explanation is given how the fundamental entities, properties, or states of the quantum gravity theory fill these roles. It is this last step, specific to each theory of quantum gravity, which delivers a necessary part of the justification for a candidate theory of quantum gravity.

We will not systematically and exhaustively list every aspect of spacetime required at step (SF1). Even without that, it seems clear that recovering significant aspects of the geometry of emergent spacetime from the fundamental quantum gravity entities will go a long way toward satisfying (SF1). In this spirit, we will focus on the geometric quantities of area and volume, which have a particularly straightforward connection to LQG and which will serve as our exemplar case of

178 OUT OF NOWHERE

how spacetime geometry emerges in LQG.[27] It is clear that (measurements of) area and volume play central roles in capturing at least the spatial geometry of spacetimes. As for (SF2), this is where the concepts offered by Butterfield and Isham can be applied. We will show that it is at least plausible that something like relativistic spacetimes emerge from the fundamental structures of LQG if a suitable combination of approximations and limiting procedures is applied: such a combination of procedures will thus give us the general scheme to establish how the fundamental structures fill the relevant geometric roles even though they are not straightforwardly spatiotemporal.

Our claim is twofold. First, we will argue that although existing work in LQG has not been related to the Butterfield-Isham scheme (apart from in our own work), it naturally fits into this conception. Second, the Butterfield-Isham scheme together with the more general framework of spacetime functionalism presented in this section outline *all* the remaining work that needs to be done in order to establish the emergence of spacetime.

Two caveats. First, what follows remains, unfortunately, vague and tentative. It is vague because work on the research program of LQG is ongoing and many details remain to be worked out—as of course they are on other research programs in quantum gravity. It is tentative because the particular proposal of how to understand semi-classical states we will study may turn out to be unfruitful (or indeed because the entire research program may be abandoned). It should be emphasized, however, that both the vagueness and the tentativeness affect all philosophy of quantum gravity. Second, apart from a few remarks about dynamics at the end, we will restrict ourselves largely to the much better understood kinematical level and the spin network states. This restriction results from the fact that most work on semi-classical states to date focuses on the kinematical level. Of course, it imposes another form of tentativeness on the argument in this section, as it might turn out that states in the physical Hilbert space relate to semi-classical states in importantly different ways from how kinematic states relate to them. We accept these limitations, also in the hope that the development of the full theory will benefit from a fuller understanding of how spacetime might emerge from kinematical states.

The key to the construction of semi-classical states from states in the kinematical Hilbert space $\mathcal{H}_K$ is to identify those states in $\mathcal{H}_K$ where the quantum fluctuations are considered to be negligibly small. These states are assumed to correspond to nearly flat three-metrics. There are various approaches to execute this construction of semi-classical states. Following Thomas Thiemann and Oliver Winkler, one ansatz is to use coherent states.[28] Alternatively, Madvahan Varandarajan has proposed to use (generalized) 'photon Fock states' (Varadarajan 2000; Ashtekar

---

[27] For more details, see Wüthrich (2017); Lam and Wüthrich (2018) focuses on relative localization, and its connection to geometric quantities such as area and volume.

[28] For a review of this approach, see Sahlmann et al. (2001) and Thiemann (2007, §11.2). Thiemann's book also discusses weave states in §11.1 and the photon Fock states in §11.3.

THE DISAPPEARANCE AND EMERGENCE OF SPACETIME IN LOOP 179

and Lewandowski 2001), and Abhay Ashtekar and collaborators have developed the 'shadow states' program (Ashtekar et al. 2003).[29] Perhaps the most prominent approach uses so-called 'weave states'. We will confine our discussion to this approach.

Introduced by Ashtekar, Rovelli, and Smolin (1992), a *weave state* is an eigenstate of the volume operator $\hat{V}$ with an eigenvalue which approximates the corresponding classical values for the volume of a spatial region $\mathcal{R} \subset \mathcal{M}$ of a spacetime, as determined by the classical gravitational field.[30] These spin network states are simultaneously eigenstates of the geometrical area operator for a surface S. It is quite obvious that the very characterization of weave states is hostage to the geometric interpretation of spin networks introduced and discussed in §6.1.1.

Let us introduce the concept of a weave state and its status as a semi-classical approximation a bit more technically.[31] Starting out from a macroscopic spacetime, let us consider a 3-dimensional region $\mathcal{R} \subset \mathcal{M}$ with a 2-dimensional surface $S \subset \mathcal{M}$ as its boundary and the 3-dimensional triad field or gravitational field $e_a^i(\vec{x})$ defined for all $\vec{x} \in \mathcal{R}$.[32] The gravitational field has an associated metric field $q_{ab}(\mathbf{x}) = e_a^i(\mathbf{x}) e_b^j(\mathbf{x}) \delta_{ij}(\mathbf{x})$, where δ_{ij} is the Kronecker delta. It is possible to construct a spin network state $|S\rangle$, which approximates the metric at sufficiently large scales Δ much larger than the Planck length ℓ_{Pl}.[33] For a classical spacetime, the area of a 2-dimensional surface S and the volume of a 3-dimensional region $\mathcal{R}$ with respect to a fiducial gravitational field $^0 e_a^i$ are given by

$$\mathbf{A}[^0 e, S] = \int |d^2 S|, \tag{6.2}$$

$$\mathbf{V}[^0 e, \mathcal{R}] = \int |d^3 \mathcal{R}|, \tag{6.3}$$

---

[29] As Thiemann (2007, §11) points out, there are deep connections between the various approaches.

[30] Rovelli (2004, §6.7.1) gives an intuitive account of weave states and of their interpretation. He analogizes these states of the gravitational field to a fabric of weaves, which appears in some sense smooth if seen from afar, but turn out to be discrete when examined from close up. Hence their name.

[31] Readers who are not interested in these technicalities should feel free to skip this paragraph.

[32] Here, we follow the choice of Rovelli (2004, §2.1) in referring to the triad field as the 'gravitational field'.

[33] Strictly speaking, $|S\rangle$ really is a spin network state, as opposed to an *s*-knot state as defined in §6.1.1. In other words, it is a 'pre-kinematic' state which solves the Gauss constraints, but not necessarily the spatial diffeomorphism constraints. As such, it is not represented by abstract graphs, but as a graph embedded in a background manifold. This choice is convenient because it allows us to directly relate the weave states to three-metrics, rather than to equivalence classes of three-metrics under spatial diffeomorphisms, and for this reason is made in most of the relevant literature. It does not imply any interpretive difficulty, as we could also characterize weave states as fully kinematic spin network states (i.e., to be precise, *s*-knot states) invariant under spatial diffeomorphisms. This can be achieved by introducing a map P_{diff} which projects states in the 'pre-kinematic' Hilbert space related by a spatial diffeomorphism unto the same element of $\mathcal{H}_K$. Then, the state $|s\rangle = P_{diff}|S\rangle$, which is a state in $\mathcal{H}_K$, is a *weave state* of the classical three-geometry $[q_{ab}]$, i.e., of the equivalence class of three-metrics q_{ab} under spatial diffeomorphisms, if and only if $|S\rangle$ is a weave state of the classical three-metric q_{ab} as defined in the text. We emphasize and elaborate this point in order to make clear that we are not reintroducing a background manifold.

such that the measures for the integrals are given by $^0e_a^i$ (Rovelli 2004, §2.1.4). Although this is not necessary, the fiducial metric is often chosen to be flat. *Weave states* are defined by two conditions. First, they are spin network states, i.e., simultaneous eigenstates of the area operator $\hat{\mathbf{A}}$ and the volume operator $\hat{\mathbf{V}}$. Second, their eigenvalues are equal to the classical values given in (6.2) and (6.3), up to small corrections of order of $\ell_{\mathrm{Pl}}/\Delta$ per dimension:

$$\hat{\mathbf{A}}(S)|S\rangle = \left(\mathbf{A}[^0e, S] + \mathcal{O}(\ell_{\mathrm{Pl}}^2/\Delta^2)\right)|S\rangle, \tag{6.4}$$

$$\hat{\mathbf{V}}(\mathcal{R})|S\rangle = \left(\mathbf{V}[^0e, \mathcal{R}] + \mathcal{O}(\ell_{\mathrm{Pl}}^3/\Delta^3)\right)|S\rangle. \tag{6.5}$$

It is this second condition which captures the requirement that weave states approximate the classical geometry at sufficiently large scales. In fact, the length scale Δ *characterizes* weave states, which we thus denote $|\Delta\rangle$. Weave states are by construction semi-classical approximations, in that above scales of order Δ, they closely approximate the corresponding classical geometry in delivering the same areas and volumes as does the metric q_{ab}. At scales clearly smaller than Δ, however, we should expect quantum features to dominate and thus the approximations to break down.

Weave states thus realize the functions of supplying a measure for areas and volumes, which is central for recovering physical space. They offer a clear example of the usefulness of the functionalist perspective in which geometrical, and so spatiotemporal, properties emerge from a non-geometrical, and so to a substantial degree non-spatiotemporal, foundation. However, it is important to note that this geometrical emergence only applies to weave states and cannot in general be extended to generic states.

The constraints (6.4) and (6.5) on weave states secure the functional realization and so the emergence. They do not, however, uniquely determine the weave states for a given 3-metric q_{ab}, as the constraint is only on properties smeared over surfaces and regions, which are large compared to the Planck length. Thus, the correspondence between weave states and classical geometries is many-to-one, and many distinct spin network states in general play the same 'averaged' metric role. In physics, this is a common situation, also encountered for instance in thermodynamics, where many microscopic states give rise to the same averaged, macroscopic state. In the context of functionalism, the situation seems reminiscent of 'multiple realizability', where distinct lower-level properties or states play the same functional role at the higher level. A standard conception of multiple realizability in philosophy demands that different *kinds* or types of states or properties all realize the same higher-level property or state (Bickle 2020). This is not the case for different weave states, so we have here at best a rather weak example of multiple realizability.

Glossing over some technical details, the notions both of approximation (definition 18) and of a limiting procedure (definition 20) prove useful in understanding this emergence. In order to arrive at weave states, which are eigenstates of the area and volume operator, all those operators defined in terms of connection operators had to be ignored, since the 'geometrical' eigenstates are maximally spread in these operators and the weave states are those which are peaked around the geometrical values as determined by the fiducial metric. In this sense, some physical quantities have been neglected. There remains an open question as to whether the neglect of connection-based operators can ultimately be justified. If if cannot, then it seems as if weave states ought to be replaced by semi-classical states which are peaked in both the connection and the triad basis such as to approximate classical states. This would still amount to a selection of states, though perhaps not of physical quantities. Either way, the procedure, if well justified, satisfies the conditions of definition 18 and thus qualifies as an approximation.

In order to give a complete account of this approximation, a physical mechanism which drives generic kinematical states to the semi-classical weave states must be provided. Decoherence is the primary candidate for such a mechanism. As we have seen above, we conceive of decoherence either in the cosmological setting with a physically meaningful separation into cosmological and environmental degrees of freedom, or in the astrophysical context in a regional perspective either in the same sense or as the interaction between a localizable system and its external environment. Either way, the idea would be that the generic quantum state would decohere into an eigenstate of the geometric operators with approximately classical values, i.e., into a weave state. Thus, decoherence would be the physical mechanism underwriting the approximation, ascertaining the emergence's physical salience.

It would be important for decoherence to also provide a mechanism to suppress the non-spatiotemporality of generic quantum states. Consider the issue of disordered locality, introduced in §6.1.3. Generally speaking, the conditions (6.4) and (6.5) do not rule out a mismatch between the adjacency structure of the spin network at the LQG level and the locality structure of the emergent relativistic spacetime. It is in principle possible that a weave state satisfying the constraints (6.4) and (6.5) and thus filling the geometric roles of area and volume, nevertheless diverges in its adjacency structure rather drastically from the locality structure of the apparently corresponding emergent spacetime. In this case, the weave state would be playing some metrical roles, but would not instantiate any standard 'localizing function'. Given that standard spatiotemporal locality would not be instantiated in this case, the threat of empirical incoherence would loom large once again. For this threat to be dispelled, it is necessary that all relevant spacetime roles are in fact realized by the fundamental structure. These relevant roles include, to repeat, all functions required to ground our empirical evidence. As for the case of disordered locality, it can be reasonably expected that in the actual world at most a mild version of it is present. In the context of semi-classical LQG,

182   OUT OF NOWHERE

this expectation translates into an assumption that (single) weave states actually do execute the correct localizing function at the emergent level. Since our world is apparently spatiotemporal and the locality structure of GR seems to be tracking its locality structure, if the world is fundamentally in a weave state then that latter better be well behaved in this sense. Perhaps there is a physical mechanism, such as decoherence, which is 'enforcing' this, or whether it is just by sheer luck (and thus explainable ex post facto by anthropic reasoning), we cannot tell at this stage.

However, weave states, even with the right locality structure, are not classical spacetime just yet. In order to relate them to classical spacetime, a limiting procedure in the sense of definition 20 can be applied. This procedure will include taking the limit $\ell_{Pl}/\Delta \to 0$, which will make the small corrections in (6.4) and (6.5) disappear and so relate the weave states to classical geometry. This limit can be accomplished either by letting Δ go to infinity, or ℓ_{Pl} go to zero (or both). Letting Δ go to infinity amounts to increasing the size of the spatial region $\mathcal{R}$ to grow beyond limits, and thus corresponds to the 'factual' limit $N \to \infty$ discussed above. In contrast, letting ℓ_{Pl} go to zero amounts to the 'counterfactual' case of $\hbar \to 0$. The first choice gives a physical explanation of how a quantum world can appear to be classical, while the second provides a mathematically rigorous, but physically questionable, relation between a quantum theory and a strictly classical description of the spatial geometry.

Let us step back and consider the scope of our functionalist claim. First, we have used area and volume as natural geometric roles that the fundamental structures ought to be playing. As it turns out, this is a convenient focus for (SF1) in the case of LQG, and of weave states in particular. However, area and volume only serve as exemplars for the functionalist work we have in mind. We emphasize that these roles are not sufficient for the program to succeed; in general, all spatiotemporal information necessary to do physics, and ultimately to underwrite ordinary experience, will have to be shown to be recoverable from the fundamental QG structures. Second, the functionalist account for area and volume as outlined only works for weave states, i.e., rather special states in LQG. There is no reason to think that it generalizes to all spin network states, let alone to generic states in LQG. It remains an open question to what extent other states in LQG give rise to spacetime-like effective behavior. For states other than weave states, we expect that some will exhibit an even more partial spatiotemporality and that some will not do so at all. This expectation, of course, could be corrected by future research in LQG. Finally, as we already flagged in §6.1.6, the functionalist account depends on the geometric interpretation of these weave states.

According to the standard conception of LQG, our considerations regarding the kinematical weave states only concern the emergence of *space*, rather than *spacetime*. Let us add a few words about how this might play out in the dynamical completion of LQG and thus say something about what might additionally come into play in the case of full spacetime emergence.

The spinfoams encountered in the dynamical context in the previous chapter suffer from limitations of their spatiotemporality corresponding to the quantum indeterminacy (§6.1.2) and disordered locality (§6.1.3) of spin network states, mutatis mutandis. We contend that the functionalist perspective serves equally well to conceptualize the classical limit and thus the emergence of full relativistic spacetime in this case. Without going into the details (those can be found in Rovelli and Vidotto 2015, ch. 8), spinfoams turn out to play the spacetime role of a Regge simplicial discretization in the classical limit understood using coherent states rather than weave states. Ultimately, this means that they are responsible for localization by playing the role of a spacetime lattice as in a lattice field theory. Unlike in standard lattice field theory, however, localization in the spacetime lattice is dynamically and functionally instantiated, rather than given as a fixed background. It will play this role in serendipitous, or at least contingent, circumstances when the right functional relationships among the dynamical entities are exemplified. Two limits have to be distinguished: the classical and the continuum limit. The classical limit involves an approximation procedure and relates the coherent states to the Regge complexes. The continuum limit involves a limiting procedure and relates these discrete complexes to classical relativistic spacetimes.

We should note that the functionalist emergence discussed in this section does not by itself solve the quantum measurement problem. Ultimately, a full account of the emergence of spacetime and thus of classicality will also have to solve the measurement problem. Solving it is non-trivial in non-relativistic quantum mechanics, harder in the context of special-relativistic physics, and even more deeply mystifying once we move to quantum gravity. In light of these difficulties, it would already be a significant accomplishment to arrive at a detailed, complete, and consistent quantum theory of gravity with a well-understood approximation to semi-classical physics and a rigorous limiting procedure connecting the semi-classical regime to classical states of the gravitational field.

In sum, spacetime functionalism offers a valuable perspective on the problem of the emergence of spacetime in LQG, as it does for other theories in quantum gravity. Its strength derives from a flexibility about what it is that instantiates the relevant spacetime roles, in two ways. First, functionalism naturally incorporates the fact that different spin oams may instantiate a particular relevant spacetime role. These different spinfoams may differ with respect to other relevant spacetime functions, which leaves room for novel predictions, such as any empirical consequence of disordered locality. Second, it strips spacetime (including relativistic spacetime) of any ineffable metaphysical natures and reduces it to its relevant roles in physics, or more widely in our manifest world. Furthermore, the functionalist does not expect that spacetime's functional roles exhaust the nature of the fundamental entities or structures which instantiate these roles, and so leaves open the metaphysical interpretation and so future developments of the fundamental

184   OUT OF NOWHERE

theory at stake. The fundamental spin networks or spinfoams may well possess other properties beyond being spacetime realizers which may turn out to be physically relevant.

6.5 Eternalism, presentism, or neither?

Since, as we argued above, in LQG the fundamental ontology is non-spatiotemporal and as you might already have expected from the fact that there was a 'problem of time', it is not surprising that the philosophy of LQG has implications for the metaphysics of time. The main debate in standard philosophy of time rages between those who endow time fundamentally with a dynamical aspect, a sense of 'becoming' or 'passage', and those who do not. Often motivated by the apparent phenomenology of temporality, such dynamical views necessarily award a metaphysically privileged status to an ever advancing 'present'. These dynamical views most prominently include presentism and the growing block theory. *Presentism* asserts that all and only present entities exist—in other words, the sum total of existence only includes present entities.[34] The dynamical aspect finds expression in the fact that the present is continually updated as it changes. The *growing block theory* also includes past entities into the sum total of existence, which thus consists in an ever growing 'block' to which new layers are continually being added as they become. In either theory, the present is special either in that it is the only thing that exists or in that it forms the cusp of what has become and thus exists.

On the other side of the debate are those who favor a metaphysics of time which does not include becoming at the fundamental level. For them, to the extent to which it is part of objective reality, this dynamical aspect emerges at some scale, possibly from interactions between fundamental physics and the cognitive and perceptual apparatus of agents, such as humans. Often motivated by physics, views in this camp do not accept the present to be metaphysically special. The main representative is *eternalism*, which maintains that the sum total of existence contains what presentists would label past, present, and future entities. However, a characterization of eternalism as holding that past, present, and future entities all exist on a par is problematic insofar as the fundamentality of such a distinction is denied by the eternalist. Having said that, it is clear that sum total of existence according to eternalism is a strict superset of that of the growing block view, which is a strict superset of that of presentism.

When we discussed the philosophy of causal set theory in §4.3, we considered the theory's potential interpretation in terms of a metaphysics of a growing block and found that while in some sense possible, the eternalist interpretation remains to be preferred. How does this debate play out in the context of LQG?

---

[34] Again, our use of 'entities' is meant to convey our liberal ontological attitude.

If we take seriously the apparent result in canonical quantum gravity that time has altogether disappeared from the fundamental level, then it remains prima facie puzzling how to even make sense of the debate in that context. For a fundamentally timeless universe, only an atemporal metaphysics would seem appropriate. Although eternalism rejects a fundamental notion of 'present', it nevertheless seems to accept a fundamental temporal ordering of entities,[35] as long as the temporal relations do not invoke some distinction between past, present, and future.

Le Bihan (2020) argues that LQG favors what he dubs 'atemporal eternalism', according to which "any proper part of the natural world exists *simpliciter* and that the material content of the natural world does not depend on any particular location in it"(2, 12), in contrast to "standard eternalism", which still requires, in Le Bihan's conception, a fundamental causal ordering of events. Although the two views will arguably coincide in most traditional contexts, we agree that the core of eternalism better be defined to maintain that what exists does not depend on any "particular location" in the natural world, including on any temporal or causal 'location'. Le Bihan then produces a two-step argument to the conclusion that LQG may empirically settle the debate in the metaphysics of time in favor of eternalism.

The first step seeks to establish that at the level of LQG itself, eternalism is the only viable option. As a result of the phenomenon of disordered locality, which Le Bihan supposes extended from spatially interpreted spin networks to relations in fundamental structures which are analogous to timelike separations, there exists no partition of the fundamental structure into past, present, and future parts. Since presentism and the growing block view depend on such a partition, they are in conflict with structures as they appear in LQG, and only eternalism, according to which the sum total of existence does not depend on such a partition, remains in play. Although we believe that the extension of disordered locality from spatially interpreted structures to in some sense spatiotemporally interpreted ones deserves further scrutiny, we find it overwhelmingly plausible that the quantum nature of the fundamental structures will disallow any clean and ontologically robust partition into past, present, and future parts. Thus, we agree that presentism and the growing block view find little hope in LQG.

In the second step, which seeks to extend eternalism from the fundamental level to higher levels, Le Bihan concedes that in principle there could be what he calls "scale fragmentalism", i.e., different scales of description might deliver different ontologies. The second step therefore amounts to a rebuttal of this possibility at least for "the particular case of temporal ontology" (18). In other words, we could not combine an eternalist metaphysics at the fundamental level with a presentist one at human scales. Let us call such a combination *emergent presentism*. Using the example of a macroscopic past event E, such as Caesar's crossing of the Rubicon,

---

[35] Of course, this order will in general not be total, but at best a partial order.

186 OUT OF NOWHERE

Le Bihan displays the unpalatable consequences of emergent presentism: while presentism would deem E as non-existing at the macroscopic level, the parts of the fundamental structure which ground E, or give rise to E, would exist from the point of view of the eternalist fundamental structure. Although Le Bihan does not describe this situation in these terms, emergent presentism implies a kind of supervenience failure, and with it an analogous kind of failure of the functionalist reductionism we have advocated. In fact, Le Bihan's case against emergent presentism can be strengthened by the observation that a present macroscopic event P, such as your reading of these lines, would have fundamental grounds which we would expect to be somewhat 'distributed' across the fundamental structure, since there exists no fundamental temporal (or causal) ordering. More generally, even if we managed to partition macroscopic events into past, present, and future ones, there exists no map of such an ordering into the fundamental realm. Any form of emergent presentism simply has no basis in the fundamental. Overall, we thus arrive at the same conclusion as Le Bihan.

Recently, Rovelli (2019) has offered arguments rejecting not just presentism, but also eternalism. His arguments are based on relativity (and McTaggart!), not quantum gravity, but we assume that his views are informed by his work on quantum gravity, and in particular on LQG. His rejection of presentism follows well-rehearsed standard lines, but his rejection of eternalism is based on a particular conception of eternalism and worth discussing in the present context. For him, eternalism endorses two theses:

> [T]he entire four-dimensional spacetime is 'equally real now', and becoming is illusory. (1328)

In our view, both parts are problematic. The formulation of the first part is misleading as it leads people (including Rovelli, cf. 1329) to assert that eternalists maintain that dinosaurs exist now and similar nonsense. The first part is unsuitable because it characterizes the position in a way that implies its endorsement of trivially false propositions. As stated above, although the sum total of existence includes all events in spacetime, this does not mean that they exist *now*.

The second part of the characterization of eternalism is Rovelli's prime target. Although some eternalists have indeed defended that becoming is illusory, this is not a necessary part of an eternalist position. What is essential to eternalism is that there exists no partition of entities into past, present, and future at the fundamental level. Consequently, if becoming relies on such a fundamental partition, it is doomed. But there is no reason to think that it does, and so becoming need not be considered illusory. Given the relevance of motion, change, and becoming for human phenomenology, denying all these dynamical aspects of our experience as illusory seems a prohibitive price to pay for a metaphysics, which takes seriously our best science as well as our experience. What is needed is a scientifically-based account of our phenomenology of temporality which does not rely on a fundamental notion of present or of a 'flow'. Such an account is possible, as relevant work e.g., in Huggett (2014, 2023) and Callender (2017, see particularly ch. 11) testifies, but this is not the place to enter into its details.

Rovelli adamantly rejects the mainstream view according to which becoming is not fundamental in contemporary physics. He argues that given its centrality to human experience, fundamental physics should be interpreted in ways which bring this dynamical aspect of the world to the fore. For instance, he asserts that Einstein's field equation describes the vivid "unfolding of happenings" (1331) in the world, although of course the resulting becoming is local, "complex and multilayered" (1332), rather than a global progression of objective coming into being. We will not repeat here the discussion of §4.3, where we argued for our eternalist preference, but only add that Einstein's field equation, which relates geometry and energy-matter content of the universe locally at each point of spacetime, is a dynamical equation only in the rather minimal sense that it includes derivatives with respects to spacetime directions. Such an equation is perfectly consistent with an eternalist reading.

6.6 Conclusion

Just as in causal set theory, it turns out that the fundamental structures postulated in LQG are at least not fully spatiotemporal and diverge quite radically from relativistic spacetimes (§6.1). This again raises the concern that LQG may be empirically incoherent. With the aid of understanding emergence in terms of approximating and limiting procedures (§6.3) and of the resources of spacetime functionalism (§6.4), this threat was averted as we sketched how spacetime may emerge from LQG. These general lessons of spacetime emergence remain valid in covariant LQG (§6.2). Not only did this framework organize the mathematical aspects of spacetime emergence, but its two parts, together with the central assumption of what we called the 'geometric interpretation', jointly establish its physical salience. Although many of the details of the account of spacetime emergence wait to be filled in as the theory gets more fully articulated, we are confident that the general templates offered here contribute to a fuller understanding of the situation and thus will remain relevant in future developments of the theory.

Finally, we considered the implications of our interpretation of LQG for the metaphysics of time (§6.5), where we concluded that on balance LQG does not override relativity's verdict in favor of eternalism. On all these accounts, the general conclusions we draw from our consideration of LQG are remarkably parallel to those we drew from causal set theory in chapter 4.

Chapter 7
A string theory primer

These days there are a number of textbooks[1] available to those who seek an introduction to the physics and formalism of string theory, and while these all have their limitations, it would be an unnecessary duplication of effort to reproduce them wholesale here. Instead this chapter will draw on them to give a somewhat heuristic overview of the subject, emphasizing physical intuitions, taking a slightly different trajectory to that often followed. We especially want to emphasize from the start how the basic physics of string theory is largely a new application of standard physics; this is especially true of the classical string, but considerable continuity remains even when more advanced ideas arise on quantization.

For the record, all of the physics of this chapter is drawn from the existing literature (especially the texts listed in footnote 1). All that is original is (some of) the manner of presentation, and in particular the audience to which it is addressed—philosophers of physics. We take it that this group is familiar with (at least) the basic ideas of Langrangian mechanics for classical fields, quantum mechanics (including some quantum field theory), and relativity (including some general relativity)—yet not necessarily familiar with advanced concepts involving quantum fields. In other words the physics described generally assumes less than graduate level texts (explaining some important background ideas), but (much) more than popular discussions (giving a deeper understanding). It is thus aimed at filling a gap in the literature, a gap which has previously slowed string theory's assimilation into the standard topics of philosophy of physics. (We hope that it is also useful to others who already understand the mechanics of string theory, but

---

[1] Here are some that we have found most useful. First, there are Susskind's amazing Stanford Continuing Studies lectures, available online (starting with Susskind 2011); these are beautifully intuitive, but still formally honest. With luck he will eventually write these up in a book; Susskind and Lindesay (2005) already overlaps. Second, Zwiebach (2004) is highly recommended. It provides a very helpful introduction to string theory, including advanced concepts, without assuming substantial graduate training, often through simplified models that allow the reader to see basic mechanisms, without all the details. (It is perhaps a little slow in the early chapters for the purposes of philosophy of physics.) Then the classic texts, of the first and second string revolutions respectively, are Green et al. (1987) and Polchinski (1998); these certainly assume more understanding of quantum field theory, but are extremely good in presenting the logic of string theory, illuminating the important physical principles. In addition, Becker et al. (2006) and Kiritsis (2011) are very useful, emphasizing applications through worked problems and problem sets. And finally, we have found McMahon (2008) very useful as a companion to the other texts, explaining some of the steps in calculations when they exceeded our understanding of quantum field theory! In what follows we have attempted to give the first, or most authoritative, or most useful citation for particular concepts that we discuss, although typically multiple of the above sources contributed to our understanding, or might be consulted.

Out of Nowhere. Nick Huggett and Christian Wüthrich, Oxford University Press. © Nick Huggett and Christian Wüthrich (2025). DOI: 10.1093/oso/9780198758501.003.0007

wish to see an overview that emphasizes some important conceptual aspects over more formal ones.)

Finally, two remarks about the relation of this chapter to the rest of the book. First, this chapter and chapter 9 are technical, providing more rigorous formal background and arguments; but the remaining two chapters on string theory (8 and 10) can be read, and the arguments and conclusions followed, without reading them. The two technical chapters provide a deeper background, and more precise account of the physics involved; this one especially is more like a textbook than the rest of the book. Readers who find this chapter hard going can therefore either skim through it, or skip it, and still understand the main philosophical conclusions of the following ones. Second, this chapter strictly goes beyond what is needed for the remainder of the book, aiming also to provide interested readers with the conceptual tools to pursue further reading and research, and with an awareness of where the important concepts for the book fit into the broader framework of string theory.

We will start, in §7.1, with a classical relativistic string in spacetime, an object with one spatial dimension, and hence a worldsheet of one spatial dimension and one temporal dimension. (As we proceed it will be important to have the worldsheet picture in mind.) It's actually a rather fun object to play with, using the standard tools of classical mechanics in an unfamiliar way (for instance, see Zwiebach 2004, problem 6.5 for a quick example of how a string can accelerate to the speed of light in a finite time!). We will first develop some physical intuition about the classical string, and its equations of motion; and then develop a more sophisticated framework for its physics (§7.2). At that point, in §7.3 we will see what happens when the classical theory is canonically quantized: especially, the existence of string excitations with the defining characteristics of quantum particles. Indeed, understanding that conclusion is the main lesson of this chapter.

The attentive reader may be puzzled by this approach after reading the introductory chapters: isn't the point of the book that theories of quantum gravity, including string theory, do *not* posit spacetime as part of the fundamental furniture? Yet here we apparently lay out theories of classical and quantum strings *in* spacetime! But in this chapter 'spacetime' is to be taken in a formal sense, referring to a Lorentzian signature differential manifold as a mathematical object; a 'background' in which the abstract theory is defined. In subsequent chapters we will argue that, despite appearances, properly understood this theory does not posit classical spacetime as a fundamental concrete physical structure; rather, spacetime emerges from fundamental non-spatiotemporal structure.

The chief limitation of the chapter is its focus on the bosonic string; since fermions exist, realistic physics needs to include them too, leading to supersymmetry. This large subject lies outside the scope of the book, aside from a brief

190   OUT OF NOWHERE

introduction in §7.4, and a few remarks in later chapters. However, the physical ideas developed here, and the conceptual issues raised in this book, do carry over to superstrings: their omission will not undermine what we aim to do.

7.1 Classical string basics

7.1.1 Equation of motion

Our string is free, subject to internal tension, but under the influence of no external forces (adding background fields adds important new features); let's suppose too that there is no gravitational field, so that it lives in Minkowski spacetime, of arbitrary dimension D. So how do we expect this object to behave? Well, like a spring tossed in the air (ignoring gravity) it can have an overall velocity, and it can wobble, stretch, and compress. So let's specify that the string satisfies a simple wave equation.[2]

For now, consider the open string, with two separate ends, pictured in figure 7.1. Assign inertial coordinates X^μ ($\mu = 0, 1, \ldots, D - 1$) to the spacetime in which the string lives; and label the points of the string worldsheet itself with two (dimensionless) 'worldsheet coordinates', $0 \leq \sigma \leq \pi$ and τ, spacelike and timelike with respect to spacetime, respectively.[3] Now we can describe the string in spacetime by mapping each worldsheet point (τ, σ) to the coordinate $(X^0, X^1, \ldots, X^{D-1})$ of the corresponding spacetime point: a worldsheet is represented by a function from the (τ, σ) to the $(X^0, X^1, \ldots, X^{D-1})$. It's natural to think of this function as a prescription for embedding the string in spacetime, but formally it is a D-component *field* on the string worldsheet, $X^\mu(\tau, \sigma)$. (The latter way of thinking will be central in what follows, so make sure that you have both conceptions straight before moving on.[4])

To say that the string is 'wobbling' is to say that this field—and hence individual components of the field—oscillate on the worldsheet, with respect to the internal components. Defining $\dot{X}^\mu \equiv \partial X^\mu/\partial\tau$ and $X'^\mu \equiv \partial X^\mu/\partial\sigma$, the simplest wave equation takes the form

$$-\ddot{X}^\mu + c^2 X''^\mu = 0 \qquad \text{or with } c = 1 \qquad -\ddot{X}^\mu + X''^\mu = 0, \tag{7.1}$$

---

[2] This section draws heavily on Becker et al. (2006) for the formal results; the reader is referred there for more details. However, the theory is physically motivated somewhat differently here.

[3] Note the switch from 'coordinate free' indices $a, b, \ldots$ of earlier chapters, to Greek indices $\mu, \nu, \ldots$, indicating a particular coordinate choice. The range for σ keeps equations tidy later.

[4] As we shall soon see, X^μ is treated as a vector in the spacetime tangent space. But note that it is obviously not a vector in the worldsheet's tangent space, as it has too many components. In that sense X^μ is not a *vector* on the string worldsheet; rather, when we think of the worldsheet as the base manifold, X^μ is an 'internal' vector, with the spacetime tangent space as the fiber. (Thanks to Tushar Menon for this observation.) The conception of the string in terms of a worldsheet field is discussed in Huggett and Weingard (1994).

and assumes that the string has no intrinsic mass—though because it is extended, only the *ends* of an open string actually move at the speed of light (Zwiebach 2004, §6.9). So let's suppose that there are worldsheet coordinates in which the equations of motion indeed take this form.

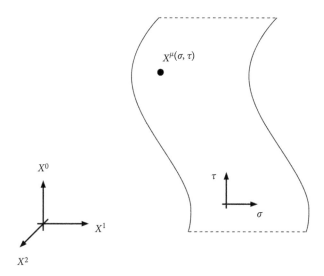

Figure 7.1 An open string (τ, σ), in an external spacetime with coordinates X^μ.

As usual when solving wave equations, a generic solution has right-to-left and left-to-right moving parts: the X^μ are functions of $\tau \pm \sigma$. These in turn can be decomposed into a sum of oscillatory modes moving in each direction (plus a constant x^μ specifying the 'initial' location of the string):

$$X^\mu = x^\mu + \ell_s \left(\alpha_0^\mu \cdot (\tau - \sigma) + \tilde{\alpha}_0^\mu \cdot (\tau + \sigma) \right) + i \frac{\ell_s}{2} \sum_{k \neq 0} \left(\alpha_k^\mu e^{-i2k(\tau - \sigma)} + \tilde{\alpha}_k^\mu e^{-i2k(\tau + \sigma)} \right). \quad (7.2)$$

Here we have used a natural length parameter, ℓ_s to give a solution with units of length; as we shall see, this 'characteristic string length' is related to the tension in the string. Since the coordinates, X^μ, are always real numbers it is easy to show that the dimensionless coefficients satisfy

$$(\alpha_k^\mu)^* = \alpha_{-k}^\mu \quad (7.3)$$

and similarly for the $\tilde{\alpha}$.

Now, (7.2) still needs some interpretation: first, are there any physical boundary conditions that we need to impose? Second, we know that the X^μ represent positions in spacetime in an arbitrary inertial frame, but we haven't said much about how the (τ, σ) label worldsheet points (only specifying the range of σ, that they are

192 OUT OF NOWHERE

timelike and spacelike, and that the wave equation holds). We can address the first point next, and return to the second below.

We should consider two possible kinds of string: closed strings, in which the string forms a loop, as well as the open strings in which there are two ends, which we have already seen. For a closed string the σ coordinate can take on any value, but we require periodic boundary conditions

$$X^\mu(\tau, \sigma + \pi) = X^\mu(\tau, \sigma), \tag{7.4}$$

since (τ, σ) and $(\tau, \sigma + \pi)$ represent the same point on the string. This condition entails that k take on integer values and that $\alpha_0^\mu = \tilde{\alpha}_0^\mu$. Thinking physically about the string it is clear that the α_0 terms don't describe vibrations, but a linear motion of the string through spacetime; thus it makes sense to write (remembering that $c = 1$)

$$\ell_s p^\mu \equiv \alpha_0^\mu + \tilde{\alpha}_0^\mu = 2\alpha_0^\mu, \tag{7.5}$$

so that the general solution for a closed string is given by:

$$X^\mu(\tau, \sigma) = x^\mu + \ell_s^2 p^\mu \tau + i\frac{\ell_s}{2} \sum_{n \neq 0} \frac{1}{n} \left(\alpha_n^\mu e^{-i2n(\tau-\sigma)} + \tilde{\alpha}_n^\mu e^{-i2n(\tau+\sigma)} \right). \tag{7.6}$$

This describes an initial position and linear velocity, and then the relative oscillatory motions of the points in terms of left- and right-moving oscillations. (Note that if there are no oscillations—all $\alpha_n^\mu = \tilde{\alpha}_n^\mu = 0$—then X^μ is independent of σ, which is to say that the string is contracted to a point under its own tension.)

For the free open string, our boundary condition should be that momentum cannot 'escape' from the ends. Figuring out the momentum of the string requires an action, which we will introduce below, but for now let's guess that for fixed σ it behaves like a typical momentum and goes like $\dot{X}^\mu$, the field's rate of change along a timelike curve in the string. (We shall see that this guess is correct.) We then expect the total momentum along the string, $P^\mu \sim \int_0^\pi d\sigma \dot{X}^\mu$, to be conserved:

$$0 = \frac{dP^\mu}{d\tau} \sim \int d\sigma \frac{\partial^2 X^\mu}{\partial \tau^2} = \int d\sigma \frac{\partial^2 X^\mu}{\partial \sigma^2} = \left[\frac{\partial X^\mu}{\partial \sigma} \right]_{\sigma=0}^\pi, \tag{7.7}$$

where we used our wave equation (7.1) to change the derivative in the integrand. Thus the *Neumann* boundary condition[5]

$$\frac{\partial X^\mu}{\partial \sigma}\Big|_{\sigma=0,\pi} = 0, \tag{7.8}$$

[5] As we shall see later, a different *Dirichlet* boundary condition is also possible, namely that the ends be fixed in space: $\dot{x}^\mu|_{\sigma=0,\pi} = 0$.

implies momentum conservation for an open string, which in turn implies (as the reader can easily check) that the general solution for an open string is

$$X^\mu = x^\mu + \ell_s^2 p^\mu \tau + i\ell_s \sum_{n\neq 0} \frac{1}{n} \alpha_n^\mu e^{-in\tau} \cos 2n\sigma, \tag{7.9}$$

where we have defined for the open string

$$\ell_s p^\mu \equiv \alpha_0^\mu. \tag{7.10}$$

7.1.2 String actions and symmetries

To get further with the physics of the string—and eventually to quantize—we need to switch to the Lagrangian formulation. To tidy up expressions at this point we define $\sigma^0 \equiv \tau$ and $\sigma^1 \equiv \sigma$ (and let $\alpha, \beta = 0, 1$). Then there are two standard, equivalent formulations of the relativistic string action for the classical equations of motion that we have just been studying. First, there is the *Nambu-Goto action*:

$$S_{NG} = -T \int d\sigma^2 \sqrt{-\det\left(g_{\mu\nu} \frac{\partial X^\mu}{\partial \sigma^\alpha} \frac{\partial X^\nu}{\partial \sigma^\beta}\right)}. \tag{7.11}$$

Here $g_{\mu\nu}$ is the metric of the spacetime in which the string lives, expressed in the chosen coordinates. It is very often taken to be Minkowski, especially in pedagogical treatments. We will make this assumption here too, but for now we will leave it as a general Lorentzian metric, with a $(-1, +1\ldots, +1)$ signature (we shall study the consequences of non-Minkowski metrics in detail in chapters 9 and 10). T is a constant, whose significance will become apparent momentarily. Finally, the determinant is obviously with respect to worldsheet indices α and β.

This action has a nice geometrical interpretation. $g_{\mu\nu}$ of course transforms as a tensor, in particular in a change from spacetime to worldsheet coordinates

$$g_{\mu\nu} \rightarrow g_{\alpha\beta} = g_{\mu\nu} \frac{\partial X^\mu}{\partial \sigma^\alpha} \frac{\partial X^\nu}{\partial \sigma^\beta}. \tag{7.12}$$

Of course $g_{\alpha\beta}$—the 'induced metric'—is restricted to the worldsheet, since its indices are α and β. But it gives a spacetime scalar product and causal structure for vectors tangent to the worldsheet, the spacetime length of curves on the worldsheet, and the invariant spacetime area of infinitesimal regions of the worldsheet. For such a region, with sides dX^μ and dX^ν, the area is

194   OUT OF NOWHERE

given by $dA = \sqrt{-\det g_{\mu\nu}} \cdot dX^\mu dX^\nu$; so the area of an infinitesimal region of the string is

$$dA = d\sigma^2 \sqrt{-\det g_{\alpha\beta}} = d\sigma^2 \sqrt{-\det\left(g_{\mu\nu}\frac{\partial X^\mu}{\partial\sigma^\alpha}\frac{\partial X^\nu}{\partial\sigma^\beta}\right)}. \qquad (7.13)$$

Therefore, minimizing the action (7.11) according to *Hamilton's least action principle* here means extremizing the spacetime area of the string! This action can be compared to the action for a free point particle: the geodesic principle says that the trajectory will extremize the length, or proper time.[6] So clearly the action should simply be the length of the worldline:

$$S \sim \int ds = \int d\lambda \sqrt{-g_{\mu\nu}\frac{\partial X^\mu}{\partial\lambda}\frac{\partial X^\nu}{\partial\lambda}}, \qquad (7.14)$$

for any worldline parameter λ. Thus the free dynamics of one and two dimensional surfaces have mutatis mutandis the same geometric interpretation—which can indeed be generalized to objects of arbitrary dimension, p (known as 'p-branes').

The constant T in (7.11) is interpreted as a fundamental, string tension. To see why, consider a static string fixed between two spatial points in Minkowski spacetime a distance L apart along the x-axis[7], and select worldsheet coordinates in which $\tau = t$ and $\sigma = \pi x/L$: thus $X^\mu(\tau,\sigma) = (t, L\sigma/\pi, 0, 0)$. Then (7.11) becomes

$$S_{NG} = -T\int d\sigma^2 \sqrt{-\det\begin{pmatrix} -1 & 0 \\ 0 & L^2/\pi^2 \end{pmatrix}} = -T\int d\tau \int_0^\pi d\sigma\frac{L}{\pi} = \int d\tau\,(-TL). \qquad (7.15)$$

But of course, since there is no kinetic energy, the action will simply be the time integral of (minus) the potential, $-V$: hence $V = TL$, where T is the tension and L the length. Note that unlike a Hooke's law spring, the tension is a constant, and not proportional to the length: of course it couldn't be because Hooke's law is not relativistic—the frame-independent tension cannot depend on the frame-dependent length! Instead the energy depends directly on the length, and T must be a constant *tension*, independent of how long the string is stretched. It follows that relativistic string points lack physically meaningful synchronic identities in the following sense. Because the tension increases with extension in a non-relativistic, Hooke's law string, the potential energy is not determined by its length, but by how the individual elements are displaced. For instance, such a string has

---

[6] Note that we need to extremize, not necessarily minimize in relativity: straight paths can be locally *longest*! This fact is familiar from the twins paradox.

[7] Instead of the Neumann boundary condition that we considered above, such a string satisfies the Dirichlet conditions that are discussed below.

its lowest potential energy when it is uniformly extended (hence this is the stationary solution). That is, it makes a physical difference if the separation between two points is changed, while keeping the overall string length the same, and so the reidentification of the points under such changes (even if merely counterfactual) is physically meaningful. But since the total length does determine the potential energy of our *relativistic* string, mere changes of the relative positions of its points are unphysical, and they have no identities across changes in separation.

The second, classically equivalent form of the action is the *sigma action* (also often called, rather controversially, the 'Polyakov action'). It lacks the immediate geometric understanding of the Nambu-Goto action, but has a well (or better) defined path integral, and so is needed to define a quantum theory: put another way, the Hamiltonian for the action S_{NG} contains square roots of observables, and so leads to square roots of operators on quantization, which are not well defined. Even classically, it is generally easier to work with the sigma action (we will see the classical equivalence of the actions at the end of this section).

The trick is to introduce a new tensor on the worldsheet: $h_{\alpha\beta}$. $h_{\alpha\beta}$ behaves as a metric on the worldsheet, but it is not (repeat *not*) the induced metric $g_{\alpha\beta}$, the restriction of $g_{\mu\nu}$, that we considered above (7.12). It is an 'auxiliary' metric, used for convenience, but without fundamental significance for the classical string—in the sense that S_{NG} already contains all the true physical degrees of freedom. We will come to understand it better as we unpack the action in this section.

To keep our formulae clean we will now generally avoid explicit appearance of both the h and g metrics, and so write, for instance, $g_{\mu\nu}V^\mu W^\nu = V \cdot W$ or $g_{\mu\nu}V^\mu V^\nu = V^2$: i.e., spacetime dot products are understood as given by the spacetime metric. Then (with $h \equiv \det h_{\alpha\beta}$) the sigma action is introduced as

$$S_\sigma = -\frac{T}{2} \int d^2\sigma \sqrt{-h} h^{\alpha\beta} \frac{\partial X}{\partial \sigma^\alpha} \cdot \frac{\partial X}{\partial \sigma^\beta}. \tag{7.16}$$

'At face value', this action describes a D-dimensional vector field X^μ (whose inner product is given by $g_{\mu\nu}$) living on a 2-dimensional 'spacetime' with metric $h^{\alpha\beta}$. Indeed, as we'll discuss in chapter 10, such an interpretation is sometimes proposed. (As noted, in this chapter 'spacetime' will always denote the D-dimensional Lorentzian 'background' formally appearing in the action; its ontological status is of course a primary question of the following chapters.)

S_σ will be symmetric under any isometries of $g_{\mu\nu}$, so it will be Poincaré invariant with respect to transformations of the X^μ if spacetime is Minkowski. It is also invariant under arbitrary changes of worldsheet coordinates, $(\tau, \sigma) \to (\tau', \sigma')$: such diffeomorphism invariance is to be expected because the action is written generally covariantly, with the metric appearing explicitly. (Of course it is also invariant with respect to diffeomorphisms of the X^μ.) Finally, crucially, the

196   OUT OF NOWHERE

action also satisfies *Weyl invariance*: it is unchanged under multiplication by a *conformal factor,* $h_{\alpha\beta} \rightarrow e^{\phi(\tau,\sigma)} h_{\alpha\beta}$. (Unfortunately, the standard terminology is confusing at this point. As we will explain below, there are also diffeomorphisms whose effect is to rescale the metric, but which should be distinguished from Weyl transformations: however, in either case one speaks of a 'conformal factor'.) All of these symmetries are straightforward to check (e.g., Becker et al. 2006, 30).[8]

Weyl symmetry arises because the auxiliary string metric, $h_{\alpha\beta}$, is only a formal, auxiliary device introduced for computational convenience, not a physical quantity. The only physical significance of $h_{\alpha\beta}$ concerns causal structure on the worldsheet, which is preserved by Weyl transformations: if v is a tangent vector in the worldsheet, then $e^{\phi} h_{\alpha\beta} v^{\alpha} v^{\beta}$ has the same sign as $h_{\alpha\beta} v^{\alpha} v^{\beta}$, so the classification into spacelike, timelike, and lightlike is invariant. But we shall see at the end of this section that this causal structure is fixed by the spacetime metric, and hence is not a separate dynamical freedom. So, *the Weyl transformation does not alter the string in spacetime at all, because it only acts on $h_{\alpha\beta}$.* However, the importance of Weyl symmetry for the physics of strings, classical and quantum, cannot be overstressed, as we shall see repeatedly in our discussion of strings. It is relatively unfamiliar of course, because it is not normally a spacetime symmetry (except for source-free fields)—and neither is it here, holding only for the worldsheet metric, not the spacetime metric.

Worldsheet diffeomorphism and Weyl symmetries can be used to simplify the action considerably: if the Euler characteristic[9] of the worldsheet $\chi = 0$, then one can use the freedom in coordinates and a Weyl transformation to set[10]

$$h_{\alpha\beta} = \eta_{\alpha\beta} = \begin{pmatrix} -1 & 0 \\ 0 & 1 \end{pmatrix}, \tag{7.17}$$

in which case (7.16) becomes

$$S_{\sigma} = \frac{T}{2} \int d^2\sigma(\dot{X}^2 - X'^2). \tag{7.18}$$

---

[8] While we have motivated the Nambu-Goto and sigma actions via the simple wave equation to which they lead, it is important to realize that they are almost unique given the symmetries listed: see Polchinski (1998, 15f), who notes "simplicity of the action is not the right criterion in quantum field theory. Symmetry not simplicity is the key idea ..."

[9] Recall that the Euler characteristic is a topological invariant, which for polyhedra is given by $\chi \equiv$ Faces - Edges + Vertices. (For metric spaces the Euler number is given by the integral of the scalar curvature.) The cylinder, torus, Möbius loop, and Klein bottle are all examples of surfaces with $\chi = 0$, while the double torus has $\chi = -2$.

[10] The quick argument: in two dimensions, the metric, being symmetric, has three independent parameters, so can be chosen uniquely at any point, given the two independent infinitesimal diffeomorphisms plus Weyl factor at each point. Of course, the quick argument breaks down once there is a topological obstruction preventing this fact about each point from being applied to all points at once: it will, however, hold for some finite region about any point.

A STRING THEORY PRIMER   197

This choice of $h_{\alpha\beta}$ amounts to *picking a gauge*: choosing a particular form for some invariant object, in this case by a choice of (τ, σ) coordinates and a conformal factor. So the symmetries no longer remain explicitly in the expression of the theory, although of course they remain for physical quantities.

We can quickly confirm that the momentum is as we guessed in (7.7). According to Lagrangian mechanics $P^\mu = \delta\mathcal{L}/\delta\dot{X}_\mu$, which in $h_{\alpha\beta} = \eta_{\alpha\beta}$ coordinates yields

$$P^\mu = \frac{T}{2}\frac{\delta\left(\dot{X}^2 - X'^2\right)}{\delta\dot{X}_\mu} = T\dot{X}^\mu \qquad (7.19)$$

using (7.18)—our earlier guess is vindicated, and the Neumann condition (7.8) does entail momentum conservation. We can now also write down the Hamiltonian for the string in our chosen coordinates:

$$H = \int d\sigma(\dot{X}\cdot P - \mathcal{L}) = \int d\sigma\left(\dot{X}\cdot T\dot{X} - \frac{T}{2}(\dot{X}^2 - X'^2)\right) = \frac{T}{2}\int (\dot{X}^2 + X'^2)\, d\sigma. \quad (7.20)$$

It's also straightforward to confirm that in this gauge, minimizing the action with respect to X^μ yields our wave equation (7.1) and the Neumann boundary condition (7.8). Quickly,

$$\delta S_\sigma = T\int d\sigma^2\ \dot{X}\delta\dot{X} - X'\delta X'$$

$$= T\int d\sigma^2\ \partial_\tau(\dot{X}\delta X) - \partial_\sigma(X'\delta X) - (\ddot{X} - X'')\delta X$$

$$= T\int d\sigma[\dot{X}\delta X]_\tau - T\int d\tau[X'\delta X]_\sigma - T\int d\sigma^2(\ddot{X} - X'')\delta X. \qquad (7.21)$$

In the first step we simply used the product rule to rewrite the action; in the second step we integrated out total derivatives with respect to τ or σ in the first two terms. In applying Hamilton's principle, the variation vanishes at the start and end of the path, so $\delta X = 0$ at the limits of the τ integration, and the first term is zero. To minimize with respect to arbitrary variations δX, the third term must also vanish, giving the wave equation (7.1). (As we shall discuss later, it does not follow that *every* solution to the wave equation is physical: the gauge choice gives rise to constraints.)

That leaves the second term, which must also vanish to minimize the action. There are three ways that this can happen: (*i*), the string may be closed, in which case the end points 'coincide' and $[X'\delta X]_\sigma = 0$. If the string is open, however, we can have either (*ii*) that $X' = 0$ at the end points—i.e., the Neumann condition introduced above—or (*iii*) that $\delta X = 0$ at the end points. But how can this final condition be realized for arbitrary variations? Only by literally fixing the ends of

198 OUT OF NOWHERE

the string to something. In this case we must demand $\dot{X} = 0$ at the end points: this is a distinct, second boundary condition, known as the *Dirichlet* condition. Of course, if the ends are so attached, first we have a fixed sink/source of momentum, and so no longer expect momentum conservation in the string; second, we effectively have additional spacetime background structure, and so may lose Poincaré symmetry.

Finally to the equivalence of Nabu-Goto and sigma actions.[11] One proceeds by minimizing the *gauge-invariant* sigma action (7.16) with respect to $h^{\alpha\beta}$

$$0 = -\frac{2}{T}\frac{1}{\sqrt{-h}}\frac{\delta S_\sigma}{\delta h^{\alpha\beta}} = \partial_\alpha X \cdot \partial_\beta X - \frac{1}{2}h_{\alpha\beta}h^{\gamma\delta}\partial_\gamma X \cdot \partial_\delta X, \qquad (7.22)$$

whose solution is

$$h_{\alpha\beta} = e^{\phi(\tau,\sigma)}\partial_\alpha X \cdot \partial_\beta X = e^{\phi(\tau,\sigma)} \cdot g_{\mu\nu}\frac{\partial X^\mu}{\partial\sigma^\alpha}\frac{\partial X^\nu}{\partial\sigma^\beta}, \qquad (7.23)$$

where $\phi(\tau, \sigma)$ is an arbitrary scalar function arising from Weyl symmetry. If this expression is substituted into (7.16) with $\phi = 0$ one obtains the Nabu-Goto action (7.11): classically, S_{NG} and S_σ are equivalent, and describe the physics of the string that we intuitively sought.

We have written out the RHS of (7.23) explicitly to make clear that in classical solutions the auxiliary metric is, up to a conformal factor, nothing but the restriction of the spacetime metric to the string worldsheet (7.12). In the first place, the identification shows that at most the conformal factor could be an additional degree of freedom, since $h_{\alpha\beta}$ is otherwise determined by $g_{\mu\nu}$; Weyl symmetry means that $\phi(\tau, \sigma)$ is not a degree of freedom either. In the second place, because a conformal factor preserves causal relations, we see that the causal structure of the auxiliary metric must agree with that of spacetime in any classical model of the sigma action.

7.2 Physics in the string

So far we have often appealed to a conception of a string as a 1-dimensional object moving in D-dimensional spacetime, in order to use our elementary knowledge of such a system to give an intuitive development of the basic equations. However, as we noted at the start and when we introduced the sigma action (7.16), formally we are describing the physics of a D-component vector field, $X^\mu(\tau, \sigma)$, on a 2-dimensional 'spacetime', namely the string worldsheet. In this conception, the spacetime metric, $g_{\mu\nu}$ is an inner product for an *internal* vector space, acting on

---

[11] See Becker et al. (2006, exercise 2.6).

the vector field at each point of the worldsheet (see footnote 4). To develop string theory further, more of the tools of field theory need to be brought to bear, and so we now switch to this alternative point of view. In a later chapter (§10.3.1) we will discuss whether the 2-dimensional field description points to the correct interpretation of string theory: that the string worldsheet is fundamental—the 'container' of the X^μ field—and spacetime as we know it is merely a field on the string. For now we only adopt the worldsheet interpretation formally: we now continue developing string theory as a particular kind of field, to which standard field theory applies.

7.2.1 Conformal symmetry

A highly significant feature of this field theory is that even though we gauge fixed Weyl and diffeomorphism symmetries by our choice (7.17), symmetry remains. As this symmetry is at the root of many of the most important physical and mathematical features of string theory, we will unpack it carefully.

The simplest way to see the symmetry is to 'rotate' the worldsheet coordinates into the complex plane, by taking time to be imaginary: $\tau \to -i\tau$. In such complex coordinates our wave equation (7.1)—which comes from the *gauge fixed* action (7.18)—becomes the Laplace equation:

$$\ddot{X}^\mu + X''^\mu = 0. \tag{7.24}$$

This equation has 'conformal symmetry': any solution will again be a solution under any diffeomorphism whose only effect is to change the metric by a conformal factor. A simple example is a uniform expansion of the complex plane.[12] Before we move on, it is worth clarifying this significant fact.

---

[12] To see the conformal symmetry of the Laplace equation, recall some basic facts about complex analysis, which are important in string theory. Consider a complex valued function $f(z) = f(x, y) = u(x, y) + iv(x, y)$, where $z = x + iy$. If $\frac{df(z)}{dz}$ is to be well defined—in which case f is *holomorphic*—then it must be the case that the direction in $\mathbb{C}$ in which the derivative is evaluated is irrelevant. Equivalently, the Cauchy-Riemann equations hold:

$$\frac{\partial u}{\partial x} = \frac{\partial v}{\partial y} \quad \text{and} \quad \frac{\partial v}{\partial x} = -\frac{\partial u}{\partial y}. \tag{7.25}$$

Differentiating the first equation with respect to x, the second with respect to y, and adding yields

$$\frac{\partial^2 u}{\partial x^2} + \frac{\partial^2 u}{\partial y^2} = 0, \tag{7.26}$$

so that the real part of any holomorphic function satisfies the Laplace equation (and similarly for the imaginary part). The Cauchy-Riemann equations relate the real and imaginary parts, so it's clear that they are not automatically satisfied by a pair of functions satisfying the Laplace equation, so the converse does not hold. However, since X^μ satisfies the Laplace equation, it is possible to find a

200 OUT OF NOWHERE

First, it may not be clear why the system under discussion manifests conformal symmetry. Earlier we said that the string has diffeomorphism and Weyl symmetry with respect to worldsheet coordinates and metric, but we also gauge fixed these symmetries to set $h_{\alpha\beta} = \eta_{\alpha\beta}$ in (7.18): so how does any symmetry remain in the expressions derived from it? Consider what happens when we use worldsheet diffeomorphism and Weyl symmetries to choose coordinates and a conformal factor to gauge fix—to set an arbitrary $h_{\alpha\beta}$ to a flat metric, $\eta_{\alpha\beta}$. There are two possible cases: either $h_{\alpha\beta}$ is conformally flat ($h_{\alpha\beta} = e^{\psi(\tau,\sigma)}\eta_{\alpha\beta}$) or it is not. If it is conformally flat, then *no* diffeomorphism is needed, as a Weyl transformation will suffice. If it is not, then to set the metric flat we first apply a passive diffeomorphism d (a change of coordinates) such that $d^* h_{\alpha\beta} = e^{\phi(\tau,\sigma)}\eta_{\alpha\beta}$, and then apply a Weyl transformation. By supposition, for all $\psi(\alpha,\beta)$, $h_{\alpha\beta} \neq e^{\psi(\tau,\sigma)}\eta_{\alpha\beta}$: so $d^* h_{\alpha\beta} \neq e^{\psi(\tau,\sigma)}d^*\eta_{\alpha\beta}$. Hence, for all $\psi(\alpha,\beta)$, $e^{\phi(\tau,\sigma)}\eta_{\alpha\beta} \neq e^{\psi(\tau,\sigma)}d^*\eta_{\alpha\beta}$, which is to say that d is not a diffeomorphism whose only action on $\eta_{\alpha\beta}$ is to introduce a conformal factor. Therefore, whether or not $h_{\alpha\beta}$ is conformally flat, gauge fixing did not involve a diffeomorphism whose only effect on $\eta_{\alpha\beta}$ is to introduce a conformal factor; and so symmetry with respect to the subgroup of conformal transformations remains. (This subgroup has measure zero in the group of diffeomorphisms: see Polchinski 1998, 85f on which our discussion draws.) Bear in mind that Weyl transformations of the worldsheet are with respect to auxiliary metric $h_{\alpha\beta}$, not with respect to spacetime—they do not change the size or embedding of the string worldsheet in spacetime.

Second, this argument helps distinguish Weyl and conformal symmetries. *Neither acts on spacetime*, but the former is a rescaling freedom of the worldsheet auxiliary metric $h_{\alpha\beta}$ in the sigma action (7.16), arising because it is not a true degree of freedom. The latter is a residue of diffeomorphism freedom, left after gauge fixing (7.17), and so can be understood as a freedom in the choice of worldsheet coordinates. Both symmetries are important in string theory. Weyl symmetry leads to 26 dimensions (as we shall discuss below), and to gravity (as we shall discuss in chapters 9 and 10). Conformal symmetry has very important consequences for the mathematical treatment of strings, as we shall soon see; in

holomorphic function for which it is the real (or imaginary) part: $\Psi(z) = X(z) + i\Phi(z)$ (hiding the μ index).

Now, the important property of holomorphic functions $f(z)$ is that they realize conformal transformations in the complex plane according to

$$\Psi(z) \rightarrow \Psi \circ f(z) = X \circ f(z) + i\Phi \circ f(z) \tag{7.27}$$

(more specifically, $f(x)$ must be *biholomorphic*, with an inverse that is also holomorphic). But, as is easily checked using the Cauchy-Riemann equations, the holomorphic property is preserved under composition. Putting everything together, if $X(z)$ satisfies the Laplace equation, then it is the real part of holomorphic $\Psi(x)$; and so $X \circ f(z)$ is the real part of holomorphic $\Psi \circ f(z)$, and so also satisfies the Laplace equation—which is to say that the Laplace equation is conformally symmetric.

chapter 9 we shall also see that conformal symmetry plays an important role in the path integral formulation of quantum string theory.

Finally, conformal symmetry indicates that there is no physically meaningful diachronic identity for points of the string (we already saw an absence of synchronic identity in §7.1.2). Any timelike direction tangent to the worldsheet through a point can be transformed into any other by a conformal transformation, without changing the gauge-fixed equations of motion (7.21), and so none is a preferred direction in which the point can be said to move. Put another way, lines of constant σ can be invariantly transformed into one another, and so do not represent the worldlines of persisting points of the string. It is in that sense a truly 2-dimensional object, not a 1-dimensional one propagating through time—its only points are spacetime ones, not spatial points at different times.

In this regard conformal symmetry can be compared with more familiar spacetime symmetries. Lorentz invariance in Minkowski spacetime—and Galilean invariance in Galilean spacetime—is similar in the sense that it implies that no inertial trajectory represents the 'true' trajectory of a point of space, so that we have spacetime rather than space-at-different-times. However, in the natural view of string theory, it is now the points of the string, not of spacetime—of the 'contained' not the 'container'—that have no diachronic identity.

7.2.2 String mass

The key idea of string theory is that what we take to be different quanta of quantum fields (including gravitons) are in fact quantized strings in different states of internal vibration. But even before quantization we can start to see this account emerge, because the mass of the classical string also depends on its mode of vibration. Seeing how will also develop some of the formal concepts we will need.

Taking the point of view in which $X^\mu(\tau, \sigma)$ is a (classical) worldsheet field, the quantity in (7.22) is the standard expression for the Hilbert stress-energy tensor, $T_{\alpha\beta}$ of the string, due to its internal metric. Of course, since the metric is purely formal, no physical quantity can depend on it. $T_{\alpha\beta}$ does not describe quantities like the tension, energy, or momentum of the string as a system in spacetime; rather, at least for our purposes, 'stress-energy tensor' is just a conventional name for the quantity in (7.22).

Using $h_{\alpha\beta} = \eta_{\alpha\beta}$—our gauge choice (7.17)—we can write (7.22) as

$$T_{\alpha\beta} = \partial_\alpha X \cdot \partial_\beta X + \frac{1}{2} \begin{pmatrix} -1 & 0 \\ 0 & 1 \end{pmatrix}_{\alpha\beta} (\dot{X}^2 - X'^2). \qquad (7.28)$$

202 OUT OF NOWHERE

In which case we simply find that

$$T_{\tau\sigma} = T_{\sigma\tau} = \dot{X} \cdot X'$$

$$T_{\tau\tau} = T_{\sigma\sigma} = \frac{1}{2}(\dot{X}^2 + X'^2). \tag{7.29}$$

Things simplify further if we change coordinates to $\sigma^\pm = \tau \pm \sigma$, in which case $\partial_\pm = \frac{1}{2}(\partial_\tau \pm \partial_\sigma)$.[13] Because lightlike trajectories in the worldsheet have constant values of σ^+ or σ^- (depending on whether they are left or right moving), these are known as worldsheet 'lightcone' coordinates (later we will also use so-called spacetime 'lightcone' coordinates). They are useful for exploring conformal symmetry because they correspond to the complex coordinates obtained by making time imaginary: $\sigma_z = \sigma - i\tau$. As we indicated earlier (footnote 12), conformal symmetry has special connections to complex analysis, and these will also be seen in worldsheet lightcone coordinates.

We start by transforming the worldsheet stress-energy tensor into the new coordinates, using the standard formula for transforming tensors. For instance,

$$T_{++} = \frac{\partial\sigma^\alpha}{\partial\sigma^+}\frac{\partial\sigma^\beta}{\partial\sigma^+}T_{\alpha\beta} = \partial_+X \cdot \partial_+X \tag{7.30}$$

$$T_{--} = \frac{\partial\sigma^\alpha}{\partial\sigma^-}\frac{\partial\sigma^\beta}{\partial\sigma^-}T_{\alpha\beta} = \partial_-X \cdot \partial_-X.$$

Similarly, $T_{+-} = T_{-+} = 0$. Now we substitute the mode expansion (7.6) of the closed string, to obtain:

$$T_{++} = 2\ell_s^2 \sum_{m=-\infty}^{\infty}\left(\frac{1}{2}\sum_{n=-\infty}^{\infty}\tilde{\alpha}_{m-n}\cdot\tilde{\alpha}_n\right)e^{-2im(\tau+\sigma)} \equiv 2\ell_s^2\sum_{m=-\infty}^{\infty}\tilde{L}_m e^{-2im(\tau+\sigma)} \tag{7.31}$$

$$T_{--} = 2\ell_s^2\sum_{m=-\infty}^{\infty}\left(\frac{1}{2}\sum_{n=-\infty}^{\infty}\alpha_{m-n}\cdot\alpha_n\right)e^{-2im(\tau-\sigma)} \equiv 2\ell_s^2\sum_{m=-\infty}^{\infty}L_m e^{-2im(\tau-\sigma)}$$

so that L_m and $\tilde{L}_m$ denote the left- and right-moving modes of the stress-energy tensor in lightcone coordinates.

Using the canonical Poisson brackets—for instance, $\{P^\mu(\tau,\sigma),X^\nu(\tau,\sigma')\} = \eta^{\mu\nu}\delta(\sigma - \sigma')$ (remembering that $P^\mu = T\dot{X}^\mu$, from (7.19))—one can first find the Poisson brackets for the αs, and hence these modes[14]:

$$\{L_m, L_n\} = i(m - n)L_{m+n}. \tag{7.32}$$

Although a full discussion is outside the scope of this introduction, this is the algebra of conformal transformations (biholomorphic functions) in the complex

---

[13] $\tau = \frac{1}{2}(\sigma^+ + \sigma^-), \sigma = \frac{1}{2}(\sigma^+ - \sigma^-)$ and, for example, $\partial_+ = \frac{\partial\tau}{\partial\sigma^+}\frac{\partial}{\partial\tau} + \frac{\partial\sigma}{\partial\sigma^+}\frac{\partial}{\partial\sigma}$.
[14] For instance, Becker et al. (2006, 34f).

plane: the algebra of *Virasoro generators*.[15] This agreement is no coincidence, but represents the fact that the modes generate worldsheet conformal transformations. It follows, from the usual connections between symmetries and conserved quantities that they are the conserved charges arising from conformal symmetry.[16] Formally, since they are biholomorphic functions, the powerful constraints of complex analysis apply: the residue theorem holds and integrals involving them or the stress-energy tensor depend solely on their poles, and functions of them have unique analytic continuations. In fact, *a great deal of the mathematical power of string theory is that it captures the physics of a D-dimensional world in terms of a 2-dimensional conformal theory.*

Moving on, the Virasoro generators came from the expression for the stress-energy tensor (7.22), but this describes the variation of the sigma action with respect to the metric $h^{\alpha\beta}$, so the equation of motion for the metric requires that it vanish in physical solutions. We have chosen $h^{\alpha\beta} = \eta^{\alpha\beta}$: therefore we need to impose $T_{\alpha\beta} = 0$ as a gauge constraint on physical solutions, in addition to the wave equation (7.1). Since the modes in (7.31) are linearly independent, we therefore have the 'Virasoro constraints':

$$\forall m \in \mathbb{Z} \quad L_m = \frac{1}{2} \sum_{n=-\infty}^{\infty} \alpha_{m-n} \cdot \alpha_n = 0, \qquad (7.33)$$

$$\forall m \in \mathbb{Z} \quad \tilde{L}_m = \frac{1}{2} \sum_{n=-\infty}^{\infty} \tilde{\alpha}_{m-n} \cdot \tilde{\alpha}_n = 0.$$

Quantum versions of these constraints will have to hold.

The Virasoso constraints play an important role in understanding how 'elementary' particles appear as excitations of strings. Observed at scales well above their characteristic length, intuitively strings will appear as (spatially) point-like objects—particles—since their extension 'can't be seen'. Viewed as a particle in this way, a string's linear four-momentum must satisfy the usual relation: $p^2 = -M^2$. But recall (from (7.5) and (7.10)) that the linear momentum of the string is just the zeroth mode, α_0, which (7.33) relates to the vibrational modes, as follows. First, the $m = 0$ Virasoro constraint (7.33) can be rewritten

$$L_0 = \frac{1}{2} \sum_{n=-\infty}^{\infty} \alpha_{-n} \cdot \alpha_n = \frac{1}{2} \alpha_0^2 + \sum_{n=1}^{\infty} \alpha_{-n} \cdot \alpha_n = 0 \qquad (7.34)$$

and similarly for $\tilde{L}_0$. Then, for the open string, substitute $\alpha_0^2 = \ell_s^2 p^2$ from (7.10); for the closed string use $L_0 + \tilde{L}_0 = 0$ and substitute $\alpha_0^2 = \tilde{\alpha}_0^2 = \ell_s^2 p^2/4$ from (7.5) to obtain

---

[15] Briefly, a typical infinitesimal conformal transformation has the form $z \to z' = z - \epsilon_n z^{n+1}$ (with ϵ_n infinitesimal). The generator l_n of the transformation is given by $\delta z = \epsilon_n[l_n, z] = -\epsilon_n z^{n+1}$: hence $l_n = -z^{n+1} \partial_z$. The l_n can readily be seen to satisfy $[l_m, l_n] = i(m - n)l_{m+n}$.

[16] See Green et al. (1987, §2.1.3) for further discussion, especially how conformal symmetry leads to the vanishing of the trace of the stress-energy tensor, $\mathrm{Tr}(T_{\alpha\beta}) = 0$.

204   OUT OF NOWHERE

$$\text{Open string} \qquad M^2 = \frac{2}{\ell_s^2} \sum_{n=1}^{\infty} \alpha_{-n} \cdot \alpha_n.$$

$$\text{Closed string} \qquad M^2 = \frac{4}{\ell_s^2} \sum_{n=1}^{\infty} (\alpha_{-n} \cdot \alpha_n + \tilde{\alpha}_{-n} \cdot \tilde{\alpha}_n) \qquad (7.35)$$

since $-p^2 = M^2$. In other words, it follows from the constraint that the 'particle-mass' of a string depends on its vibrational modes—different vibrations give 'particles' of different masses. Not surprisingly, as we shall see in the next section, when the string is quantized, the possible modes and hence masses become discrete, so that we can properly talk of the different modes as distinct particle (or quanta) species.

In the 1960s it was found that hadrons in the 'particle zoo' then being discovered in high energy physics could be grouped in families, whose squared masses were proportional to their spins, J: $M^2 \propto J/\alpha'$, with α' the *Regge slope* of the family. Looking at (7.35) it is natural to identify $\alpha' = \ell_s^2/2$: indeed, as we shall see, on quantization masses indeed increase in jumps proportional to $\ell_s^2/2$. This fact explains early interest in string theory—as a potential theory of the strong interaction (Rickles 2014, part 1). Moreover, a concrete calculation of the relation between spin and energy (which we will not perform—see Zwiebach 2004, §8.6) shows that $\alpha' = 1/(2\pi T)$, thereby relating all three of the parameters we have used. (It is standard in the physics literature to use the slope instead of the equivalent length, but we will stick with ℓ_s in this chapter, to maintain contact with our intuitive picture of the string; in chapter 9 we will switch, because our picture will also have shifted.)

Note, however, that from (7.35) the energy gaps between the particles depends on $1/\ell_s$. This quantity is typically taken to be a couple of orders greater than the Planck length, say $\ell_s \sim 10^{-33}$m, in which case the mass gap is of the order 10^{-7}g, or 17 orders of magnitude greater than the proton mass—far too wide a gap to accommodate observed particles. Contemporary string theory obtains the mass spectrum of the standard model by other means: typically involving 'Dp-branes' (see chapter 8).

Finally, the zeroth Virasoro generator is related to the Hamiltonian of the string. To see this, we expand the Hamiltonian (7.20) in terms of the modes of X^μ. For the open string (7.9) a simple calculation shows that in fact,

$$H = \frac{1}{2} \sum_{n=-\infty}^{+\infty} \alpha_{-n} \cdot \alpha_n$$

$$= \frac{1}{2} (\alpha_0^2 + 2 \cdot \sum_{n=1}^{+\infty} \alpha_{-n} \cdot \alpha_n)$$

$$= \frac{\ell_s^2}{2} (p^2 - M^2). \qquad (7.36)$$

A STRING THEORY PRIMER 205

(7.36) may be a surprise, if one expected the relativistic energy-momentum relation to hold: $E^2 = H^2 = p^2 + m^2$. The point is that the string Hamiltonian is *not* the relativistic energy of the string *viewed as a particle*, whose mass comes from vibrations of the string. Instead, as (7.34) and (7.36) show, the Hamiltonian is the zeroth Virasoro generator: $H = L_0$.

7.3 Quantization

We have developed enough of classical string theory now to turn to quantization, and ultimately to the string particle spectrum, the heart of the theory and the route to spacetime emergence. Before we do that we need to formally develop the theory of the quantized string, and in particular explore the crucial issue of gauge symmetry in quantum string theory, which ultimately (notoriously) requires 'extra dimensions' to spacetime.

7.3.1 Canonical quantization

String theory is, as we have emphasized, a field theory, and so the standard techniques of quantum field theory (QFT) can be applied when we quantize the string.[17] That is, we start in the usual way by treating the X^μ 'field' as now operator-valued, and imposing equal-'time' commutation relations:

$$[X^\mu(\tau, \sigma), P^\nu(\tau, \sigma')] = i\eta^{\mu\nu}\delta(\sigma - \sigma'). \tag{7.37}$$

Here of course we apply *bosonic* commutation relations because our classical string was described by real valued, commuting variables. (Note that one speaks even of the classical string physics that we have discussed as 'bosonic'.) With knowledge of QFT, one will wonder about the possibility of *fermionic* anticommutation relations for the string; these will be discussed in the final section.

It follows from the explicit form of the field, (7.6) and (7.9), and of the canonical momentum (7.19) that the modes satisfy the relations

$$[\alpha_m^\mu, \alpha_{-n}^\nu] = m\eta^{\mu\nu}\delta_{m,n}. \tag{7.38}$$

Because X^μ is Hermitian,

$$\alpha_{-n}^\mu = \alpha_n^{\mu\dagger}. \tag{7.39}$$

---

[17] We will continue merely to sketch the development, following the more detailed treatments in the textbooks cited at the start of the chapter: especially the formal treatments of Green et al. (1987, §2.2–2.3) and Becker et al. (2006, §2.4–2.5). Zwiebach (2004, chs. 12–13) is also useful. Interpretational comments draw on Polchinski (1998, §1.3–1.4).

206 OUT OF NOWHERE

Therefore, for $m > 0$, $\frac{1}{\sqrt{m}}\alpha_m^{\mu\dagger}$ and $\frac{1}{\sqrt{m}}\alpha_m^{\mu}$ have the algebra of raising and lowering operators:

$$\left[\frac{1}{\sqrt{m}}\alpha_m^{\mu}, \frac{1}{\sqrt{n}}\alpha_n^{\nu\dagger}\right] = \eta^{\mu\nu}\delta_{m,n} \qquad m > 0. \tag{7.40}$$

This creation/annihilation algebra has the familiar Fock representation for a bosonic field, with basis states labeled by mode occupation numbers, *except for one important difference*:

$$\left[\frac{1}{\sqrt{m}}\alpha_m^0, \frac{1}{\sqrt{m}}\alpha_m^{0\dagger}\right] = -1, \tag{7.41}$$

not +1, because the Lorentz-signature metric (with $\eta^{00} = -1$) determines the commutator. It follows that there are negative norm states—*ghosts*. For instance, letting Ω be the (normalized) vacuum state, annihilated by all the α_m^{μ}:

$$|\alpha_1^{0\dagger}\Omega|^2 = \langle\alpha_1^{0\dagger}\Omega|\alpha_1^{0\dagger}\Omega\rangle = \langle\Omega|\alpha_1^0\alpha_1^{0\dagger}|\Omega\rangle = \langle\Omega|(\alpha_1^{0\dagger}\alpha_1^0) - 1|\Omega\rangle = -1, \tag{7.42}$$

where we have used the commutator. Since this is the norm squared of a state, this result is not compatible with the usual probabilistic interpretation of QM, so what has gone wrong?[18]

The answer is that the Virasoro constraints (7.33) have not yet been applied; so on the one hand, the algebra assumes a 'gauge' in which the string metric $h_{\mu\nu} = \eta_{\mu\nu}$ (since the mode expansion follows from a wave equation that only follows from making that choice in (7.18)), but we have not actually applied the constraints (7.33) which follow from that choice. Without the constraints the system has unwanted, unphysical gauge degrees of freedom in it; specifically, the conserved charges L_m of the conformal symmetry can take on unphysical values. The cure should be to impose a quantum version of $L_m = 0$—i.e., only quantum states for which the constraints are satisfied are physical, and the rest (which we hope includes those of negative norm) should be discarded as an artifact of the quantization procedure.

This procedure—known as 'string theory covariant quantization'—can be pursued, but for various reasons, physicists prefer to develop the theory in a different way: by applying enough constraints to prevent negative norm states from appearing in the first place. So doing involves *lightcone quantization*.

---

[18] What if we chose $\eta^{00} = +1$ instead? Then $\eta^{ii} = -1$, so now the spatial excitations have negative norm. Whatever we choose, not all states have norms of the same sign, frustrating a probability interpretation.

The first step is to change to so-called *lightcone coordinates*, this time for spacetime rather than the worldsheet:

$$x^\pm = \frac{1}{\sqrt{2}}(x^0 \pm x^1), \qquad x^i \text{ unchanged for } i = 2, 3, \ldots, D-1. \qquad (7.43)$$

Again, the reason for the name is clear: a particle moving at the speed of light along the x^1-direction has a constant coordinate along its worldline. But the name is also a little misleading, for the coordinates only live on the lightcone in the $x^1 - x^0$-plane, not the full $(D-1)$-dimensional lightcone: 'lightlike' coordinates might have been a better name, but that battle is long lost. Since such a change of coordinates amounts to boosting an inertial frame to the speed of light, it is not implemented by a Lorentz transformation: the effect of the change has to be determined by a general coordinate transformation. Thus, while they are not accelerating, lightcone coordinates are not inertial in the strict sense, and standard invariant relations will fail to hold.

7.3.2 Lightcone coordinates

Before we proceed with deriving some of the consequences of this coordinate change, we will pause for a brief heuristic overview of the physics of lightcone coordinates (based on Susskind's 2011 wonderful Stanford lectures), which will help develop some intuitions about this possibly unfamiliar approach. Consider a composite relativistic particle (in D dimensions) whose momentum in the x_1 direction $p_1 \approx 0$: as a particle it satisfies $E^2 = p^2 + m^2$, but its mass comes from the potential energy, V, of its constituents (ignoring their masses), so we have $E^2 = p^2 + V^2$. Now suppose that we approach lightcone coordinates by boosting the frame in the x_1 direction such that $p_1 \gg p_i$ for $i = 2, 3, \ldots D-1$. Then the energy in the boosted frame is given by

$$E = p_1 \cdot \sqrt{1 + \frac{\sum_i p_i^2 + V^2}{p_1^2}} = p_1 \cdot \left(1 + \frac{1}{2}\frac{\sum_i p_i^2 + V^2}{p_1^2} + \ldots\right) \approx p_1 + \bar{p}^2/2p_1 + V^2/2p_1,$$

$$(7.44)$$

where $\bar{p}$ is the momentum in the directions transverse to x_1. Consider the second two terms on the RHS of (7.44). Naively, up to a constant factor, $\bar{p}^2/2p_1$ is a *non-relativistic* kinetic energy coming from motion only in the directions transverse to the direction of the boost, x_1. Similarly for $V^2/2p_1$: it (not something proportional to V) is a *non-relativistic* potential energy. And because the boost can be chosen to contract distances Δx_1 to arbitrarily small values, the potential can typically be made to again depend only on displacements transverse to x_1. Because the energy in these two terms comes from positions and motions transverse to the boost that

208 OUT OF NOWHERE

relates the frames, it is not changed by the transformation. The upshot is that all the interesting relativistic physics can be found in a non-relativistic theory in a space of one fewer spatial dimensions: results in this simpler context will also apply to the original relativistic theory.

Applying this correspondence to the string, we can immediately deduce an important consequence. In the case of a classical string, the non-relativistic potential energy is given by integrating Hooke's law along its length, $\frac{1}{2}kL^2$ for uniform linear extension. Identifying this quantity with the final, potential term in (7.44) means that $V^2 \propto L^2$: hence the relativistic potential energy of the string is given by $V = TL$, where T is a constant tension (violating Hooke's law). Of course, in (7.15) we saw that T appears correctly in the action to produce just this potential energy. That is, the correspondence between relativistic physics in an inertial frame, and non-relativistic physics in the transverse dimensions of the lightcone frame holds for the string. This correspondence is exploited by quantizing in lightcone coordinates, as we shall now see.

7.3.3 Lightcone gauge

Now that we have a better understanding of lightcone coordinates, we can return to the problem of ghosts in quantum string theory. As a first step, we need to recast some of the important features of the *classical* string in the new lightcone framework. First, the Minkowski spacetime metric in lightcone coordinates is

$$\eta'_{\mu\nu} = \begin{pmatrix} 0 & -1 & 0 \\ -1 & 0 & 0 \\ 0 & 0 & I \end{pmatrix} \tag{7.45}$$

where I is a $(D-2)$-dimensional identity matrix.[19] The fact that $\eta' \neq \eta$ makes clear that the lightcone frame is not a standard inertial frame, that it is not obtained by a Lorentz transformation, which by definition preserves the metric. Thus the string is no longer *manifestly* Lorentz invariant, although a choice of coordinates alone cannot—in classical physics—break the symmetry: one just has to take care in making Lorentz transformations to see it. However, symmetries are more delicate in quantum mechanics and, as we shall see, the Lorentz symmetry now may be broken unless we are careful.

Since (as we discussed above) worldsheet conformal symmetry remained after we gauge fixed (7.17), an additional gauge choice can be made to further simplify matters: the 'lightcone gauge', in which one sets

---

[19] A vector with inertial frame components v^μ has lightcone components $v^\pm = \frac{1}{\sqrt{2}}(v^0 \pm v^1)$, so it is easy to check that $v \cdot w$ is invariant under the change to lightcone coordinates.

$$X^+(\tau, \sigma) = x^+ + \ell_s^2 p^+ \tau. \tag{7.46}$$

According to this choice, the string has no vibrations along the x^+-axis, only linear 'momentum' $p^+ = \alpha_0^+/\ell_s$ for the open string (or with a factor of 2 for the closed string—see (7.5) and (7.10)). Of course that is because the gauge fixes τ and σ that way: for instance, so that worldsheet curves of constant τ have constant X^+. (It takes a little work to show that conformal symmetry makes this choice possible.)

The virtue of this gauge choice is that it leaves *all* the dynamics in the transverse direction, as anticipated by our discussion of lightcone coordinates. To see this explicitly, first from (7.29) we have that $(\dot{X} \pm X')^2 = 2(T_{\tau\sigma} \pm T_{\tau\tau})$. But again, from (7.22) $T_{\alpha\beta}$ is the variation of the action with respect to $h_{\alpha\beta}$, and so vanishes for physical solutions (imposed as a gauge constraint). Hence, using the lightcone metric (7.45)

$$0 = (\dot{X} \pm X')^2$$

$$= -(\dot{X}^+ \pm X'^+)(\dot{X}^- \pm X'^-) - (\dot{X}^- \pm X'^-)(\dot{X}^+ \pm X'^+) + \sum_{i=2}^{D-1}(\dot{X}^i \pm X'^i)^2, \tag{7.47}$$

so by (7.46)

$$\dot{X}^- \pm X'^- = \frac{1}{2p^+\ell_s^2} \sum_{i=2}^{D-1}(\dot{X}^i \pm X'^i)^2. \tag{7.48}$$

If X^- is expanded in modes in the usual way, then this equation implies that

$$\alpha_n^- = \frac{1}{2p^+\ell_s} \sum_{i=2}^{D-1} \sum_{m=-\infty}^{\infty} \alpha_{n-m}^i \alpha_m^i. \tag{7.49}$$

Thus we see explicitly that solving for the X^i completely fixes the vibrational motion of the string: there is none along the x^+-axis, and that along the x^--axis is fixed by the transverse modes. Choosing lightcone coordinates and gauge gives an effective system in one fewer spacetime dimension.

Using lightcone coordinates and choosing the lightcone gauge solves the problem of negative norm states for the quantum string. Such ghosts arose because, on quantization, the α^0 modes satisfy the wrong commutation relations (7.41); these (and the α^1 modes) have been replaced by $\alpha^\pm$, which in turn have been given in terms of the α^is. But these latter—now the only modes in the theory—satisfy the correct commutation relations, and so act in a standard Fock space with no negative norm states. Again, the price paid is the loss of manifest Lorentz invariance; with consequences for quantization to be explored.[20]

---

[20] Note: (as we shall see next) the strategy in lightcone quantization is to impose *some* gauge constraints classically, but others as operator constraints on quantum states. If one instead pursued covariant quantization, then all the constraints would be quantum.

210   OUT OF NOWHERE

Before moving on to the crucial result of this chapter, we should emphasize that although we have been using QFT, what has been presented is the quantization of a single string; 'first quantization', analogous to the quantization of a *particle*. In contrast, 'second quantization' means quantum *field* theory, and the possibility of creating and annihilating particles—or better, 'quanta'—and we have not yet introduced that for strings. (Though it will become an increasingly important topic of subsequent chapters.) Of course, as we said at the start of quantization, the reason that we can use QFT techniques to first quantize the string is that its dynamics of motion in space can be interpreted (formally at least) as the dynamics of a field (of coordinates) on the worldsheet. Second quantizing that field is equivalent to first quantizing the motion of the string. So, looking ahead, in a sense then, stringy QFT—unlike particle QFT—involves second quantization *twice over*: once to promote the classical string to a quantum string, and then again to promote quantum strings to the quanta of a string 'field' (a point that will be stressed in chapters 9 and 10).

We will return to this point below, but for now we will continue our discussion, not of string field theory, but of the quantum string.

7.3.4 Quantization and stringy particles

In §7.2.2, we derived a relation between the mass and vibrational modes of a classical string—when the string was viewed 'from a distance' as a particle. We have now developed enough theory of the quantized string to perform the analogous analysis; though in this context we will find a much stronger result, a discrete spectrum of (quantum) particles. To proceed, we thus again consider the string at scales at which a string's extension cannot be resolved; though still subject to the relativistic mass-shell condition $M^2 = -p^2$. For the open string $p = \alpha_0/\ell_S$, by (7.10), so we can rewrite

$$M^2 = -\frac{\alpha_0^2}{\ell_s^2} = \frac{-1}{\ell_s^2} \cdot \left(\sum_{n=-\infty}^{\infty} \alpha_{-n} \cdot \alpha_n - \sum_{n=-\infty}^{-1} \alpha_{-n} \cdot \alpha_n - \sum_{n=1}^{\infty} \alpha_{-n} \cdot \alpha_n \right). \qquad (7.50)$$

Then we use (7.34) to obtain a form of the mass that is useful for quantization:

$$M^2 = \frac{-1}{\ell_s^2} \cdot \left(2L_0 - 2\sum_{n=1}^{\infty} \alpha_{-n} \cdot \alpha_n \right)$$

$$\equiv \frac{2}{\ell_s^2} \cdot (N - L_0). \qquad (7.51)$$

Of course, the classical constraints mean that $L_0 = 0$, but we are going to impose Virasoro constraints *after* quantization, so we will not assume that. Instead, we consider the quantum string.

The canonical quantization relations (7.38) for the transverse modes are unchanged by our choice of lightcone coordinates (and gauge), and it is easy to check that they entail that on quantization N will be a number operator (assuming normal ordering), with

$$N \, \alpha^i_{-m}|0\rangle \equiv \left(\sum_{n=1}^{\infty} \alpha_{-n} \cdot \alpha_n \right) \alpha^i_{-m}|0\rangle = m\alpha^i_{-m}|0\rangle. \tag{7.52}$$

(Recall from (7.39) that the commutation relations mean that α_m is a raising operator when $m < 0$.)

Next, consider the classical Virasoro generators (7.33): $L_m = \frac{1}{2} \sum_{n=-\infty}^{\infty} \alpha_{m-n} \cdot \alpha_n$. What form should they take in the quantum theory? As always, the issue is that of operator order. First we note that from (7.38) α_m and α_n commute, unless $m = -n$; so clearly for $m \neq 0$ we can freely normal order L_m, with all lowering operators to the right. However, the order in which the αs appear in L_0 will lead to different operators. The ambiguity is represented by choosing the operator L_0 to be normal ordered, but defining the quantum Virasoro generator to be $L_0 - a$, where a is the constant which would be picked up by normal ordering whatever generator is in fact physically correct. Hence the zeroth quantum Virasoro constraint says that for all physical states $|\psi\rangle$, $(L_0 - a)|\psi\rangle = 0$, or

$$L_0|\psi\rangle = a|\psi\rangle. \tag{7.53}$$

Now we can use our mass operator to fix a. Consider the $N = 1$ first excited states, which are obtained by the action of a single mode on the vacuum: $\alpha^i_{-1}|0\rangle$.

$$M^2 \, \alpha^i_{-1}|0\rangle = \frac{2}{\ell_s^2} \cdot (N - L_0) \, \alpha^i_{-1}|0\rangle$$

$$= \frac{2}{\ell_s^2} \cdot (1 - a) \, \alpha^i_{-1}|0\rangle, \tag{7.54}$$

using (7.51–7.53). (Strictly, these states should also carry a quantum number for the linear momentum of the string, but we will suppress it here.)

Bear in mind that M represents the mass operator for a quantum particle that is, in reality, an excitation of the string. Now, there are $D - 2$ transverse modes α^i_{-1}, and our gauge means that there are only transverse modes. So these states form a representation of $SO(D - 2)$, although the string lives in D-dimensional Minkowski spacetime: the correct state space for a massless particle according to quantum representation theory (e.g., Weinberg 1995, §2.5), which identifies particle kinds according to their representations of the spacetime symmetries.[21] This

---

[21] Heuristically, we can always pick a frame in which a massive particle is at rest, and then rotate its polarization in any direction we like; but a massless particle travels at the speed of light and has no rest frame, so its polarization can never be set parallel with its motion.

212 OUT OF NOWHERE

massless particle, the first excited state of the open string, has the correct symmetry properties of the photon, but is not sufficient to give a stringy account of quantum electrodynamics, for the simple reason that none of the string excitations carry any charge—so this 'photon' does not mediate an electromagnetic force. Charges have to be added to the theory by other devices, beyond the scope of this discussion.

That the first excited state is massless means that $M^2 \alpha^i_{-1}|0\rangle = 0$, and hence from (7.54) $a = 1$. More generally, for physical states the mass operator (7.51) is therefore

$$M^2 = \frac{2}{\ell_s^2}(N - 1).$$
(7.55)

It is worth pausing to emphasize something rather important, concerning the role of Lorentz symmetry in the argument; namely in the assumption that stringy particles are 'on mass shell', so that we can infer the masslessness of the first excited states from the fact that they form a representation of $SO(D - 2)$. It may seem an additional postulate, motivated empirically by the desire to recover relativistic particle physics. What, mathematically speaking, would happen if we tolerated the breaking of Lorentz invariance on quantization—a Lorentz 'anomaly'? Answering this question requires considerations beyond the framework of lightcone quantization, specifically the path integral formulation, to be discussed in chapter 9. In short though, it is Weyl symmetry that determines that $a = 1$, and hence that the first excited state is massless. And, as we shall discuss, because Weyl symmetry is a local symmetry (unlike global Lorentz invariance) it cannot be broken on quantization: if such symmetry is broken on quantization, the anomaly leads to pathologies. Hence it is not an additional postulate that stringy particles are on mass shell, it is a logical consequence of the quantum theory. That the Weyl anomaly would manifest itself as a Lorentz anomaly is an artifact of lightcone quantization, which obscures the strength of the result.

Unfortunately, the mass operator (7.55) thus entails that $M^2|0\rangle = -2/\ell_s^2|0\rangle$, so the string ground state has a negative mass squared—it is a tachyon, with a spacelike 4-velocity! Despite popular wisdom, faster than light particles are not necessarily incompatible with relativity (as Maudlin 1994, 72ff explains). However, a negative mass is a problem in a quantum theory, because it means that the vacuum is unstable. The problem is that we have only included bosonic degrees of freedom, satisfying commutation relations, and have neglected the other possibility in a quantum theory, to have fermionic degrees of freedom, with *anti*commutation relations. Once both are included, in 'supersymmetric string theory', the tachyon disappears (moreover, fermions appear in the string spectrum, as they must). The point of studying the bosonic string, rather than the full supersymmetric theory, is that it is simpler, yet many of the key ideas are the same. (Note that the tachyon is the simplest particle in the spectrum, a spinless, scalar particle: so in fact we will use it later to illustrate some ideas in string theory, even though it is unphysical. In those situations its negative norm will not be relevant.)

A STRING THEORY PRIMER   213

The mass spectrum of the closed string is treated similarly, but taking into account that we have left and right moving modes. Normal ordering considerations are symmetric between the two, so for normal ordered L_0 and $\tilde{L}_0$ (7.33) physical states must satisfy the quantum Virasoro constraints:

$$(L_0 - a)|\psi\rangle = (\tilde{L}_0 - a)|\psi\rangle = 0. \tag{7.56}$$

It immediately follows that $L_0 - \tilde{L}_0 = 0$ holds as an operator equation. From (7.34) and the definition of the number operator (7.52) (mutatis mutandis for $\tilde{N}$)

$$L_0 - \tilde{L}_0 = \frac{1}{2}\alpha_0^2 + N - \frac{1}{2}\tilde{\alpha}_0^2 - \tilde{N}, \tag{7.57}$$

But the closed string periodic boundary conditions entail that $\alpha_0^\mu = \tilde{\alpha}_0^\mu$, so we have that

$$N|\psi\rangle = \tilde{N}|\psi\rangle, \tag{7.58}$$

for physical states—the 'level matching condition' for closed strings. In other words, left- and right-moving modes must be excited to the same level. In particular, while the ground state has $N = \tilde{N} = 0$, the lowest excited state has $N = \tilde{N} = 1$, so both a left- and a right-moving mode—the state in which only one of the modes is excited is unphysical.

The quantum mass operator for closed strings is derived as for open strings (7.51), but starting with $M^2 = -p^2 = 4\alpha_0^2/\ell_s^2$ from (7.5):

$$M^2 = \frac{8}{\ell_s^2}(N - L_0) = \frac{8}{\ell_s^2}(N - 1). \tag{7.59}$$

Once again the first excited state is therefore massless, and the ground state is a tachyon (which once again is absent in supersymmetric string theory). For both open and closed strings the very small value of ℓ_s means that higher excited states have masses far greater than those of observed particles. As mentioned in §7.2.2, a more realistic spectrum requires a more complicated model than we have developed here.

The first excited states have a basis $|\Omega^{ij}; k\rangle = \alpha_{-1}^i \tilde{\alpha}_{-1}^j |0; k\rangle$; where we have made explicit the linear momentum dependence of the state (i.e., the α_0^i mode) in addition to the polarization. Since i and j range over the transverse (lightcone) coordinates, there are $(D-2)^2$ such states. A superposition thus has the form $\epsilon^{ij}|\Omega^{ij}; k\rangle$. It transforms as a second rank tensor under rotations in $(D-2)$-space, which can be decomposed into three irreducible representations of $SO(D-2)$, scalar, antisymmetric, and traceless-symmetric:

214 OUT OF NOWHERE

$$\epsilon^{ij} = \frac{1}{D-2}\delta^{ij}\epsilon^{kk} + \frac{1}{2}(\epsilon^{ij} - \epsilon^{ji}) + \frac{1}{2}(\epsilon^{ij} + \epsilon^{ji} - \frac{2}{D-2}\delta^{ij}\epsilon^{kk}). \tag{7.60}$$

Again, these string transformation properties are identified with the transformation properties of relativistic particles: representations of the Poincaré group. The massless scalar particle is called the *dilaton*; the massless antisymmetric representation of $SO(D-2)$ describes a spin-1 'Kalb-Ramond' particle; and the traceless, symmetric representation is that of a massless spin-2 particle, on which we will focus. Thus the closed, bosonic, quantum string realizes three kinds of bosonic particles at the lowest level above the ground state.

Note that there are two senses of 'bosonic' here: we introduced 'worldsheet bosons' $X^\mu(\tau, \sigma)$ by imposing commutation relations with respect to the worldsheet lightcone structure. We have now seen that their excitations appear as quanta with integer spin with respect to spacetime transformations, 'spacetime bosons' according to the spin-statistics theorem. That worldsheet bosons are spacetime bosons is perhaps not surprising, but needed to be shown, not assumed.

All three of the particles found in the first excited state are physically interesting, but it is the third—traceless-symmetric, hence massless spin-2 particle—that is of greatest significance.[22] The weak field approach to GR decomposes the metric field into a flat part, plus a symmetric tensor: formally describing a tensor gravitational field in Minkowski spacetime. That is, we write: $g_{\mu\nu} = \eta_{\mu\nu} + h_{\mu\nu}$. For weak $h_{\mu\nu}$, the Einstein Field Equation (EFE) to first order is:

$$\Box h_{\mu\nu} = -16\pi G(T_{\mu\nu} - \frac{1}{2}\eta_{\mu\nu}T). \tag{7.61}$$

This linear equation is symmetric under infinitesimal transformations $x^\mu \to x^\mu + \epsilon\lambda^\mu(x)$, corresponding to diffeomorphism invariance of the full theory. Exploiting this gauge symmetry, coordinates can be found in which there are plane wave solutions $h_{\mu\nu} = \epsilon_{\mu\nu}e^{ikX}$, propagating at the speed of light, with symmetry under 180° rotations: or with 'helicity' 2, since $180^\circ = 360^\circ/2$. When such a tensor field is quantized, such plane waves correspond to the massless spin-2 quanta of the resulting quantum field, 'gravitons'.

Thus the existence of massless spin-2 states in the spectrum of the closed string offers hope that string theory is a quantum theory of gravity—that gravitons are stringy excitations—but does not yet show it. First, isn't it possible that the graviton is just one kind of massless spin-2 particle, while there are others coupling to matter in a different way? Actually, things are much more tightly constrained

---

[22] See Kiefer (2004, §2.1) for a comprehensive survey of the following; and Misner et al. (1973, ch. 18) for a review of weak fields.

than one might anticipate. Weinberg (1964) showed that the principles of 'S-matrix theory' (e.g., Cushing 1990), including Lorentz invariance, entail that all infrared spin-2 massless quanta have the same coupling constant: universal coupling for the graviton. While Weinberg (1965) shows that in Dyson-Feynman perturbation theory, Lorentz invariance entails that any field with massless spin-2 quanta has the operator form of the EFE as its Heisenberg equations of motion, plus possible terms coupling matter to curvature (i.e., minimal coupling is not entailed).[23] In other words, a QFT of massless spin-2 quanta is—at low energy—a theory of gravity, and so any such quanta deserve the name 'graviton'. We will discuss at much greater length in chapters 9 and 10 how the corresponding string excitations give rise to gravity.

However, we should (again) distinguish quantum *particles* from field *quanta*: the latter can be created and annihilated, and form states of (superpositions of) arbitrary numbers, but quantum particles just are. This is the difference between first and second quantization. Our treatment of string theory thus far has been of a *single* string, which corresponds to a single first quantized particle at low energy, not a second quantized field quantum. It is true that string excitations can be created and annihilated, and superposed, but from the spacetime perspective higher excitations of the string do not produce *additional* particles, but rather excite a single string to 'become' a *different* particle, with a greater mass. (Excitations can also be viewed as 'quanta' of the $X^\mu(\tau, \sigma)$ field from the worldsheet perspective; but when recovering the particle spectrum we are concerned with how strings appear as spacetime objects.) And while a superposition of different number states is a state of the worldsheet quantum field, it is not a state of some spacetime quantum field, but rather a superposition of single particles of different kinds. Spacetime fields are recovered from spacetime quanta of course, and so we must take that perspective.

Hence to recover spacetime fields, and in particular for Weinberg's results to apply, there has to be more than the states of a single string: string states corresponding to many particles, particle creation and annihilation, and indefinite particle number, and interactions that dynamically cause such events and states. Most naturally one anticipates that there must be states of *many* strings, each representing a single quantum; that these strings can be created and annihilated, so that quanta are created and annihilated; that there can be an indefinite number of strings, so an indefinite number of quanta; and that there are string-string interactions. This is the kind of framework that we will describe and investigate in

---

[23] There is a small lacuna: the stress-energy tensor is not explicitly constructed. The paper does prove explicitly that massless spin-1 quanta satisfy the operator form of the Maxwell equations, up to non-minimal coupling. See Salimkhani (2018) for a philosophical discussion of Weinberg's results, and Weinberg (1995, 595ff) or Kiefer (2004, 34ff) for pedagogical proofs of universal coupling.

216   OUT OF NOWHERE

chapters 9 and 10. We shall see that it leads—by a route different from Weinberg's—to the classical EFE (at low energy). This result of course shows that string theory is a theory of gravity, and it also provides evidence for the quanta interpretation of strings: that full string theory really is well approximated by QFT at scales above ℓ_S, that quanta of physical fields can indeed be understood in terms of different string states.

7.3.5   26 dimensions

Finally, at this point, pedagogical presentations give a heuristic argument that bosonic string theory requires $D = 26$ dimensions. The argument is typically accompanied with apologies, as it is motivated by giving the right answer, drawing on the right considerations; but it is not rigorous, and should be thought of as a plausibility argument. We will discuss it here mainly to clarify this common but puzzling line of thought. A more rigorous derivation uses the path integral approach described in chapter 9.[24] But nothing in the rest of the book really hangs on this discussion, so the tired reader could safely skip this section!

The basic idea is that the normal ordering constant a can be calculated directly, and then compared with the value found earlier. Start with the zeroth quantum Virasoro generator

$$L_0 - a = \frac{1}{2} \sum_{n=-\infty}^{\infty} \alpha_{-n} \cdot \alpha_n, \tag{7.62}$$

understood as an operator equation: L_0 is normal ordered. The $n \geq 0$ terms already have lowering operators to the right, so we can split up the sum and apply the commutation relations (7.38) to normal order the $n > 0$ terms as follows:

$$\frac{1}{2} \sum_{n=-\infty}^{\infty} \alpha_{-n} \cdot \alpha_n = \frac{1}{2} \left(\sum_{n=0}^{\infty} \alpha_{-n} \cdot \alpha_n + \sum_{n=-\infty}^{-1} \alpha_{-n} \cdot \alpha_n \right)$$

$$= \frac{1}{2} \left(\sum_{n=0}^{\infty} \alpha_{-n} \cdot \alpha_n + \sum_{n=-\infty}^{-1} (\alpha_n \cdot \alpha_{-n} - n\eta_\mu^\mu) \right)$$

$$= L_0 + \frac{D-2}{2} \cdot \sum_{n=1}^{\infty} n, \tag{7.63}$$

because $\eta_\mu^\mu = D - 2$ (and note that the limits of the sum have been changed). (Note that this is how dependence on the number of dimensions enters the argument.)

---

[24] See Polchinski (1998, §3.4); find a somewhat more physically motivated version of the following on p.22 of the same book.

Hence we have an alternative expression for a, which we already know is equal to 1: hence

$$\frac{D-2}{2} \cdot \sum_{n=1}^{\infty} n = -1. \tag{7.64}$$

The embarrassment in the presentations has the remaining—divergent—sum as its focus. The most straightforward approach is to consider the Riemann zeta-function,

$$\zeta(s) = \sum_{n=1}^{\infty} n^{-s}, \tag{7.65}$$

which is holomorphic in the complex plane, except at its poles. It follows that its analytic continuation to $s = -1$ is unique: it's a remarkable and important feature of complex analysis that if a function is holomorphic in some finite region, and if any two other functions are holomorphic in that region, then the latter agree over any larger region over which they are both holomorphic. That unique continuation is $\zeta(-1) = -1/12$. The situation here is akin to that found in bosonic field theory, in which each mode gives a finite contribution to the energy even in the ground state, and so an infinite constant must subtracted. In neither case is the derivation analytically rigorous, but they are reasonable in the circumstances. And of course it's worth stressing again that the same result can be obtained by more rigorous means. It should also be stressed that zeta-function regularization, introduced in Hawking (1977), is a standard and well-understood tool of quantum field theory, as physically motivated as other regularization schemes. Our point here refers to the specific use in the heuristic, pedagogical argument just given, not against more careful use.

But using this regularization, we can then conclude from (7.64) that $D = 26$. When Claude Lovelace discovered this result (by other means, in a related model), he remarked that '$D = 26$ is obviously otherworldly' (1971, 502)! (See also Lovelace 2012.) But remarkably, extra dimensions 'too small to observe' have come to be accepted as a consequence of string theory: in part, as this book aims to explain, spacetime is not (necessarily) a background in the theory. For now we simply stress that this heuristic result depended on $a = 1$ ($D = 4$ would have required $a = 12$), which in turn is, as we saw, required by the Lorentz invariance of the theory—specifically that the first excited state must be massless. Again, a more rigorous derivation of $D = 26$ (by the techniques described in chapter 9), like that of $a = 1$, reveals that the source is in fact Weyl invariance.[25]

Thus ends our technical introduction to elementary bosonic string theory. Starting from an intuitive conception of a relativistic string, we have sketched the

---

[25] While in terms of covariant quantization, the result is that when $D = 26$ and $a = 1$ the theory is free of ghosts: the 'no ghost theorem'. See Rickles (2014, ch. 4).

218 OUT OF NOWHERE

outline of the resulting theory, with special attention to the resulting mass spectrum of the quantized theory, and the symmetries of the theory. Of special interest going forward are the worldsheet's Weyl symmetry, and the (distinct) conformal symmetry that remains after gauge fixing the worldsheet metric. The former we have already indicated has important consequences for string theory, and we shall see in chapter 9 that it also leads to the field equations of GR! The latter will also be important, because it considerably simplifies the mathematics of string theory.

7.4 Fermionic strings and supersymmetry

Bosonic strings involve coordinates X^μ that commute classically, and have non-trivial commutation relations on quantization. But QM permits *anti*symmetric degrees of freedom too, so there's no reason that the quantized string should not also have these; indeed, it seems to be a requirement, since (as we saw) with only bosonic degrees of freedom there is a tachyon in the string spectrum, indicating a relativistic pathology. Now, one[26] might suggest that adding fermionic degrees of freedom is an ad hoc maneuver, made to deal with the tachyon. However, this would be to think about strings the wrong way round, as derivative objects: the-quantization-of-classical-strings. Of course, that is the way we (as most texts) present things, but that is simply a pedagogical matter, to develop the formalism and concepts in a more intuitive physical context. But the quantum string is the fundamental object, and the classical string derivative: the classical-limit-of-a-quantum-string. (A similar point applies to any quantum system, of course.) So one should not think that strings 'should' only have bosonic degrees of freedom, since that is the result of quantizing our classical string. No, the string is a quantum object, and both bosonic and fermionic degrees of freedom are equally valid, though only the former have an interpretation as spatiotemporal degrees of freedom, and play that role in the classical limit.

Once we have thus included both symmetric and antisymmetric degrees of freedom, we find that the string spectrum includes internal states corresponding to both bosonic and fermionic quanta, and so is capable (in principle) of accounting for all the quanta that we observe. That is to say, it promises a remarkable unification of our current physics of particles and forces, including gravity. However, the key conceptual ideas regarding the emergence of spacetime are already contained in the simpler bosonic subtheory, so we will focus on that in the following chapters. In this section we will then briefly review the basic ideas of fermionic string theory for completeness, and to develop a couple of ideas that will be relevant later.

---

[26] The 'one' that we have in mind is John Norton, who made some very helpful comments on this chapter, and in particular raised this criticism (amongst others).

We again adopt a worldsheet point of view, and proceed in parallel with the bosonic case, now adding anticommuting, Grassman number-valued fields to the commuting $X^\mu(\tau,\sigma)$ fields to the classical string; and then quantizing by imposing anticommutation and commutation relations respectively.[27] This procedure produces a QFT of bosons and fermions in a standard field theoretic way, in 1+1-dimensional spacetime.

From a spacetime point of view it is less clear what is going on: the fermionic and bosonic degrees of freedom seem to have formally similar roles in the theory, so do the former represent *anticommuting* spacetime dimensions? What would that mean? The question becomes even more pressing when one considers an action with supersymmetry, a transformation in which bosonic and fermionic excitations are exchanged—a symmetry under a simultaneous change of boson and fermion number. Not only does supersymmetry imply an equivalence between the commuting spatial and new anticommuting degrees of freedom, it also has a natural interpretation as an extension of familiar spatial transformations, leading to the conception of 'superspace', combining commuting and anticommuting dimensions. Yet how can *space* truly be anticommuting? We will not take up this issue here, referring the reader to relevant discussions in Weingard (1988), Baker (2020), and Menon (2021). Our only goal here is to indicate how the treatment of such fermionic fields develops in parallel with that of bosonic fields, to reassure us that things will not completely change conceptually, so that we can extrapolate the following investigations from the bosonic context to superstrings.

A suitable (gauge-fixed) action for string theory is given by:

$$S_{\text{SUSY}} = \int d^2\sigma \frac{T}{2}(\dot{X}^2 - X'^2) + \frac{1}{2\pi}[\Psi_1(\partial_\tau + \partial_\sigma)\Psi_2 + \Psi_2(\partial_\tau - \partial_\sigma)\Psi_1], \qquad (7.66)$$

where $\Psi_\lambda(\tau,\sigma)$ are 'Marjorana' spinors rather than Grassmann fields (λ indexes the different fields). This action simply describes a bosonic field (unchanged from (7.18)) and a fermionic field in a 2-dimensional spacetime—physically the worldsheet. S_{SUSY} exhibits classical supersymmetry, under an infinitesimal mixing of the bosonic and fermionic fields.[28]

---

[27] We won't take a stance on the classicality (or on the physical possibility) of such Grassman fields; we speak only formally here. If they are allowed, then even if one does think of strings as the-quantization-of-classical-strings, fermionic degrees of freedom are again just as valid as bosonic. We did not make this argument earlier to avoid objections from those who think that Grassman fields are merely formal.

[28] See Green et al. (1987, §4.1) for more details. As they explain, in this formulation it is in fact the worldsheet rather than spacetime that gains anticommuting dimensions, which is made manifest by expanding (7.66) in terms of a pair of worldsheet Grassmann coordinates and considering the effect of the supersymmetry (§4.1.2). To see the full spacetime supersymmetry described above, one has to consider the equivalent action that they describe in (§5.1).

220   OUT OF NOWHERE

Minimizing the fermionic part of the action in parallel with (7.21) leads to the equations of motion[29]

$$(\partial_\tau + \partial_\sigma)\Psi_1 = 0 \qquad \text{and} \qquad (\partial_\tau - \partial_\sigma)\Psi_2 = 0. \tag{7.67}$$

Thus the two components of the spinor $\begin{pmatrix} \Psi_1 \\ \Psi_2 \end{pmatrix}$ are right and left moving respectively: $\Psi_1 = \Psi_1(\tau - \sigma)$ and $\Psi_2 = \Psi_2(\tau + \sigma)$. One also obtains the boundary conditions

$$\Psi_1(\tau, 0) = \Psi_2(\tau, 0) \qquad \text{and} \qquad \Psi_1(\tau, \pi) = \pm\Psi_2(\tau, \pi) \tag{7.68}$$

(after fixing an arbitrary phase). The choice of $+$ at $\sigma = \pi$ leads to *Ramond* boundary conditions, while that of $-$ leads to *Neveu-Schwarz* boundary conditions, in comparison to the Neumann boundary condition of the bosonic string.

The different choices lead to different Fourier expansions, and hence on quantization to different raising and lowering operators—reflected in physically different particle spectra (including tachyons), and sectors of state space. The operators will also contribute to the zeroth Virasoro generator in (7.62), and hence change the number of dimensions in the quantum theory: from $D = 26$ to $D = 10$. Although the operators satisfy anticommutation relations, different combinations with each other and with the bosonic operators give rise to both fermions and bosons (with respect to the worldsheet) within each sector. Thus quantum supersymmetry can be understood simply in terms of a transformation within a sector, a rotation in Hilbert space from one excitation to another (of course it is a property of the action that such a transformation is a dynamical symmetry). Moreover, these particles form subspaces that transform correctly as spacetime bosons and fermions, providing a more physically realistic basis for quantum field theory (though matching the observed particle spectrum is a further goal).

Then different consistent quantized forms of string theory (inter alia, those without a tachyon) are obtained by combining different subspaces of the Ramond and Neveu-Schwarz sectors: type I includes unoriented open and closed strings, while types IIA and IIB are for oriented closed strings. (In addition there are 'heterotic' theories, in which left- and right-moving modes are bosonic and fermionic, respectively.) As the reader may well be aware, these theories are in fact not independent, but related (in appropriate limits) by a network of symmetries known as 'dualities' (a simple version will be studied in chapter 8), suggesting that they are in fact limiting cases of some as yet unknown underlying theory, often referred to a 'M-theory'. While this idea will be referred to on occasion in what follows, we

---

[29] This remainder of this section follows the nice elementary treatment found in Zwiebach (2004, ch. 14).

emphasize that the main goal of this work is to analyze what already follows from bosonic string theory, without supersymmetry and without speculations about M-theory.

We will thus not develop fermionic and supersymmetric string theory further here. What we have indicated already in this short section is how the general line of development, from classical modes, to quantized modes and particle spectrum, and constraints on dimensions, runs similarly for bosonic and fermionic strings. But the math is different because the fields are anticommuting rather than commuting.

As we explained at the start, this chapter was designed to summarize some of the key ideas of string theory, in a way that we hope will make the subject more accessible to philosophers of physics. At this stage the reader should be equipped, not only to understand in more detail the arguments of later chapters, but also be in a position to tackle the kind of textbooks which we have cited. There is much more to learn!

Chapter 8
Duality

Having developed string theory in the previous chapter, now we will see how it contains some surprising symmetries—'dualities'—which, we will argue, put pressure on the view that the spacetime in which strings are described can be literally identified with classical, physical spacetime—instead it is 'emergent' from the theory. While the discussion here stands on the previous chapter, and exemplifies its physics, it can be read on its own to understand the essential conclusions.

First, in §8.1, we will explicate the simplest duality, 'T-duality', which holds between models with reciprocal radii for space: a symmetry between a very large and a very small universe! In §8.2 we review the different interpretational questions that face one given such a symmetry: is a duality merely a 'translation manual', relating two languages stating the same facts differently? If, instead, the models represent factually different worlds, what is the significance of the symmetry? Is it epistemic: we can't tell which of two possible worlds we inhabit? Or ontological: the models have 'surplus structure', representing facts that go beyond those that make up the world? We will argue for the latter option in §8.3–8.4, claiming that the space in which strings live is indeterminate in size. But classical space, in the sense we use that term, to refer to the space of astrophysics, particle physics, and indeed the everyday, certainly has a determinate radius—it's very big! It is for this reason that we conclude that the space assumed in string theory is not, after all, classical space; hence the latter must be emergent.

With that important conclusion obtained for T-duality, in §8.5 we review additional dualities, for which we conjecture analogous reasoning holds, strengthening the claim that spacetime is emergent in string theory. Showing how is then the task of subsequent chapters.

8.1 T-duality

Consider a closed, classical bosonic string in Minkowski spacetime with a compact spatial dimension, x, of radius R.[1] See figure 8.1. As we saw in the previous chapter, its state is a function $X(\tau, \sigma)$ describing the x-coordinate of the point of the string worldsheet with worldsheet coordinates (τ, σ): hence the state of a string specifies an embedding of the worldsheet into spacetime.

---

[1] Our technical presentation follows, amongst others, Brandenberger and Vafa (1989), Greene (1999, 237), Zwiebach (2004, ch. 17), and Zaslow (2008).

Out of Nowhere. Nick Huggett and Christian Wüthrich, Oxford University Press. © Nick Huggett and Christian Wüthrich (2025). DOI: 10.1093/oso/9780198758501.003.0008

DUALITY 223

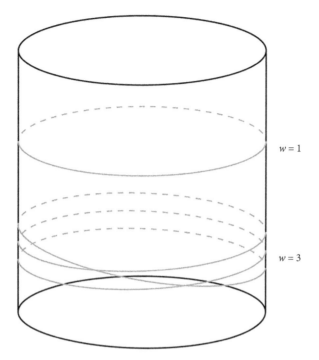

Figure 8.1 Two closed strings in a 2-dimensional space with one compact dimension. One string is wrapped once around x, and the other three times—with winding numbers $w = 1$ and $w = 3$, respectively.

We adopt the conventions that the spacelike string coordinate σ is periodic with period π, while the compact spatial coordinate x is periodic with period 2π (so that σ and $\sigma + n\pi$ are labels for the same point of the string worldsheet, and x and $x + 2n\pi$ are labels for the same point of space). Then the state of the string with respect to x is

$$X(\tau, \sigma) = 2w\sigma R + 2\ell_s^2 p\tau + \text{vibrational terms.} \qquad (8.1)$$

This expression differs from equation (7.6) by the addition of the first term, which describes the w-fold winding of the string: for instance, if the string is wound once around x, so $w = 1$, then X ranges from 0 to 2π as σ ranges from 0 to π. The second term represents the linear momentum of the string; the constant ℓ_s is the 'characteristic string length'. For simplicity, we shall ignore vibrations, since they do not change the substance of this chapter.[2]

---

[2] Important: to clean up expressions in this chapter, we rescale ℓ_s by a factor of $1/\sqrt{2}$, and, in (8.2), T by a factor of $\pi/4$, relative to chapter 7. Though there is a small risk of confusion arising, we believe this decision makes this chapter more accessible to those reading it without studying the previous chapter. These unconventional changes have no physical significance; indeed, one could have achieved most of the same result by trivially rescaling the worldsheet coordinate.

224 OUT OF NOWHERE

In wisely chosen string coordinates (and suitable units), substituting X in the Hamiltonian (7.20) gives

$$H = \frac{T}{8\pi} \int \dot{X}^2 + X'^2 \, d\sigma = \frac{T}{2\pi} \int (\ell_s^2 p)^2 + (wR)^2 \, d\sigma, \tag{8.2}$$

where T is another constant, the string tension. Not surprisingly there is a kinetic term, plus a term from the winding, hence stretching, of the string around the closed dimension. The next step is to quantize.[3]

Momentum first. The closed dimension implies a periodic boundary condition for momentum eigenstates $\Psi_k(x) = e^{ikx}$ (ignoring normalization, and with $\hbar = 1$)

$$\Psi_k(0) = \Psi_k(2\pi R) \Rightarrow e^{ik\cdot 0} = 1 = e^{ik\cdot 2\pi R} \Rightarrow k = 0, \pm 1/R, \pm 2/R \dots. \tag{8.3}$$

In other words, momentum is quantized: $|k| = n/R$, with 'wave number' n. Substituting into the Hamiltonian (8.2), we obtain the spectrum

$$E_{n,w} = \frac{T}{2\pi} \int (\ell_s^2 n/R)^2 + (wR)^2 \, d\sigma. \tag{8.4}$$

Now winding. When we quantize, w is not a constant, classical 'c-number' of the system: first, we can have superpositions of states of different winding, and second, assuming interactions, a string can change the number of times it is wound around a closed dimension. In other words, w is now a dynamical *quantum* quantity, described by a wavefunction. Pay close attention to this point, as it is *crucial*: without dynamical winding superpositions there is no T-duality—in this sense it is a quantum phenomenon.

The winding term in (8.2) depends on $l = wR$, which must have a discrete spectrum since w does. Thus much as before, these eigenstates have the form $\Phi_l(y) = e^{ily}$ around a circle with coordinate y, but with radius $1/R$.[4] In that case the periodic boundary condition $\Phi_l(0) = \Phi_l(2\pi/R)$ yields

$$e^{il\cdot 0} = 1 = e^{il\cdot 2\pi/R} \Rightarrow l = 0, \pm R, \pm 2R, \cdots = wR, \tag{8.5}$$

as required. Overall then, the state of a quantum string involves (the tensor product of) two wavefunctions, one representing its position/momentum, and another representing its winding.

---

[3] This time we do not follow the quantization procedure of chapter 7, but considerably simplify matters by using an approximation based on the 'double field theory' approach (see e.g., Aldazabal et al. 2013). Rest assured that the duality that results is *not* an artifact of the approximation, but also holds in a full treatment.

[4] This quantity seems to have units of *length*$^{-1}$, but the numerator can be taken as an *area* to give overall correct units. A similar point applies everywhere that quantities appear to have the wrong units.

DUALITY 225

The question is of course, 'where is the circle on which the $\Phi_l(y)$ wavefunction lives?' It can't just be in space, because then $\Phi_l(y)$ describes a second object which we could expect to find somewhere. Instead, there must be a new 'internal' dimension associated with each closed dimension of space; hence Edward Witten calls y 'another "direction" peculiar to string theory' Witten (1996, 29). His proposal is not that *space* has an extra dimension for every dimension a string can wrap around, but rather that treating winding as a quantum observable means that it can be represented like momentum on a *non-spatial* circle. Or more precisely, when we consider the space of all states of any momentum or winding, we find two quantum 'position' operators, x and y, respectively corresponding to position in physical space (radius R) and in a new 'winding space' (radius $1/R$). But observables represent physical quantities, so we have to take both 'positions' and spaces equally seriously, even if only one is physical space; let's call the other 'winding space'. But remember, the string winds around physical space, while the winding number wave lives in winding space. (Take the term 'physical space' with a grain of salt here: it is the space in which the string moves, but precisely because of T-duality we will have to clarify below exactly how it relates to ordinary, observable, classical space.)

Semi-technical aside: as usual, x and y are 'position' operators for physical and winding space. Moreover, as $p = -i\partial/\partial x$ is the momentum observable with eigenvalues $k = n/R$ in the periodic plane wave states $e^{inx/R}$, so $w = -i\partial/\partial y$ is the winding observable with eigenvalues $l = wR$ in the winding states e^{iwRy}. Thus each space is associated with identical canonical commutation relations, $[x, p] = i$ and $[y, w] = i$ (observables from different spaces commute). Therefore, since position and momentum generate the algebra of observables, each space has, formally speaking, exactly the same observables, individuated as functions of x and p or y and w.[5]

Such 'internal' spaces are familiar—for spin states and gauge field states, for instance—so there is nothing new yet. But look again at (8.4), the spectrum of the Hamiltonian. It's easy to check that a string with wave number n and winding number w in a space with radius R has the same energy eigenvalue as a string with *winding* number n and *wave* number w, *but which lives in a space of radius* ℓ_s^2/R:

$$n \leftrightarrow w \quad \text{and} \quad R \to \ell_s^2/R. \tag{8.6}$$

The second string has a spatial wavefunction in a compact dimension of radius ℓ_s^2/R, and hence—by the same reasoning as before—a winding wavefunction that

---

[5] Of course there are observables involving operators from both the spaces, but since the latter commute, such observables are always the commutative product of a pair of observables, one from each space. So all the points we need go through trivially, and we will ignore them.

lives in a compact dimension of reciprocal radius, namely R/ℓ_s^2. If the first string lives in a space with radius $R > \ell_s$, then the second string lives in a space of radius $\ell_s^2/R < \ell_s$: the strings are 'reflected' through ℓ_s. See figure 8.2.

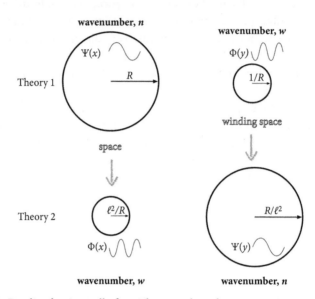

Figure 8.2 Reading horizontally first, Theory 1 describes a string moving in a space x of radius R with wavenumber n—more generally, with wavefunction $\Psi(x)$; and winding described by the wave number w—more generally, wavefunction $\Phi(y)$—in a 'winding space' y with reciprocal radius, $1/R$. Now reading vertically, a second, 'dual' theory is obtained by simultaneously taking $R \to R/\ell_s^2/R$ and $n \leftrightarrow w$ (equivalently, $\Psi \leftrightarrow \Phi$).

Now, n and w label eigenstates of momentum and winding, and so in terms of wavefunctions, $n \leftrightarrow w$ corresponds to $\Psi_n(x) \to \Phi_w(x)$ and $\Phi_w(y) \to \Psi_n(y)$: the wavefunctions are exchanged between space and winding space. Thus for *general* spatial and winding wavefunctions (i.e., superpositions of momentum or winding eigenstates) $\Psi(x)$ and $\Phi(y)$, respectively, let $\Psi(x) \to \Phi(x)$ and $\Phi(y) \to \Psi(y)$: the 'dual' string has the same—but exchanged—wavefunctions. (And if we took account of the vibrations of the string, these would also be exchanged between physical and winding spaces.)

Then, (*i*) because they have the same Hamiltonian, both strings will have the same mass spectrum (because in string theory the Hamiltonian determines the mass: see equation 7.36). Moreover, (*ii*) because the roles of momentum and winding are reversed in the Hamiltonian by (8.6), the dynamics of the spatial wavefunction in one string become the dynamics of the winding wavefunction in the other, and vice versa: in other words, the strings remain dual over time under the exchange of momentum and winding.

Further, (*iii*) because momentum and winding are exchanged by (8.6), every observable pertaining to physical space is exchanged with a corresponding observable pertaining to winding space; and because the wavefunctions are also exchanged, the expectation value of the new observable for the dual string will equal that of the original observable for the original string. (And vice versa.) In short, the *pattern* of observable quantities will be preserved by (8.6); what changes is whether the quantity is understood to pertain to physical or to winding space.[6] (And similarly once one includes the vibrations of the string.)

Continuation of the semi-technical aside: a little more formally, the point is that the algebra of observables on spatial wavefunctions for one string is mapped onto the identical (as we saw above) algebra of observables on winding wave-functions of the other—with $x \leftrightarrow y$ and $p \leftrightarrow w$. Since the wavefunctions are also exchanged, the values associated with *all* corresponding elements of the algebra of observables are preserved by (8.6)—the entire pattern of expectation values. (But generally not the expectation values of specific operators.)

Thus the systems are equivalent in the following sense: the Hamiltonian and hence dynamics are the same, and the pattern of physical quantities (formally represented by observables) agree. This equivalence, and others comparably strong, are known as 'dualities', and (as implicitly anticipated) the two theories related by it are 'dual' to each other, or 'duals'.[7] In particular, (8.6) is known as 'T-duality', where—depending on whom you ask—'T' either stands for 'target space' (i.e., the space in which the string lives), or for 'torus'. T-duality holds not only for the bosonic string just considered, but also for supersymmetric string theory, in which case it also changes the character of the string: for instance, type IIA strings are T-dual to type IIB strings.

There is a lot to unpack and justify in these statements, which will be the work of the following section, but for now a concrete example, taken from Branden-berger and Vafa (1989), will help illustrate some of its implications. Take T-duals, T_1 and T_2, which differ in the radius that they postulate for a closed spatial dimension: a circumference of 10^{12} light years (two orders of magnitude bigger than the visible universe) on the one hand, and 10^{-94}m on the other (assuming a value for the characteristic string length of 10^{-33}m, two orders of magnitude above the Planck length). Thus T_1 and T_2 (apparently) make radically different assertions about the size of a spatial dimension. Before T-duality, one would assume that

---

[6] The mapping introduces a ℓ_s^2 factor, but these can be absorbed in a trivial rescaling of observables, so we will ignore it.

[7] There are competing accounts of what exactly makes a symmetry a 'duality' (or even a 'symmetry'). We have in mind an account along the lines of that given in de Haro (2020), which aligns with our interpretative goals. This sense may not align perfectly with traditional use of the term, nor (according to Dawid 2017) its novel methodological meaning for string theorists. See also de Haro (2019).

228   OUT OF NOWHERE

simple observations would rather readily choose between them, but that can't be right if the duals are physically equivalent.

To understand how the equivalence manifests itself, in a beautiful conceptual analysis, Brandenberger and Vafa consider an archetypical measurement of the radius: fire off a particle of known velocity—a photon, say—and time its journey around space. Suppose the result is a trillion years: that seems pretty conclusive evidence for the large radius story, T_1. But, in terms of T_1, how is the measurement described? The photon has a spatial wavefunction $\psi(x, t)$, which evolves, according to the Hamiltonian, from being localized nearby, via a journey away of 10^{12} years, to being localized nearby again. However, T_2 can also account for this result.

Note that the photon is a low energy mode of the string, the easiest thing to excite. Indeed, using the Planck tension of about 10^{44}N as an estimate of the string tension, we find that in T_1 the first excited winding state corresponds to the mass-energy of 10^{15} supermassive black holes; so 'low' energy here is relative to almost inconceivable scales! On the other other hand $k = n/R$, so momentum is inverse to the radius of space, and the very large radius of space in T_1 allows states of very low momentum, hence of low energy. In other words, in T_1 the low (by any sensible measure) energy photon used to measure the circumference has a state involving *only* momentum excitation, and the lowest, $w = 0$, winding state.

But in T_2, with reciprocal spatial radius, even the smallest excitation has tiny wavelength and huge momentum, while the energy of stretching a string around a dimension of radius 10^{-94}m is tiny (about a tenth of the electron mass). Thus in T_2 the (same) low energy photon in the experiment is described by winding modes, and the lowest, $k = 0$ momentum state. And of course the states are indeed dual: a T_1 momentum state maps to a T_2 winding state, the former with wave number, and the latter with winding number, n. Thus, according to T_2 the photon has the dual state represented by the same wavefunction $\psi(y, t)$ *but in winding space.* Then, because the Hamiltonian is the same, but with the roles of physical and winding space reversed, the photon evolves in exactly the same way—namely a 'journey' around *winding* space, taking 10^{12} years, as observed. No surprise there: in T_2 physical space is tiny, hence winding space with reciprocal radius is huge. (Of course, from the point of view of physical space, the evolving winding wavefunction describes some changing superposition of states in which the string is wound different numbers of times around the tiny radius.)

This analysis shows that because the experiment is characterized as timing a low energy particle (of a given type), by construction it involves a process in the larger of physical and winding space, and so is guaranteed to take a long time, guaranteed to produce the phenomena of a large radius space in either dual. Indeed, because they are dual, the process is guaranteed to take the same time and hence be an observation of the same large radius.

And the equivalence generalizes. Any process in *physical and winding* space according to T_1, corresponds to a process in T_2 in *winding and physical* space, and so no measurements or observations of even the most hypothetical kind will distinguish them. And so, the question comes up of which we should take to be correct; or indeed, whether the difference between tiny and huge is a true physical one at all. In the following discussion we will review some responses to this situation, and give reasons to favor one.

But first, we should briefly acknowledge that assuming that physical states supervene on expectation values is to take a strong stance on the interpretation of QM. For instance, that assumption is clearly false according to Bohm's theory: indeed, Nikolić (2007) has shown that Bohmian string theory breaks T-duality as a symmetry. While we take the Bohmian view very seriously, in this discussion we will explore the consequences of duality for interpretations in which there are no 'hidden' variables.

8.2 Two interpretive forks

Having explained T-duality in the previous section, we now turn to unpacking its significance for the nature of space in string theory (and by extension, the possible significance of other dualities). In this section we will lay out different ways in which one might understand a duality. We will consider two interpretational forks: first, whether, despite appearances, duals are physically equivalent, in fact representing the same physical states of affairs; or whether they are inequivalent, but equally compatible with any observation. Second, there is the question of what common physical state of affairs the two duals should be taken to represent; in particular, do their apparent disagreements merely reflect different linguistic conventions, or are the duals compatible because where they make incompatible assertions, they do not state any facts at all?[8] Subsequently we will turn to deciding which interpretational path to take at each fork: we argue in §8.3 for the physical equivalence of the duals, and in §8.4 that the duals describe strings living in a space of no determinate radius—the main conclusion of this chapter.

8.2.1 The first fork: physical equivalence?

Consider the following statement, summarizing the previous section:

T-duality provides a mapping (8.6) between a pair of theories that agree (under the mapping) on the expectation values of all observables in all states, and on the evolutions of all states.

---

[8] Note that Matsubara (2013), Read (2014), and Le Bihan and Read (2018) provide alternative complementary categorizations of the stances on duality, useful in the context of different background philosophical issues. Dieks et al. (2015) propose a distinction similar to our second taxonomic fork.

230 OUT OF NOWHERE

To lay out the interpretive choices, we need to be careful about several terms here.

First, note that there is a mismatch between the philosophers' and the physicists' use of 'theory' at this point. Roughly, physicists distinguish theories by the mathematical form of the laws (or action), while philosophers often further distinguish them by the values of their constants: in physicists' but not philosophers' terms, T-duals are the same theory. However, there is no substantive disagreement about the facts, and we have chosen to follow philosophers' sense of 'theory' here, because it will facilitate the following discussion to emphasize the differences between the duals. We, however, follow the physicists in describing T-duality, and other dualities in which the laws take the same form in the duals, as 'self-dualities'. (Later we will describe 'holographic duality', which is not a self-duality.)

It's also important to emphasize that 'observables' here does not have any narrow philosophical empiricist meaning: it denotes the collection of Hermitian operators (subject to any selection rules), *not* some 'special' collection of properties to which we have especially 'direct' access. Indeed, the observables are thus those operators normally thought of as representing the totality of physical, quantum mechanical quantities, including those far from immediate 'observation' by humans. And with respect to those quantities the theories are—under the mapping—in perfect agreement.

Finally, the qualifier 'under the mapping' is crucial. *Prima facie*, in one system a string has momentum n/R, and is wound w times around a dimension of radius R. In the other, it has momentum Rw/ℓ_s^2 and is wound n times around a dimension of radius $R' \equiv \ell_s^2/R$. And in the quantum mechanical treatment spatial and winding-spatial parts of the wavefunction are interchanged: $\Psi(x) \otimes \Phi(y) \to \Phi(x) \otimes \Psi(y)$ (which, in the case of simultaneous momentum and winding *eigen*states, entails $n \leftrightarrow w$). If the physical interpretation of the operators is held fixed, then the theories are inequivalent. So the crucial question will be that of their interpretation.

Normally one thinks of c-numbers such as R as physical, in which case duals again describe different physical situations. But normally, c-number parameters can be determined by the values of quantum quantities: the charge on the electron, say, by scattering probabilities. A duality arguably means they cannot be so determined by the values of the observables of the theory: the pattern of expectation values is preserved. So we should at least leave open that such differences in the c-numbers do not, after all, represent physical differences. In §8.3, invoking a simple duality, we will argue that indeed they do not. But for now we have our *first interpretive decision*: either the T-duals agree on the physical world or they do not. If they do, then for the purposes of this enquiry they say the same, *all in*; we will not be interested in any putative non-physical differences.

Most commentators have agreed that the T-duals should indeed be taken as giving the same physical description: especially, see Dawid (2007), Rickles (2011) and

DUALITY   231

(2013), Matsubara (2013), and Vistarini (2019, ch. 3). However, there have been recent skeptical or dissenting voices: especially Read and Møller-Nielsen (2020), Weatherall (2020), and Butterfield (2021). String theorists often seem to endorse physical equivalence of duals (for instance, Greene 1999, 247). However their words sometimes seem ambiguous on the point: Teh (2013) identifies remarks suggesting that one dual may be more fundamental than another.[9] We will argue below (§8.3) that physical equivalence is a reasonable conclusion, but we will also clarify how this should be understood, and modify the position taken in Huggett (2017), in the light of more recent work.

8.2.2 Three senses of 'space'

If—as we do—one takes the view that duals are physically equivalent, a *second interpretive decision* awaits.[10] To describe the options now facing us, it is first necessary to be more careful in distinguishing the different conceptions of space that have entered the discussion. So far we have already distinguished 'physical' and 'winding' spaces, but to proceed we need to revise this division. First, the former concept will be bifurcated. Second, it should be clear by now that winding space is every bit as 'physical' as 'physical space', so we will drop that terminology as misleading.

Instead we start from an analytical distinction between 'theory' and 'phenomena'. That is, when a new, more fundamental scientific theory explains an established, less fundamental theory, which has stood the test of experiment, then we can speak of the latter as the 'phenomena' *relative* to the former. Being relative, the distinction is suited for the historical process by which today's novel 'theory' becomes experimentally vindicated, and eventually becomes tomorrow's bedrock empirical given: Kepler's laws were phenomena for Newton's, but the laws Newton inferred from them were themselves phenomena for general relativity (GR).[11] In the present case, the more fundamental theory is string theory, which aims to explain, amongst other things, our current account of space. This account is expressed in our current best scientific theories, quantum field theory (QFT) and GR, in the small (high energy physics) and large (cosmology). All of these are the 'phenomena' relative to string theory. In Huggett (2017) the space that they describe was thus called 'phenomenal'; however, this term has suggested to some

---

[9] See also Read and Møller-Nielsen. We will discuss in §8.5.2 a sense in which relative fundamentality does play a role in holographic duality.

[10] We want to thank Dave Baker for emphasizing that there are distinct options here.

[11] It is not essential for the reader to accept this historical picture. It's helpful to accept the theory-phenomenon distinction that has its origins in Cartwright (1983, 1ff) and Bogen and Woodward (1988). What we take from them is the idea that 'phenomena' are abstracted from direct observation events, and so have a 'theoretical' structure themselves: we then add that a theory might therefore be phenomenal relative to a more fundamental theory.

232 OUT OF NOWHERE

the mental content of spatial experience rather than a physical structure, so in this book we have used the terms 'classical', or 'relativistic', or 'observed', or plain 'space' instead. Under this concept we also include the looser prescientific geometrical space we take ourselves to observe in the everyday, including the experience of three large dimensions. (Of course, QFT, GR, and everyday experience describe space in strictly incompatible ways: flat and curved, relativistic and not. But the relations between these descriptions, especially as limits of each other, are well-enough understood to make the notion of a single classical space, described by these phenomenal theories, clear enough for our purposes.)

In contrast, any 'space' that appears in the formulation of string theory is 'theoretical' in the sense described. There are in fact two theoretical spaces in (quantum) string theory: the first, of course, is winding space. The second is the space in which strings (and their momentum wavefunctions) live, which in the previous section we equated with classical space, under the concept of 'physical space'. But, according to the theory-phenomena distinction, classical space is a 'phenomenal' physical space, while strings live in a 'theoretical' physical space. The ultimate question of this section is of the relation between these two spaces, whether they are identical or whether one is reducible in some sense to the other. To clarify that investigation we thus adopt a new term—*target space*—for the latter, the space in which strings propagate. (There are two reasons not to use 'physical space' for this theoretical notion: first, as noted all three of the concepts of space we have discussed are 'physical' in a general sense; second, on one interpretation of T-duality, target space will turn out to have novel features, and the novel name will avoid the inapt connotations that would come with a pre-existing concept.) 'Target space' is a term of art in string theory, referring to the 'background' space in which a string is embedded: by the function $X(\tau, \sigma)$. The classical string is literally located in target space, and wound around it; we will take it that it makes sense to extend this intuitive picture to the quantum string, which thus also 'lives' in a background target space. As we have formulated the theory in this chapter, this situation is represented by the string's position/momentum wavefunction being a function over target space: just as we represent a quantum particle being 'in' a region by a non-zero wavefunction over that region. It is natural when first introduced to string theory, to think that target space is simply the same space we ordinarily experience, or at least space as conceived in contemporary physics: T-duality makes this identification problematic. Hence figure 8.2 should be modified, with 'space' replaced by 'target space'.

Given that GR and QFT (and our everyday understanding) are the context of classical space, measurements of its radius are operationalized in their terms: as in Brandenberger and Vafa's thought experiment, for example, which appeals to the photons and clocks of extant physics. Thus the radius of classical space, as defined, is given by c times the duration of the photon's journey. In terms of such measurements, classical space is observed to be very large: we don't know its radius (or even whether it is compact), but we can observe 10^{10} light years of it—and even a simple glance around the room shows that it is much larger than 10^{-33}m![12] Moreover, we have also seen how the dual theories will both predict that empirical result. While giving dual descriptions of the photon experiment—one in a target space of the same radius as classical space, and one in a target space with the reciprocal radius—they will agree on its duration, and hence on the observed radius of space. Clearly we cannot immediately infer that target space and classical space are one and the same; the remainder of the chapter explores this situation.

8.2.3 The second fork: factual or indeterminate geometry?

Earlier we made our first interpretive decision (to be justified in the next section): we decided that dual theories state the same physical facts. Now that we have clearly distinguished three concepts of space—classical, target, and winding—we are in a position to describe a *second* interpretive decision that follows from that decision, in the form of a dichotomy.

Interpretation One: Suppose that the radius of space has been measured, by the Brandenberger and Vafa experiment say, and found to be very large. Consider a theory, T, that sets R, the radius of the x-dimension, equal to this observed radius. One can then understand T in a naively realist way: take x to represent target space, take the string position/momentum wavefunction $\Psi(x, t)$ to represent a string living in target space, and identify target and classical space. That's a natural way to interpret the theory. But then how is one to understand the dual theory, T', which we are taking to state the very same physical facts as T? For instance, T and T' apparently assign different radii to target space, and (for $w \neq n$) apparently assert that the string is wound a different number of times around target space: aren't these physically different states of affairs? A solution is to take the duality mapping as specifying a *translation manual*. From (8.6), in the dual theory, let 'n' denote the winding number, not wave number, and 'w' denote the wave number, so that momentum and winding are unchanged! And while in T the x-dimension represents target space and the y-dimension winding space, in T' the roles are reversed, so the same wavefunctions pertain to each space as before; and we again identify target and classical spaces so that in T' it is y, not x, that represents classical space. Finally, as we saw earlier, within each theory the x- and y-dimensions have reciprocal radii, so in T' the radius of the y-dimension is R/ℓ_s^2: if we understand T' to involve a rescaling of length units by a (dimensionless) factor of $1/\ell_s^2$, then

---

[12] In fact we will count any additional, 'small' dimensions also as 'classical': though they may be required by certain theories of quantum gravity, their possibility is not at all quantum mechanical, as the original Kaluza-Klein theories demonstrate. Even though they are microscopic relative to the ordinary dimensions, they may still have a large radius on the relevant scale, $R > \ell_s$.

234 OUT OF NOWHERE

the duals even agree on the radius of target space.[13] In short, according to this interpretation, duals only appear to be incompatible because they are written in different languages, assigning different meanings to the same words: for example, they appear to assign different radii to target space, but only because they denote different things by 'target space'.

In the framework of first-order logic, in this understanding T-duals are related by a permutation of terms that induces a different formal interpretation with respect to a domain with a fixed structure, rather than any change in the domain referred to by those terms. That is, if predicate symbols 'P' and 'Q' have extensions A and B, respectively, in one interpretation, then in the other they have extensions B and A, respectively; and no changes of any other kind.[14]

However, such a permutation is trivially possible for any theory with more than one term (of the same syntactic kind), so we are left with the question of what distinguishes a (self-)duality from an arbitrary permutation. This question was posed in Huggett (2017) (and recently pressed in Weatherall 2020, §4). The answer suggested there was, roughly, that some theorems, including those whose terms have antecedent empirical significance, are preserved by the duality. For instance, in Brandenberger and Vafa's thought experiment both duals agree on the energy of the particle observed, and the duration of the trajectory: in general, as far as the experiment is described in operational terms, the duals agree on the facts.

The significance of the invariance of claims stated in terms with antecedent empirical meaning can be brought out by the following example (from Motl 2015). Consider that if one permutes the meanings of the terms 'inside' and 'outside', then in the resulting language the Earth is hollow! That is, the solid core is 'outside' the surface, and the sky and beyond 'inside'—there may well be aliens living 'inside' the Earth! But we antecedently define the 'outside' as our direction from the ground, or by the gravitational force, or as the direction of the fixed stars. And the consequence that the moon is outside *in that sense*—in the same direction as the fixed stars, say—is not preserved in the new language. Similarly, the antecedently meaningful, including operational, claims of string theory are common to both duals, so duality is not arbitrary permutation.

Such invariance is non-trivial but, we now think, does not adequately answer the question of why dualities are physically significant, for permutations of terms are just as trivial, even if some proper part of the vocabulary is held fixed. However, we do not find such triviality as telling against this interpretation of duality. Perhaps the correct understanding is that dualities are physically trivial, and

---

[13] The factor is dimensionless because the numerator has units of *length*^{2} (footnote 4). To see that this rescaling is trivial, note that we could have simply have worked in units in which $\ell_s = 1$, in which case no rescaling is necessary.

[14] For a duality that is not a self-duality, the reinterpretation would replace old terms with new ones, through a 'translation manual', rather than permute a fixed set of terms.

their significance purely formal. So our argument against the adequacy of this interpretation of T-duality will come below, in §8.4.

Interpretation Two: The second understanding of duality—which we argue should be adopted in §8.4—also takes the dual theories as asserting all the same things about the physical world, but now under *a common interpretation* of their terms. This choice may seem at first glance paradoxical, for how can two theories that are physically equivalent (as we are assuming) disagree under the same interpretation, and so be logically inequivalent? But of course the resolution rests on the familiar point that not all logical content is physical content: frames disagree on the coordinates and velocities of particles, but do not disagree physically; similarly different choices of gauge will assign numerically different but physically equivalent values to gauge fields. Thus we resolve the 'paradox' by saying that what either dual says about the physical world must be restricted to their 'shared content', in some sense: for instance, the mass spectrum of the string is common to both and hence a physical fact. Similarly, as we saw, the duals predict the same time for a photon to circle the universe: 10^{12} years, say. Since the radius of observed, classical space is thus a shared consequence of the duals, it is a determinate, physical fact.

But the theories do not agree on the radius of target space, nor, as we saw, on what string process corresponds to the photon measurement. Since in this interpretation the terms of the duals denote all the same things, these disagreements are logical incompatibilities between them. And then, because we are taking duals to agree on the physical facts, where the duals disagree, they do not state physical facts; just as coordinate or gauge differences are not physically factual. In particular, according to T target space has radius R, while according to T', the radius is ℓ_s^2/R. Thus according to Interpretation Two there is no physical fact of the matter which is correct, and with respect to these two values *the radius of target space is indeterminate*. Similarly, it is indeterminate whether the string is wound w or n times around the dimension. And so on.[15]

By taking this interpretational path, we believe that we are following the understanding of many string theorists. If we are correct then we note that it is quite misleading (and strictly wrong) to describe dualities as giving a 'translation manual' or 'dictionary' between theories (as one sometimes hears in informal explanations). That description applies to Interpretation One, in which the disagreements between the duals are merely verbal, the result of using different terminology to express the same propositions; a dictionary, after all, relates different terms that

---

[15] Matsubara (2013, §6) argued along similar lines (as we did in Huggett and Wüthrich 2013a), proposing (with misgivings, but without elaboration) that the shared commitments of the duals be thought of as 'structure'. However, his account did not fully recognize the role of derived, classical space in the logic of the situation described in Huggett (2017). The more recent discussion of Matsubara and Johansson (2018) clarifies this point, and is in agreement with the conclusion of this chapter that (under the assumption of physical equivalence) the spaces of T-dual theories are not strictly identical with classical space, but rather give rise to it.

236   OUT OF NOWHERE

mean the same thing in two languages. On the current, second interpretational path there is only one shared language, and the so-called 'dictionary' describes which quantities in one theory correspond to which in the other, but does not say that they are the same quantities under different names. Rather, as we have just seen, if a quantity in one theory is mapped to a different quantity in the dual theory, then we should expect its value to be indeterminate. Talk of a 'dictionary' is thus loaded, and for the current interpretative path at best confused, and at worst fallacious; the reader should therefore beware, as such talk is common in physics, even when (we believe) the current interpretation is intended.

Now, it's clear that the notion of the 'shared content' of duals is crucial for this path: we reconcile the physical equivalence of duals with their incompatibilities (in a common language) by taking their physical content to be restricted to their common part. But what is this? Huggett (2017) suggests that it be understood as any common entailments of both duals. But T entails that target space has radius R, and hence that it has a determinate radius; and T' entails that target space has radius ℓ_s^2/R, and hence *also* entails that it has a determinate radius. So that 'target space has a determinate radius' is a common entailment, which contradicts the interpretation of duality that we are currently pursuing! van Fraassen (1980, 46f) makes essentially the same point about absolute motion, in an argument against the 'syntactic view' of theories, as an adequate approach to the interpretation of physical theories. Drawing the same conclusion here, the natural move is to a more 'semantic' view of shared content. Such a position has been worked out in detail in de Haro (2020) and Butterfield (2021). Leaving out the details, a theory is understood as a triple of formal states, quantities, and dynamics; such a 'bare' abstract structure will generally be realized in a more concrete mathematical framework; then symmetries in general, and duals in particular, are cases of a common bare core theory, with different mathematical representations. We endorse this account, in general terms and specifically, although (for reasons that will become clear later) view it as an idealization in at least some of the cases of interest. (That is not a criticism: we generally view philosophical theories in this way.) We shall return to the bearing of this framework on physical equivalence, but to preview that discussion, the issue for a pair of duals will be whether their differences in representation have physical content, or whether only their shared core does. In the latter but not former case they will be physically equivalent.

As argued in Huggett (2017), on this interpretation of duality, because the radius of target space is indeterminate while that of classical space is determinate, it follows that *classical space is not identical with target space*. (Similarly, it is not winding space either.) Nothing can be both determinate and indeterminate with respect to some property at once. Similarly, it follows that we cannot think naively of strings as spatial objects, since there is no fact of the matter (even in a quantum mechanical sense) of how many times they wrap around a dimension. And so on.

As Brandenberger and Vafa conclude (393), 'the invariant notions of GR ... may not be invariant notions for string theory'.

If this position seems outré, that is only because of the surprising way in which it implements perfectly ordinary considerations. Consider Newtonian mechanics: we know that the predictions of the theory are the same whatever point we choose for the origin, whatever orientation we choose for the axes, and indeed whatever constant state of motion we choose for the frame. And so we think that there is no preferred 'center', that space is isotropic, and that spacetime does not distinguish a preferred state of rest. The fact that our coordinates do distinguish a point, break isotropy, and give a notion of rest is quite clearly an artifact of the representation: inertial coordinates make distinctions beyond those we wish to represent. The same understanding can apply to string theory: T-duality shows that a definite radius for target space and a definite state of winding are not physical, but only artifacts of the representation.

Classical space in this case is therefore derived, or 'emergent' from string theory, and in particular from the common core of its dual representations. Let us be very clear that classical space is not 'unreal' or 'unphysical' for that reason. There are well-known reasons to question the existence of space, but being derived rather than fundamental is not one of them (though the sense in which it approximates a more fundamental physics may bear on the debate). Space could be perfectly real and perfectly physical, though not fundamental. In general, for all we know there is no ultimate theory of everything, so that everything is derived from something more fundamental. If one insisted that only the fundamental was 'real' then for all we know nothing is real—an absurdity, since we know of many real things! *How* space is derived will be the subject of the next two chapters, but we have already seen how one of its properties—its observed radius—can be understood in string theory.

8.3 Interpretation: physical equivalence

So we have two interpretational forks. First, do the two theories describe the same physics, or not? And second, if they do, should we take them literally, with the string living in classical space, and avoid incompatibilities by interpreting their terms differently? Or do they have the same formal interpretation, in which case only their shared consequences are physical? In this section we will work through the first fork: in §8.3.1 we will argue for the physical equivalence of duals using a simple analogy, while in §8.3.2 we will defend and clarify our position in response to some recent objections. Then we will turn to the second fork in §8.4, and argue—given physical equivalence—that some quantities—such as the radius of target space—are indeed not physically determinate.

238 OUT OF NOWHERE

8.3.1 Analogy: the harmonic oscillator

If two theories are dual then *under the duality* the expectation values of all observables are preserved. We emphasize again that 'observable' here is used in its physical, quantum mechanical sense, not its philosophical, epistemic sense. That is, observables are the correlates of the system's Hermitian operators, generally understood to encompass *all* its dynamical physical quantities, and not merely a proper subset to which we are thought to have privileged experiential access. In other words, saying that 'dual observables preserve expectation values' does *not* signify that duals agree only on the values of physical quantities visible to unaided senses, but may differ on those that are not: 'observable' means dynamical physical quantity without qualification, and certainly with no implied epistemic privilege.

That understood, systems with dual descriptions need not be physically equivalent: as has often been emphasized in the literature, a formal duality alone is not sufficient for physical equivalence.[16] To explore this issue—and better understand the equivalence of expectation values—it is helpful to look at duality in a very familiar system, a simple harmonic oscillator, such as a mass moving horizontally and frictionlessly on a spring, described by the Hamiltonian

$$H = \frac{p^2}{2m} + \frac{kx^2}{2}, \tag{8.7}$$

where p and x are momentum and displacement respectively, and m and k are the mass and spring constants respectively. This oscillator is dual to another under the duality mapping[17]

$$(m, k) \rightarrow (1/k, 1/m)$$

$$(x, p) \rightarrow (p, -x). \tag{8.8}$$

(8.8) is in close analogy to (8.6); position and momentum are the analogs of winding and momentum, and mass and spring constant the analogs of the radius of space.

As for strings, the quantum harmonic oscillator Hamiltonian and canonical commutation relations are preserved by the duality (for the latter, $[x, p] = [p, -x]$). By the same logic then, the expectation values for all pairs of dual observables agree, so that if any series of values of the quantities represented by dual observables is consistent with either oscillator, then it is compatible with both. However, the dual theories can clearly be used to describe two distinct concrete, physical

---

[16] For instance, Matsubara (2013), Huggett (2017), and de Haro (2020) all use the following harmonic oscillator example to make the point.

[17] We will generally say that position is dual to momentum and vice versa, although the sign change means that this is not quite accurate. We will pay attention to the sign when it is significant.

DUALITY 239

oscillators in our world, one with mass m, and one with mass $1/k$ (unless $m = 1/k$). No one questions that these would be dual, *but physically distinct oscillators*. Nevertheless, in this subsection we argue (with many others) that one should draw the opposite conclusion in the parallel case of T-duals differing in the radius of space. We will explain why by further unpacking our example.

In particular, we need to consider carefully the *measurements* that might distinguish the two oscillators. As with any symmetry, we are interested in the question of whether indistinguishable systems are physically identical, so we have to understand clearly what can and cannot be distinguished. In the first place, given concrete oscillators we could simply dismantle them and place their bobs on a scale to determine the masses, and thereby distinguish them. But this is not helpful to our enquiry into T-duality, because there is no analogous experiment that could determine the radius of space, only Brandenberger and Vafa's equivocal experiment, and its ilk. If we thus don't have direct empirical access to constant, classical, c-number parameters like mass, radius (or spring constant), then the question is whether duals can be distinguished by measurements of their dynamical, quantum observables. (Recalling our discussion of 'observables', we bear in mind that we are assuming the measurability of all such quantities, not a just proper subset accessible to human senses.)

For instance, the quantum harmonic oscillator energy spectrum is $E_n = \hbar\sqrt{\frac{k}{m}}\left(n + \frac{1}{2}\right)$, so observations of the energy can determine the c-number k/m. But such measurements clearly cannot determine whether $\langle$mass, spring constant$\rangle = \langle m, k\rangle$ or $\langle 1/k, 1/m\rangle$, since they agree on the ratio of mass to spring constant. In general, measurements of observables that are invariant under a duality will (obviously) not distinguish the duals. And some observables will be invariant: at least the energy, since the Hamiltonian must be invariant to preserve the duality over time.

But not all. According to (8.8) x in one dual agrees with p in the other, not (in arbitrary states) x in the other; and similarly for p and $-x$. The duality (like T-duality) does *not* assert that both theories assign the same values to the same mathematical objects (operators and their expectation values), but rather that they instantiate the same 'pattern' of values. For instance, imagine a table of pairs of measured values at a series of times; if they agree with one oscillator's x and p expectation values at those times, then they equally agree with the expectation values of p and $-x$, respectively, for the dual oscillator: at any time, the position of one is numerically equal to the momentum of the other, and the momentum of the first to minus the position of the other. The situation is exactly the same as for a string, in which the values of winding and momentum are exchanged by T-duality, as we explained and as figure 8.2 illustrates. And similarly for other dualities.

240 OUT OF NOWHERE

We agree with Weatherall (2020) that dualities are thus formally distinct from instances of empirical equivalence as usually characterized.[18] As in (8.6) and (8.8), a duality is an invariance under a *mapping* between observables; a permutation of observables if the theories are self-dual (as in T-duality), or a correspondence between distinct sets of observables if not (as in the 'holographic' duality discussed in §8.5.2). It is not simply an invariance of some quantities when others are transformed or 'translated' to new values (though some observables will be preserved by a duality). However, Weatherall's approach does not adequately recognize how the issue of empirical equivalence turns on the question of which quantities have independent physical significance, and which obtain their significance through the dual theories themselves.

So suppose again that one is given the table of oscillator observation pairs: and suppose they agree with the expectation values of x and p for some oscillator, hence for p and $-x$ for its dual. Applying the standard convention that x represents position and p momentum to *both* duals (so adopting our Interpretation Two), knowledge of which column describes position measurements and which momentum allows one to distinguish the dual oscillators. Once again, if $\langle \text{position, momentum} \rangle = \langle i, j \rangle$ for one dual, then $\langle \text{position, momentum} \rangle = \langle -j, i \rangle$ for the other, and these are generally unequal, since the oscillators move differently. This seems to be the normal case, in which of course the dual oscillators are empirically distinguishable by position and momentum measurements.

But the same pair of columns with *no* indication of which is position and which momentum do *not* distinguish the duals: $\langle i, j \rangle$ might represent $\langle \text{position, momentum} \rangle$ or $\langle \text{momentum, } -\text{position} \rangle$. In the actual world, of course there would still be a fact about which column was *really* the result of position measurements, and which momentum. The situation could only occur if, say, a careless lab assistant neglected to label the columns when recording the data; though then the inability to discern duals would only be epistemic. But what if the very meanings of 'position' and 'momentum' were called into question, raising the question of what exactly the two columns refer to?

This cannot happen in the normal case, because position and momentum are well-defined by a theoretical and experimental framework independent of the harmonic oscillator: especially, their general theoretical understanding in classical and quantum mechanics, the models of very many other physical systems in which they are dynamical quantities, and the numerous techniques for their measurement. For instance, dual oscillators can be distinguished by weighing their masses. Or by coupling them to some external system that does not respect the duality, but depends directly on the position: reflecting light off the bob, say. Or by 'looking inside' the oscillator in some other way.

---

[18] We don't agree with his suggestion that the duality literature misses this point (see footnote 16), though his critique has prompted us to explain it more carefully. He uses the simple example of source-free electromagnetic duality in much the way we and others have used the oscillator.

Calling into question the distinction between the dual quantities means ignoring all of that, something that can only be done by supposing a world in which none of that framework surrounds the oscillator, so a situation in which 'position' and 'momentum' are not 'externally' meaningful: in other words, a world in which there is nothing but a single oscillator. Then there is no weighing or shining a light on the bob, since there are no scales or light; the difference between oscillator position and momentum no longer makes a difference to other systems, because there are none. No physical operations 'look inside' the oscillator.

Equation (8.7) can be taken to describe such a world. One then reflexively imports the usual interpretations of 'x' and 'p' as position and momentum in the senses given by the actual world framework. But then one discovers (8.8), and that the pattern of values for these quantities would be the same under the opposite interpretations of 'x' and 'p'. And so the question is which of these two identifications of the quantities x and p is the correct one in the single oscillator world in which (8.7) is the *complete* physical theory? Is the property of the lone oscillator denoted 'x' position or momentum in our familiar sense? We argue that under the given circumstances there is no fact of the matter: nothing internal to the oscillator world need determine how it instantiates the qualities of our world.[19] (Clearly our free choice of the symbol 'x' to label position in our world carries no such ontological weight.) But that is to say, in theory terms, that the duals describe the same world; that they are physically equivalent, that there is just no 'inside' to see.

Further, we argue, the case is just as in string theory, taken as a theory of *everything*. Consider a world in which there is only a string in a spacetime with a closed dimension: now p (or n) and w are the duals (8.6), in analogy to x and p for the oscillator. We know that dual systems agree on the energy, and through the analysis of Brandenberger and Vafa, radius measurements. Such measurements would in principle allow one to determine the values of *all* observables, including momentum and winding. The issue is not the unobservability of momentum or winding, but rather that of determining *which measured quantity is position, and which is winding*. That is, the duality preserves the pattern of observables: some are invariant, and others permuted—if $\langle n, w \rangle = \langle i, j \rangle$ in one dual then $\langle n, w \rangle = \langle j, i \rangle$ in the other. The pattern itself will not distinguish the duals, and there is no broader theoretical framework in the lone string world to settle which value is n and which w. By stipulation, there is no broader external theory in which strings with T-dual assignments of n and w are not dual, allowing the duals to be distinguished. As with the oscillator, there is no physical way to 'look inside' the system to see which dual it is. It seems that we should draw the parallel conclusion: no preferred identification of the quantities with those of the actual world, and no physical difference

---

[19] If you like, we advocate a kind of second-order antihaecceitism for qualities: dual worlds can't differ simply in how dual qualities are instantiated. We believe that such a view is compatible with Lewis' nominalism, including some version of his natural properties view, for instance.

242   OUT OF NOWHERE

between the duals. The duals describe the same world, and there is no 'inside' to see.

But what of the 'low energy' limit of the theory in which something like quantum particle physics is found? In that framework, particle momentum is well defined, and observable. But the lesson taught by Brandenberger and Vafa is that the state of a particle, and the measurement of that state, always have dual analyses: the state of a stringy particle can be understood equally well in terms of the target space state of a string, or the dual winding space state; and any measurement of particle momentum in terms of dual target and winding space processes. The measurement of the radius of space is just one example of something general: low energy physics cannot break a duality, since any 'reduction' has a dual, hooking up a single low energy structure to either of two dual high energy structures.

We also want to head off the line of thought that we can just see—immediately experience—that the radius of space is large, and that things would seem different if it were not. Brandenberger and Vafa's argument applies here. Given that our visual experiences supervene on the physical, whatever physical process that underwrites our experience of a large dimension is realized in both duals: in one as a process involving momentum modes, say, and in the other involving winding modes. We have been arguing that we should take these to be different representations of just one process, but even on the view that counts them as distinct physical possibilities, a fairly mild assumption will guarantee the indistinguishability of the duals even in direct experience. For the two processes will only be experientially distinct if visual experiences depend on the processes grounding them involving spatial (not winding modes): that T-dual brains are not identical minds. It is, in other words, to privilege the spatial in the physical theory of mind. But we see no particular motivation for such a view: rejecting it means that dual brains have the same experiences, so that things would not appear any different at all if target space had the reciprocal radius. Hence we cannot just 'see' which of the two possibilities holds, and considerations of direct experience provide no reason to think that there are two physical possibilities at all.

We therefore conclude, in parallel with the lone oscillator, that in the lone string world there is no fact of which of p and w is momentum and which is winding, and that the duals are thus physically equivalent. Moreover, if string theory is understood as a theory of everything, then whether there is one string or many makes no difference to the argument, and so we also conclude that T-duals of full string theory are physically equivalent. Our conclusion is not a logical necessity, nor do we think there are compulsory semantic or ontological principles that can force the conclusion that dual theories of everything describe the same physical possibility. But the case of dual total theories is clearly one in which the putative differences are 'hidden' in a very strong sense—a unique mass is impossible to determine from the physical quantum quantities of the harmonic oscillator, just as a rest frame is from relativistic quantities in special relativity. And when there are quantities that

DUALITY 243

do not supervene on any of the other physical quantities, and when there is no reason to think that different values for them can be determined directly, then at least from a practical, scientific point of view, it makes sense to treat those differences as non-physical (until some new, well-supported theory shows how they are connected to physical quantities). In other words, long established, well-motivated scientific reasoning should lead us to think that dual total theories represent the same physical situation.

Of course one now wonders what physical equivalence under duality really amounts to. What exactly follows? Well, we have already described two interpretations of the claim in the previous section; in §8.4 we will turn to the question of which we favor and why. Before that we turn to some recent philosophical reflections on the claim that duals are physically equivalent, which alternatively give a more rigorous formulation of the idea, and oppose our conclusion of physical equivalence for duals.

8.3.2 On the concept of physical equivalence?

Our line of thought was presented in Huggett (2017) (Dieks et al. 2015 argue similarly), but since then other authors have clarified or questioned the conclusion of physical equivalence.[20] First, de Haro (2020) (building on a series of earlier papers cited there, including that with Dieks et al. just cited, and de Haro and Butterfield 2019) develops this idea more formally and thoroughly. Summarizing, in de Haro's terms, dual oscillators are distinct because in our world, their common core can (in a precise sense) be 'extended' to—embedded in—a larger theory that gives 'external' meaning to their terms: mass, spring constant, momentum, position. But in a world in which the common core instead describes *everything*, then the duals are nothing but different tools for computing the dynamics, with their differences (in m and k, and x and p) as nothing but empty conventions used to turn the mathematical handle. In that case there would be no larger theory of the world (without uninterpreted surplus structure) in which the core could be embedded; it is 'unextendable' to a more comprehensive theory, and hence cannot receive an external interpretation. Instead it can only have an 'internal' interpretation: possible states are fully distinguished by the value-pairs, have the same allowed histories, and are interpreted as the values of the only two physical quantities of the world. As we noted above, we endorse this picture, but we think it idealizes the situation: as we

---

[20] Philosophers have focused on the issue of physical equivalence, and the implications of the common core for emergence, but Dawid (2017) emphasizes that string theorists have found duals a useful tool to gain different perspectives on the underlying string theory. Indeed, he argues that this is their main significance; we agree with the importance of the use he describes, but claim that quotienting is the important implication for understanding the topic of this book—spacetime emergence.

244 OUT OF NOWHERE

will discuss shortly, we need not have an explicit formulation of the common core in order to know something of the shared physical content of duals.

Given de Haro's framework, the question of physical equivalence has two parts. First, could we ever reasonably believe that the common core of a pair of duals was not extendable? That it captured all the physical structure of the world in its domain, so that it was not just part of a broader (perhaps more fundamental) theory? Generally, unextendability will not be a purely formal property of a theory, but will depend on its intended application: real world oscillators are extendable, but one can arguably stipulate a world in which they are not. So a typical way to frame the question will be regarding its intended application to the actual world. Then one may want to apply methodological principles such as ontological simplicity to move from duality to unextendability, and thence to an internal interpretation. For instance, if the world constantly manifests Lorentz symmetry, why postulate some unknown physics that picks out a rest frame? Alternatively, de Haro suggests that physical principles of a theory will be used to answer the question: perhaps in this case the Lorentz symmetry of the theories. But in a sense this approach will also rest on methodological principles, for how else are we to decide the physical principles themselves?

Second, suppose that the world were such that the common core of a pair of duals indeed has no external interpretation: does it follow that the duals are physically equivalent? Perhaps instead they could describe a pair of worlds in which different physical quantities are instantiated in isomorphic patterns. This question will in part depend on considerations from philosophy of language: does an interpretation of the core provide a relation between a single world and both duals, so they refer to a single domain? De Haro shows that fairly mild assumptions about reference justify such a conclusion. However, one might ask whether it is possible for there to be two interpretations of the core, each relating the duals to different worlds. In such a case though, any one interpretation of the common core will serve as an interpretation of the duals, since only the core has physical significance. And so, de Haro points out, any one interpretation will relate all duals to the same world. Hence, even if one interpretation is 'better suited' to the mathematical representation of one dual than another, this will only be a pragmatic matter, not a semantic one that could produce inequivalence.

Positive answers to both questions (Unextendable? Unequivocal internal interpretation?) for a pair of duals means that they are physically equivalent, with their content exhausted by a single internal interpretation of their core. It should be clear that such a conclusion, therefore, does not follow simply from some formal property of theories, but is a matter of interpretation and philosophical theory, something emphasized by Butterfield (2021).

However, some recent commentators have raised substantive issues regarding physical equivalence, to which we would like to respond. Both Read and Møller-Nielsen (2020), and Butterfield (2021) discuss significant cases resembling

T-duality, in which a profound symmetry relates two theories, arguing that an inference of physical equivalence is *not* thereby justified. For on more careful consideration, there is an important difference between Lorentz invariance and T-duality, namely the *knowledge* of an underlying formal structure which unifies different frames: a common core that makes explicit the unphysical surplus structure introduced when a frame is chosen. (Unphysical in the sense that a convention is involved, even though that convention will have to refer to physical objects to pick out an origin, orientation, and so on.) In the case of T-duality there are just the duals, and no known explicit common core; that would be 'M-theory', an exact completion of string theory (something that will be a recurring topic in the remainder of this part of the book). Does this make a difference to claims of equivalence? Read and Møller-Nielsen, and Butterfield think so. Consider the related example of (full) Newtonian spacetime versus Galilean (or 'neo-Newtonian') spacetime (e.g., Earman 1989, ch. 2). Suppose one knew Galilean symmetric physics but only of Newtonian spacetime. Would one be justified—just from Galilean symmetry—in inferring the physical equivalence of two 'theories' that differed only in the standard of absolute rest? Or should one believe these to be inequivalent states of affairs? Or be agnostic? *Once* one discovers Galilean spacetime it is reasonable to take that to properly capture the geometry of spacetime, but until then? After all, that is the analog of the situation with string dualities.

It seems that everyone is agreed that things are not clear cut. But Butterfield says that inequivalence is a reasonable option (§1.2); and Read and Møller-Nielsen say that inferring equivalence is not justified (§3.3). While we think that although it is most reasonable to infer equivalence for T-duals, we acknowledge that the point is open to debate, and others could judge differently. We do not see the availability of an explicit formulation of the common core as particularly important for inferring equivalence; we believe it to be cogent to assert that the physical content of duals is 'that which they say in common', and argue that making explicit this common content is not necessary in order to accept the assertion.[21]

In particular, our intuitions about the right thing to say about the Newtonian spacetime example run in the opposite direction to those of Butterfield, Møller-Nielsen, and Read. They think that one would best assume a standard of rest until Galilean spacetime is discovered; we think that one should conclude that apparently different ascriptions of rest are in fact equivalent. (Which is not to say that we fault Newton in historical context for accepting both absolute space and Galilean relativity; we make our judgment from a historical perspective that has

---

[21] There is a recent literature debating whether Ramsifying—here that 'there are xs such that they satisfy the common commitments of the duals'—is reasonable. For instance, Dorr (2010) argues against this move in a number of instances, while Sider (2020, ch. 5) defends it (in some cases). We find that there are rather strong intuitions on both sides; it is a topic we are happy to see being explored more carefully.

246   OUT OF NOWHERE

accumulated a great deal of additional understanding of logic, semantics, mathematics, nature, and science.) And while agnosticism about equivalence is the more epistemically *cautious* course, we don't believe it to be more epistemically *virtuous* for that. What we think is that global theories—such as Newtonian gravity—do have a good track record for turning out to be unextendable, and that in particular string theory is promising as a complete unified theory in its domain, and so is reasonably thought to be unextendable. And from that we do think physical equivalence is the reasonable conclusion.

Of course, we agree that giving an explicit formulation of a common core is a significant goal even when one has accepted physical equivalence; it would be an explicit formulation of the physical content of the duals, and so crucial to fully understanding them. Indeed, the expected utility of finding such an explicit core will (all things being equal) be greater the more likely one thinks duals are equivalent; the more likely that is, the more likely it is that an explicit core exists. So, we don't find attractive Read and Møller-Nielsen's (§ 3.3) 'motivationalist' position: that one is most rational only to accept equivalence once an interpreted common core is known, and that one should seek it. As we just explained, we find sufficient reasons to accept the equivalence of T-duals, and that acceptance is (additional) motivation to seek the common core. Indeed, our discussion of Brandenberger and Vafa, and our investigations in later chapters contribute to such a project. As does (with impressive results) the work of de Haro and his collaborators (in addition to work already cited, see de Haro et al. 2016a, 2016b, and de Haro 2017).

All that said, it may be surprising that we have some skepticism that the common core of string duals can be formulated in a closed, complete formalism. The reason is that string theory as currently formulated is an essentially perturbative theory: as we shall see in detail in the following chapters, one postulates a classical limit of some as yet unknown theory, including a classical spacetime, and then studies quantum perturbations around it. As such, it is not clear that there will be some way of completely describing a shared core structure better than 'that which the duals have in common'; the duals themselves are inherently limited as descriptions of the world. In this case, de Haro's framework is just an idealization, as we suggested above. This situation is compatible with finding out specific aspects of the common core, as we have just described, but incompatible with stating a closed, complete interpretation. Instead, one hopes that the content of the string duals will be found within a theory to which they are the perturbative approximations, namely 'M-theory'. (In addition to Read and Møller-Nielsen, and Butterfield, this point is stressed as a motivation for studying duality by Le Bihan and Read 2018.) More than likely though, M-theory contains both more and less content than either of the duals, and so is not the same as simply quotienting them.

One might then ask what the value is of studying the duals, but this is a question we have addressed a number of times already: we are seeking to understand the fragments of existing quantum theories for ways in which existing concepts of space and time might be modified in a successful theory. So our attitude toward the investigation of duals has the same spirit. (However, we should acknowledge that in chapter 10 the perturbative nature of the duals will limit our investigation.)

8.4 Interpretation: indeterminate geometry

Given the arguments of the previous section, we now proceed on the understanding that T-dual theories describe the same physical situation. The question now is *what* situation that is, in particular with respect to the geometry of space. Above we described two possibilities: according to Interpretation One, it could be that the duals agree that the radius of target space is greater than ℓ_s, and the apparent inconsistency is resolved by understanding duality as a permutation of terms, a relabeling. Or, according to Interpretation Two, it could be that the duals should receive the same formal interpretation, so that only their common pronouncements describe what is physical: for instance, a unique radius to phenomenal, but not target, space. In this section we will make a couple of brief comments on the two possibilities, and then explain why we favor the second.

Talking of 'relabeling' the terms of a theory may suggest that the difference is between 'passive' and 'active' interpretations of duality (see footnote 6 in chapter 5). But that suggestion clearly isn't correct: T-duality cannot be seen as a passive transformation in the sense that the duals are descriptions of a single situation from two points of view, for the duality does not map 'observers' or concrete 'reference frames' into distinct but symmetrical observers and frames. Rather, the two interpretations that we have described are much closer to the interpretive options that arise in the case of a gauge symmetry. On the one hand, maybe there is 'one true gauge' Healey (2001): in the present context, classical space is identified with target space, and has a definite radius $R > \ell_s$. On the other, maybe apparent differences in choice of gauge are nothing but differences in 'surplus representational structure' (Redhead 1975): target space is distinguished from space, and the difference between target spaces of radii R and ℓ_s^2/R is merely a difference in representational fluff. We won't pursue this parallel to gauge symmetry in field theory at length, but a couple of points are worth making. First, duality is neither a local nor a continuous symmetry of the kind found in field theory, so much of the philosophical discussion of those theories is inapplicable. Second, that said, at $R = \ell_s$ there is a continuous SU(2) × SU(2) gauge symmetry of which T-duality is a part (e.g., Polchinski 1998, 247f). Thus, in this sense at least, T-duality is formally, and not just conceptually, a gauge symmetry.[22]

---

[22] See Healey (2007) and the responses to it for continuous gauge symmetries in general. The SU(2) × SU(2) symmetry entails that an infinitesimal increase of the radius from $R = \ell_s$ is the same

248   OUT OF NOWHERE

So, why do we advocate Interpretation Two, that R is indeterminate? After all, the definite radius view presented above is intuitive, in that it says that strings live in a space with an observed radius $R > \ell_s$; whether that space is called target or winding space. However, there is a *distinct*, indistinguishable definite radius view according to which strings live in a space whose radius is ℓ_s^2/R; whether that space is labeled target or winding space! Generally, if there is one true gauge, then there are as many distinct possibilities for it as choices of gauge: in this case two, depending on the radius of the space in which the strings literally live, move, and wind. According to one choice, the space of experience is the one in which strings live, while according to the other the space of experience is much bigger than the one in which they live: from Brandenberger and Vafa we understand that the same appearances arise from a string's momentum state in one dual, and from its winding state in the other. The bottom line is that understanding T-duality as a mere permutation of terms leaves open what underlying facts are equally described by the duals, because such an understanding is compatible with different true gauges. Hence Interpretation One does not really address the issue it was supposed to resolve: dual theories are physically equivalent on this interpretation, but there is a second pair of duals that differs physically from the first, but *only* with respect to an unobservable radius. If one is satisfied with that situation, then why was one not satisfied with physically inequivalent duals?

Moreover, these considerations point to an analogy to related cases in which we usually do accept that there is no fact of some matter (we alluded to a similar example earlier). For instance, one could claim that there is a preferred rest frame in spacetime, even though it has no physical significance in special relativity. One could even claim that it is some frame which can be picked out physically and phenomenally: for example, perhaps the fixed stars (idealized as an inertial frame) are at rest. These proposals will strike many readers as completely unmotivated. But replace 'frame' with 'radius', and the fixed stars with the observed radius, and the parallel is perfect. Looked at this way, the definite radius view appears as a reactionary attempt to preserve aspects of an old theory when it is superseded, and understood as merely effective.

However, since Huggett (2017) we have recognized a way of defending a version of Interpretation One that avoids these objections.[23] Formally, the common formal core of T-duals (in our toy model) is a pair of spaces, one big and one small. There is nothing indeterminate about the radii of the bigger and of the smaller (R and $1/R$, respectively, in $\ell_s = 1$ units); and low energy phenomena are understood in terms of states in the bigger, because it allows longer wavelength, lower energy, wavefunctions. (In figure 8.2, delete 'space' and 'winding space', and the

---

as an infinitesimal decrease. Read (2014) makes a related comparison, but between string dualities and diffeomorphism symmetry rather than conventional gauge symmetries.

[23] We are especially grateful to Neil Dewar for a useful discussion of the following.

only difference between the upper and lower figures is the trivial coordinate relabeling $x \leftrightarrow y$.) One could interpret the duality as showing that there is no more physical content to the theory than this. Especially, as showing that the *distinctions between* the terms 'target space' and 'winding space', and *between* the related 'momentum' and 'winding' of the string, are *without physical content*. For as soon as these have independent meaning, we can distinguish the two duals; even if we recognize them as different descriptions of the same state of affairs. In short, this interpretation means that talking of a 'string' in any classical spatial sense at all evaporates, because all one has is some quantum object, described by a product of wavefunctions, one in each space.

One is of course then free to adopt the *convention* that 'target space' simply means 'big space', and 'winding space' simply means 'small space', and that 'momentum' and 'winding' refer to wavefunctions in big and small space, respectively. That is to strip the terms of any of the physical content with which they were introduced at the start of this chapter. So equally, one could have adopted the opposite convention, and declared 'winding space' synonymous with 'big space'. Either way, these are just different terms one might choose for the same concept. Then one can preserve Interpretation One, understanding the permutation of terms not as a choice of 'true gauge', but as a mere linguistic convention. (Similarly, saying whether space is target space or not simply reports the convention.) But adopting either convention adds nothing to the theory, since 'big' and 'small' would do just as well; their only significance is heuristic, as reminders of a certain way to derive predictions.

In effect, this 'two space' interpretation extends the familiar idea (investigated in Earman 1989, ch. 3) that the geometry of a spacetime should have the same symmetries as the physical laws: in particular, if a spacetime has additional symmetry-breaking structure, it is preferable to find a spacetime which 'quotients' it away. The proposed new formulation of string theory quotients away the distinction between target and winding space, leaving only 'big' versus 'small' as physically meaningful, so that the symmetries of the theoretical representation match those of the dynamical physical quantities. It agrees that space is not target space, if 'target space' is supposed to denote something more than 'big space'. But it does not admit three distinct spaces, since space is not emergent, but identified with the larger of the two string spaces: the one to which we refer spatial phenomena.

While we endorse Earman's prescription in general, and so find this interpretation appealing for the model we have discussed, we do not think that it is ultimately the correct account of string dualities, but merely an artifact of the particular example. The 'double field' approximation of our toy model is based on the assumption of two wavefunctions, whose two spaces play indistinguishable roles, allowing us to quotient over them. But more realistic dualities, such those introduced in §8.5, do not share this feature: dimensionality or other global

250   OUT OF NOWHERE

topology changes, or weak couplings are exchanged with strong ones, or even the elementary with the composite, and there is no known common structure capturing the content of the theory. But without such a common core, with a structure identifiable as space, a quotienting approach breaks down.[24]

Even in the case of T-duality in a realistic perturbative string theory (as opposed to our toy system), formally a string will live in three large spatial dimensions plus six or 22 microscopic dimensions (plus time); and from the worldsheet perspective, momentum and winding will correspond to distinct sets of quantum field excitations. The microscopic dimensions are beyond normal observation, but are still 'big' on the string scale ($r \gg \ell_S$), so larger than winding space; however, small enough that at energies low on the string scale, both momentum and winding will contribute to string processes. The model we have worked with collapses all this structure into a pair of wavefunctions, but once it is put back, it is not clear even in this perturbative case how the quotienting strategy can be applied. We will return to this question briefly in §8.5.1, when we will see that open string duality makes this simple quotienting even less satisfactory.

In other words, while we agree that the proposed two space interpretation is reasonable for a system in which our double field model is the exact, complete description (as we treated it in comparison with the harmonic oscillator), when it comes to interpreting string theory we have to bear in mind that the model is an approximation. The proposed quotient depends on that approximation and, we argue, does not apply to string theory more fully; hence the two-space interpretation is not a viable account of string theory. For string theory—the object of this investigation—then, we endorse Interpretation Two instead.

To summarize our conclusions then: in §8.2 we offered two interpretational forks, and in the following two sections argued for taking one path in each case. First, in §8.3 we argued that duals indeed give physically equivalent descriptions of the world. Second, in §8.4 we argued that duals do not merely make the same statements but in different languages; rather, they make incompatible statements in a common language. How are these two claims about duals to be reconciled? Well, to avoid making inconsistent physical assertions, and thus being physically inequivalent, it must be that not every claim a dual makes is part of its physical content; indeed the physical content must be restricted to the shared content of the theories. But it is not part of the shared content that target space has radius R, nor is it part of the shared content that it has radius $1/R$, hence target space (like winding space) has a radius indeterminate between R and $1/R$. But then of course it cannot be identified with classical space, which has a determinate radius of R—in other words, space is emergent.

---

[24] We want to thank Keizo Matsubara for emphasizing especially this point to NH in their many discussions about duality.

8.5 Beyond closed string T-duality

This chapter focuses on the technically simplest example of string duality, but we should briefly survey the other important cases. The philosophical issues that arise are—we claim—the same as for closed string T-duality, though because questions of equivalence are not separable from questions of interpretation, the conclusions need not be the same. However, we will conjecture—especially in chapter 10—that the conclusions are indeed the same, so that the dualities represent further indeterminacies for spacetime, and strengthen the case for emergence. Of course, by leaving a detailed analysis for future work we also leave conclusions based on the conjecturevulnerable; but here we will focus on simply giving the formal ideas.

Below we will discuss T-duality for open strings (§8.5.1), which must be treated differently to closed strings, since they cannot enclose a dimension; and the 'holographic' duality between strings in a gravitational field and a gauge field on the boundary of spacetime (§8.5.2). Brief as those treatments are, two other dualities will only be mentioned in passing (introductions and investigations can be found in Rickles 2011).

First there is 'mirror symmetry', which states the equivalence between certain target spaces of different topologies. More specifically, this equivalence holds for supersymmetric strings, which we saw live in a 10-dimensional target space. If it takes the form of 4-dimensional Minkowski space, plus 6-dimensional closed 'compact' dimensions (as one might model our universe), then mirror symmetry holds between specific pairs of compact dimensions of different topologies. Insofar as these are taken to literally represent space then according to our preceding analysis there is no fact of the matter of the 'shape' of space. (Though one could argue instead, with Matsubara and Johansson 2018, that that such indeterminacy sometimes indicates that the compact dimensions are not spatial at all, but rather represent internal degrees of freedom.) While our conclusion that space has no definite classical topology is even more conceptually dramatic than our earlier one that spaces of definite topology have no determinate radius, mirror symmetry is in fact a mathematical generalization of T-duality (Strominger et al. 1996). Mirror symmetry is especially interesting because its discovery sparked much of the initial excitement about dualities when it was used to simplify and solve some difficult mathematical problems.

Second, there is 'S-duality', which also has applications in mathematics. This relates different 'types' of superstring theory, characterized by their different boundary conditions. In this regard S-duality is analogous to open string T-duality, to which we will shortly turn. However, it is also characterized by exchanging strong and weak values for the string coupling strength: for instance, type I string theory with a strong coupling is dual to SO(32) type with weak coupling (while type IIB has a self-duality between strong and weak couplings). S-duality has been less discussed by philosophers, but is thought significant for providing

252 OUT OF NOWHERE

an important clue that the different supersymmetric string types are different perturbations of a single underlying M-theory.[25]

8.5.1 T-Duality for open strings

On the face of things, it is hard to see how T-duality could be extended to open strings. The winding number of a closed string is conserved classically or in the absence of string interactions because the topology of target space prevents the string from being contracted to a smaller winding number. But an open string can (topologically) always be contracted to a point, in any space, so does not seem to have a winding number. But in that case, one cannot exchange momentum and winding, and (8.6) does not seem to apply at all. (Or if you prefer, w is always zero, so T-duality must fail for non-zero momentum.) However, T-duality does apply to open strings, in a surprising and very important way, which illuminates the physical content of the theory. We cannot see this at the level of the toy double field approximation we have used so far, but need to draw on the theory developed in chapter 7.

First we will work out how to implement T-duality for closed strings in those terms, and then we apply the transformations to the open string.[26] We start again with the equation for a string wound around a closed dimension (8.1), ignoring the vibrational part for simplicity for now; what follows can readily be seen to hold for that part too.

$$X(\tau, \sigma) = 2\ell_s^2 p\tau + 2Rw\sigma = 2\ell_s^2 n\tau/R + 2Rw\sigma, \tag{8.9}$$

using $p = k = n/R$ (since $\hbar = 1$). Next we expand this into left and right moving pieces:

$$X = (\tau + \sigma) \cdot \left(\frac{\ell_s^2 n}{R} + Rw\right) + (\tau - \sigma) \cdot \left(\frac{\ell_s^2 n}{R} - Rw\right) \equiv X_L(\tau, \sigma) + X_R(\tau, \sigma). \tag{8.10}$$

We observe now that if we take the rescaled reciprocal radius, $R \to \ell_S^2/R$, then:

$$X_L \to (\tau + \sigma) \cdot \left(\frac{\ell_s^2 w}{R} + Rn\right)$$

$$X_R \to -(\tau - \sigma) \cdot \left(\frac{\ell_s^2 w}{R} - Rn\right), \tag{8.11}$$

---

[25] Other interesting dualities, and some profound insights into the nature of duality in general can be found in Polchinski (2017). Especially, he discusses elementary-composite duality, also investigated by McKenzie (2017) and Castellani (2017).

[26] The following treatment follows Polchinski (1998, §8.3, 8.6), Becker et al. (2006, §6.1), and Zwiebach (2004, chs. 17–18).

so that if one also interchanges $w \leftrightarrow n$, then $X_L \to X_L$ and $X_R \to -X_R$. Then summing the transformed left and right moving parts, we find:

$$X \to 2Rw\tau + 2\ell_s^2 n\sigma/R. \tag{8.12}$$

Comparing this expression with (8.9), paying attention to the timelike and space-like coordinates, we see that it describes a closed string with wavenumber w and winding number n, but in a target space of radius ℓ_s^2/R. That is of course what we expected from our previous quantum analysis, (8.6). In other words, in this formalism we implement closed string T-duality with the transformation:

$$X_L \to X_L \qquad \text{and} \qquad X_R \to -X_R. \tag{8.13}$$

We postulate that the transformation remains unchanged for the open string.

So we start with the general equation of motion for the open string (7.9) (recalling footnote 2):

$$X = 2\ell_s^2 n\tau/R + i\sqrt{2}\ell_s \sum_{j\neq 0} \frac{1}{j} \alpha_j e^{-ij\tau} \cos 2j\sigma \tag{8.14}$$

For the open string it is illuminating to leave in the part of the solution describing vibrations, as we shall see. Break this into left and right moving parts, $X = X_L + X_R$,

$$X_L = \frac{\ell_s^2 n(\tau + \sigma)}{R} + \frac{i\ell_s}{\sqrt{2}} \sum_{j\neq 0} \frac{1}{j} \alpha_j e^{-ij\tau} e^{-i2j\sigma}$$

$$X_R = \frac{\ell_s^2 n(\tau - \sigma)}{R} + \frac{i\ell_s}{\sqrt{2}} \sum_{j\neq 0} \frac{1}{j} \alpha_j e^{-ij\tau} e^{+i2j\sigma}, \tag{8.15}$$

and apply our T-duality transformation (8.13) to obtain the dual state:

$$X_L - X_R = 2\ell_s^2 n\sigma/R + \sqrt{2}\ell_s \sum_{j\neq 0} \frac{1}{j} \alpha_j e^{-ij\tau} \sin 2j\sigma. \tag{8.16}$$

What does this T-dual state represent?

First, we notice that there is no term linear in the time, so no overall linear motion; all the string's motion is vibrational. But if there is no linear momentum term, what does n now represent? As in (8.12), the term linear in σ has the correct form for a string wound n times around a space of radius ℓ_s^2/R. But how can an open string have a meaningful winding number? To answer this question, consider the ends of the string, $\sigma = 0, \pi$. Because of the $\sin 2j\sigma$ term, they have no time dependence at all—they are fixed in space! (Or rather, they are fixed in the dimension in question.) That is, they satisfy the Dirichlet boundary conditions that we

254 OUT OF NOWHERE

noticed in (7.21). So that is why an open string can have a sensible winding number: if its ends are attached to something, it can no longer be topologically shrunk to a point! So the physical interpretation is that the state T-dual to a freely moving open string in a space radius R, is an open string wound around a space of reciprocal radius, but with its ends fixed in place. (And it's easy to see that the ends will be a distance $2\ell_S^2\pi/R$ apart.) Of course, one has to check that the transformation really is a duality, that these two very different strings do agree on expectation values (under the duality), as for closed string. But they do (as you can see in the references in footnote 26).

In a sense then, by allowing Dirichlet boundary conditions, the string theory we developed in the previous chapter already contained room for open string T-duality; we didn't have to extend the formalism. But of course, in a very important way something very new has just been discovered; for what are the ends attached to? The reason we didn't pursue Dirichlet boundary conditions to start with is that they involve invariant locations, and so violate Poincaré invariance (and momentum conservation at the ends of the string). But there is another possibility. Suppose that a more complete theory, to which our perturbative string theory approximates, contains other dynamical (properly relativistic) objects than strings; specifically multi-dimensional generalizations of strings, known as p-branes. And suppose further that open strings can attach to them, enforcing Dirichlet boundary conditions: that they are 'Dp-branes'.[27] And even further that they are non-perturbative: the approximations on which perturbative string theory is based means that they do not appear as dynamical objects in the theory. If so, then open string T-duality simply shows that perturbative string theory is not completely independent of Dp-branes, but recognizes their physical presence at the level of non-dynamical objects.

In that case, one would very much like to know whether more can be discovered about Dp-branes at the perturbative level, and of course this question has been extensively studied.[28] Those results fall outside the scope of this work, but it is worth noting that the gauge fields that arise in superstring theory naturally couple to 'charges' carried by p-branes. In other words, their presence is revealed in perturbative string theory in more ways than through T-duality. Moreover, this role opens the door to a great deal of very interesting (and physically realistic) physics.

We promised above to apply these considerations to the proposal that T-dual theories could be realistically quotiented as a pair of wavefunctions. The presence

---

[27] Where 'D' is for 'Dirichlet', and p is the dimensionality of the brane. $0 \le p \le D$ (the dimension of space); and, if you think about it, if the end of the string is attached to a brane, then it is fixed in $D - p$ dimensions, and free to move in p of them. For instance, a point attached to the z-axis in 3-space can only move in one dimension, and is fixed in the x-y-plane.

[28] For a sense of their physical significance, refer to the references of footnote 26. For philosophical discussion of their nature see Vistarini (2017) and Vistarini (2019, ch. 4).

of Dp-branes as physical objects makes this proposal even less realistic since they are not described in the two field approximation we used above. The pair of wavefunctions carry no information about branes or their locations; indeed, any quotient would have to be the same regardless of the presence or absence of branes, so insensitive to them in that sense. What one expects instead is that the duals are 'quotiented' by an underlying theory, in which spacetime concepts do not fully apply; but which can equally be approximated with and without branes, or by any pair of dual theories. But at the level of perturbation theory, around a spacetime solution, we have to give significance to spatiotemporal concepts; that they are artifacts of an approximation scheme leads to their surprising indeterminacy.

8.5.2 Gauge-gravity duality

In the early twenty-first century, perhaps the most important tool for the study of string theory has been what is variously known as 'AdS-CFT duality' (an acronym of 'Anti de Sitter-conformal field theory'), or 'gauge-gravity duality' (the gauge theory being the CFT, and the gravity theory AdS space), or 'holographic duality' (though strictly this is a more general concept). This duality was proposed by Juan Maldacena, whose argument we will briefly discuss. But it has its roots in an earlier proposal by Gerard 't Hooft (see Stephens et al. 1994) based on the observation that the entropy of a black hole is proportional to its area not volume, suggesting that the entropy counts states on the 'boundary' not 'bulk'. In other words, suggesting that the full state of the black hole can be described by microstates on its horizon. Of course, for appropriate densities of states it's always possible to match the number of states on bulk and boundary; but the point is that for a constant quantized state density the number of states will not remain equal—doubling the radius means 4 × the boundary states but 8 × the bulk states. So the idea that bulk and boundary of any system can be physically equivalent—dual—is remarkable: but it seems to be the case when one compares a theory of gravity in AdS space with a conformal gauge field on its boundary.

Maldacena's (1998) argument[29] rests on the string theoretic understanding of gauge fields on the one hand, and of gravity and curved spacetime on the other. First then, gauge quanta can be understood in terms of the modes of open strings whose ends are attached to coincident Dp-branes. Specifically, if there are N such branes then the ground states of (supersymmetric) strings connecting them will describe the quanta of an $SU(N)$ Yang-Mills gauge field—just as in the previous chapter we understood the quanta of bosonic fields as modes of bosonic strings

---

[29] See Zwiebach (2004, ch. 23) or Dawid (2017) for intuitive presentations.

256 OUT OF NOWHERE

when viewed 'close up'.[30] Moreover, the branes and strings will carry charge for this gauge field. Second, we know that string modes have mass because of their tension, and the same applies to branes. Then the energy-momentum of the gauge field, its charges, and of the string and brane masses will provide a source for a relativistic gravitational field: the matter side of the Einstein field equation.[31]

In other words both the gauge field on the boundary and gravity in the bulk can be understood in string terms if one zooms in to small enough scales; and AdS-CFT duality ultimately asserts the equivalence of apparently very different systems of strings. Maldacena's argumentative strategy for this duality was to compare a system at weak and strong string couplings; observe that a low energy equivalence holds for half of it; and conjecture (on the basis of additional evidence) that the equivalence also holds for the other half: a gauge theory at weak coupling and AdS at strong. (Note that the string coupling is a dynamic quantity, not a constant of nature, and so really will vary.)

In slightly more detail: the system in question consists of closed super-strings, and N coincident D3-branes with attached open superstrings, all in flat 10-dimensional spacetime. For weak coupling: there are no interactions between the strings, so the systems decouple; moreover, gravity is 'turned off', and so space-time is flat; and finally, at low energy the open strings will be unexcited, massless quanta of an $SU(N)$ gauge field. For strong coupling: gravity is now 'turned on', and spacetime will be curved by the energy and charge of the D3-branes; near the branes the geometry will be AdS_5 (from the 4 spacetime dimensions parallel to the branes plus the radial dimension) times S^5 from the remaining dimensions; far from the branes the geometry will be flat; then plausibly the closed strings near the branes do not interact with those far away, leaving (at low energy) a stringy gravitational field in $AdS_5 \times S^5$ and again closed strings in flat spacetime. The coin-cidence of the closed string part of the system at weak and strong couplings then supports, but doesn't prove, the claim that the other part of the system—a gauge theory at weak coupling, and $AdS_5 \times S^5$ gravity at strong—also coincides, so that these theories are equivalent, or rather *dual*.

There has been considerable philosophical discussion of AdS-CFT[32]; we have treated the general philosophical consequences of duality in terms of T-duality instead, and a discussion of the more specific features of AdS-CFT is largely beyond the scope of this work. However, we want to recommend an essay by Harlow (2020) which presents a simple but illuminating and contentful model of

---

[30] More carefully, they are quanta of a $U(N)$ field, which decouples into $SU(N)$ plus a single gauge field.

[31] In the next chapter we will explicate the nature of gravity in string theory in detail. Ultimately the curved geometry is understood in terms of graviton states of the string, so that instead of sourcing an extrinsic gravitational field (as it seems here), matter and gravitational fields are both states of strings; the theory of gravity is a theory of string-string interaction.

[32] A good cross-section is: de Haro et al. (2016a, 2016b); de Haro (2017); Matsubara (2013); Polchinski (2017); Read (2014); Rickles (2012); Teh (2013); Vistarini (2017, 2019).

DUALITY 257

AdS-CFT duality, in which the bulk contains a black hole. Briefly, three 'qutrits'[33] live on the boundary, comprising a $3^3 = 27$-dimensional Hilbert space of a boundary quantum theory. The bulk theory is represented by a single qutrit, living in a 3-dimensional subspace of the full theory; corresponding to the few degrees of freedom of a classical black hole. But the bulk theory should be dual to that on the boundary, and so also live in a 27-dimensional Hilbert space; what has happened to the other 24 dimensions? Harlow's point is that although both bulk and boundary theories are effective descriptions of more fundamental string states, the boundary *quantum* gauge theory is still more fundamental—closer to string theory—than the bulk *classical* theory. When one recognizes that point it is natural to understand the missing 24 dimensions as representing *quantum* microstates of the black hole. If so, AdS-CFT duality allows one to study the unknown quantum nature of black holes through the better understood physics of CFT. Harlow's work explores and supports this very idea.

He emphasizes that this picture also clarifies the common view that AdS-CFT duality asserts the equivalence of the two theories. This view is correct at the level of the fundamental string description, but not at the level of the effective AdS and effective CFT descriptions, for the latter is a more complete description than the former; as we just saw, it carries more information about the fundamental string state. In that sense, the classical AdS theory is not equivalent to the quantum CFT, but *derived* from it; then as we saw, the additional boundary degrees of freedom allow one to probe quantum gravity in the bulk. Moreover, this is a picture in which not all quantum degrees of freedom correspond to classical spacetime degrees of freedom, suggesting—with Brandenberger and Vafa—that the fundamental ones may not be spatiotemporal at all.[34]

8.6 Conclusions

The main conclusions of this chapter are as follows. First, T-duality is an unusually deep symmetry between theories, with respect to some very counterintuitive and surprising parameters: especially the radius of space. Gauge symmetries in field theory are similarly deep, but since they typically involve internal degrees of freedom, they are not so shocking. A touchstone of this chapter has been the analysis of Brandenberger and Vafa, which explains how there can be two theories apparently differing on the radius of space, yet predicting the same observed radius.

---

[33] A qutrit is the 3-dimensional generalization of an ordinary (2-dimensional) qubit.
[34] See Mathur (2012) for an introduction to the 'fuzzball' approach to string theoretic black holes, which proposes a specific quantum structure (and Huggett and Matsubara 2020 for further philosophical analysis). In a fuzzball model spacetime indeed 'ends' at the horizon, leaving the black hole state 'beyond spacetime'.

258   OUT OF NOWHERE

Their analysis has helped at several points to understand the physical meaning of T-duality: such a picture is crucial to understanding duality.

The symmetry is so deep—between all observables, not just empirical quantities in some superficial sense—that duals should be understood as giving physically equivalent descriptions. Since they formally disagree on some claims, we have argued (against an alternative view) that the physical commitments of dual theories are limited to their common assertions. Specifically, they disagree on the radius of target space, so that must be indeterminate between the two possible values. And in general, 'target space' is not a space in the familiar sense at all, but a 'space' with only the structures on which the duals agree (quite possibly then, a structure that appears as a formal representation of some more fundamental, as yet unknown, non-spatial object). As the analysis of Brandenberger and Vafa explains, duals do agree on the radius of observable, classical space, so that is determinate. But nothing can be both determinate and indeterminate with respect to radius, and so target space is not classical, relativistic space.

Therefore classical space, specifically as a geometric space of determinate radius, is not a fundamental object of string theory, but an appearance, arising from physical processes of the kind that Brandenberger and Vafa analyzed. That, ultimately, is the ontological significance of T-duality, and indeed of the other dualities we have described.

Chapter 9
The string theoretic account
of general relativity

The canonical and lightcone quantization treated in chapter 7 allowed us to see how modes of the string correspond to different kinds of particles, but for a more rigorous and complete understanding of string theory one needs to switch to Feynman's path integral formulation, which we sketch in §9.1. For one thing, one can then more rigorously derive the 'critical' dimension of $D = 26$. We won't describe the calculation here, but it is treated in the standard textbooks. In this chapter we will instead use the formalism to outline two results that are crucial to understanding how classical spacetime 'emerges' from string theory. These results go back to the early days of string theory, but on the one hand they have been largely neglected by philosophers (see exceptions below), while on the other physicists generally take them rather formally—an analysis of how the derivation can be considered physically salient is lacking.

The results concern the possibility of generalizing string theory so that strings propagate in a general curved target space. First, in §9.2, we will see that the generalization of the sigma action from a Minkowski target space to an arbitrary Lorentzian metric describes a string interacting with a 'coherent' state of the massless spin-2 quanta of the closed string, provisionally identified as gravitons in §7.3.4. Second, in §9.3, we will then see that the metric of target space is not arbitrary, but must (to leading order) satisfy the Einstein field equation (EFE). Taken together, these points establish that gravitons truly are the quanta of the gravitational field: they produce a metric that must satisfy the dynamics of general relativity (GR). Moreover, they offer a string theoretic explanation of generally relativistic spacetime: what is taken classically to be curved spacetime is at a more fundamental level nothing more than a state of a particular kind of string excitation, just like every other field.[1]

The discussion of this chapter is at a more technical level, for those seeking a deeper understanding of the theory behind these issues. Other readers may skip straight to the next chapter, where the key arguments are summarized in a more informal way, before the consequences are unpacked more conceptually. For those who read on, part of the work is to explicate some ideas from quantum field

---

[1] See Rickles (2014, especially §9.4) for some of the history of this line of thought, which he shows to have been a concern of string theorists since the time it was recognized as a quantum theory of gravity.

Out of Nowhere. Nick Huggett and Christian Wüthrich, Oxford University Press.
© Nick Huggett and Christian Wüthrich (2025). DOI: 10.1093/oso/9780198758501.003.0009

260   OUT OF NOWHERE

theory (QFT) that are essential for understanding the emergence of spacetime in string theory: in §9.4 we discuss quantum 'anomalies', while in appendix 9.A we investigate coherent states of a field as its classical limit. Both of these topics, while important in QFT generally, have been largely neglected by philosophers of physics, a lacuna which this chapter seeks to address.

9.1 Path integral string theory

As we saw in §7.3, string theory is, in the formal sense, a field theory: fields X^μ propagate on a 2-dimensional world sheet, with coordinates $(\tau, \sigma) \equiv (\sigma^0, \sigma^1)$. Familiar canonical quantization then leads to a Fock representation, whose excitations have the symmetry properties of spacetime field quanta; on that basis we provisionally identified string excitations with those quanta.

In this chapter we start over, looking at a quantized string from the point of view of the path integral approach. Here, *Feynman's Principle* says that the amplitude for a system to evolve from a state q_i to a state q_f is given by summing over distinct classical paths between q_i and q_f, weighted by e^{iS}, where S is the action for the path. In well-behaved cases, the canonical amplitude, in terms of the Hamiltonian, and the path integral amplitude, in terms of the action or Lagrangian, agree. Thus we start again with the sigma action from (7.16), rescaled for simplicity:[2]

$$S_\eta = \int d^2\sigma \sqrt{-h} h^{\alpha\beta} \eta_{\mu\nu} \partial_\alpha X^\mu \partial_\beta X^\nu, \qquad (9.1)$$

with an explicit Minkowski background spacetime, $\eta_{\mu\nu}$. Because we are turning our attention to general background metrics, we shall now refer to (9.1) as the 'flat sigma action', and actions with arbitrary backgrounds as 'sigma actions', S_σ. Classically, such an action will be obtained by substituting $\eta_{\mu\nu} \to g_{\mu\nu}$ in (9.1); §9.2 will explicate the quantum string theoretic significance of this substitution. Path integrals will then be functional integrals of the form

$$\int DXDh \; e^{iS_\sigma[X,h]}, \qquad (9.2)$$

taken over field configurations on the worldsheet, since these are the 'paths' in question. $X(\tau, \sigma)$ and $h(\tau, \sigma)$ describe, respectively, the embedding of the string in target space, and the auxiliary metric on the worldsheet. (We take $\hbar = 1$.)

---

[2] This section draws heavily on Polchinski (1998, ch. 3). In fact he develops these ideas after Wick rotating the theory into the complex plane $\tau \to i\tau$, yielding a theory with a Euclidean signature $h_{\alpha\beta}$. This shift takes care of singularities on non-trivial worldsheets, making the theory better defined. The discussion here is more heuristic.

THE STRING THEORETIC ACCOUNT OF GENERAL RELATIVITY   261

There is a significant problem with this expression, because the functional integral is over all embeddings X^μ, and all auxiliary worldsheet metrics $h^{\alpha\beta}$, even those related by diffeomorphisms or Weyl transformations, which are gauge symmetries for the action. That is, the integral hugely over-counts, summing over all gauge-equivalent states, even though they represent no physical difference. Of course, we said in §7.1.2 that *classically* the auxiliary metric inherits its causal structure from the target space metric: the equations of motion imply that $h^{\alpha\beta}$ is conformal to the induced metric on the worldsheet. Hence, all physically possible—'on shell'—classical auxiliary metrics are Weyl/gauge equivalent for a given solution. However, even paths that are 'off shell'—not satisfying classical equations of motion—must be summed over in the *quantum* path integral. Hence not all the $h^{\alpha\beta}$ summed over are Weyl/gauge equivalent, and gauge fixing does not mean picking a single metric, but picking one from each gauge inequivalent class. The techniques of QFT, especially the 'Faddeev-Popov' method, will take care of this problem; indeed such a calculation is what leads to the more rigorous derivation of $D = 26$ spacetime dimensions for the bosonic string. We will take this procedure as read in our discussion, and ignore the problem.[3]

Going forward, it is worth bearing in mind the place of Feynman path integrals in QFT. While they simplify perturbative methods in scattering theory, they are not intrinsically perturbative: in principle, the path integral can be taken as an exact specification of a quantum theory, via the amplitudes it determines, since these determine a QFT via the Wightman reconstruction theorem (Streater and Wightman 1964, ch. 3)—assuming that they satisfy the Wightman axioms. So it is perfectly reasonable to take the path integral as a fundamental formulation in its own right, and not necessarily as merely a part of a calculational technique for a canonically quantized theory—even if its history and pedagogy suggest that. (Of course, the formal question of the existence of the integral is open for most theories: but so is the question of the existence of most canonically quantized field theories.) That is the point of view taken here: insofar as they exist, string amplitudes are defined by the path integral (9.2). What we take from our earlier discussion of canonical and lightcone quantization is the spectrum of particles we expect the theory to contain—especially the graviton.

However, (9.2) does not fully specify 'string theory' in this sense. One must sum, not only over embeddings and metrics of the worldsheet, but *first* over every (topologically) possible way that strings might join and split, compatible with the given in- and out-states, as illustrated in figure 9.1. *Then* for each worldsheet, the path integral sums over the worldsheet metrics $h^{\alpha\beta}$. And then *finally*, there is a sum over

---

[3] The reader is invited to follow the derivation in the standard textbooks (e.g., Polchinski (1998, §3.3)). Page 89 notes that (9.2) is not really well-defined because of over-counting. The more fundamental, well-defined, action is instead that obtained at the *end* of Faddeev-Popov quantization.

X^μ fields: the different ways of embedding the string in target space. In short, (9.2) only describes the second two sums. What of the sum over topologies?

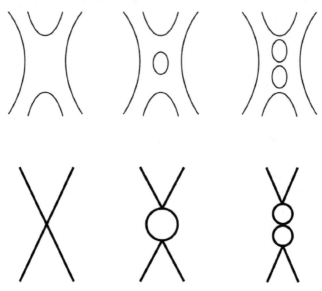

Figure 9.1 The top row shows three of the infinite number of worldsheets with two open strings scattering to produce two strings—time runs up the page. These differ by the increasing number of holes. The bottom row shows some of the corresponding ways that they could appear at low energy—as the Feynman diagrams for interacting quanta.[4]

Examining figure 9.1 it is apparent that different worldsheets represent interacting strings, splitting and joining. Assuming a coupling coefficient, g_s, at each stringy 'vertex', the diagrams then correspond to terms in a perturbative expansion, ordered by the number of vertices, or power of the coupling (equivalently, in the sum, worldsheets are distinguished by their topology). So although (9.2) may (we supposed) be taken as defining amplitudes non-perturbatively, 'string theory' as understood here also has an essential perturbative aspect. But to what theory does the sum over worldsheets then approximate? The most obvious suggestion, based on an analogy to Feynman diagrams in QFT, is that it approximates a quantum 'string field' theory, in which strings can be created and annihilated, like regular field quanta. And indeed, Witten (1986) proposed a 'cubic action' for such a theory of open strings, which produces the correct Feynman rules; related actions for closed and supersymmetric strings have also been explored (see Taylor 2009 or

---

[4] Green et al. (1987) point out that the correspondence is many-one: there are far fewer stringy diagrams at each order than ordinary, linear diagrams. Diagrams are distinguished only up to homeomorphism and, for example, compact 2-dimensional manifolds are uniquely topologically characterized by the numbers of loops, their genus.

THE STRING THEORETIC ACCOUNT OF GENERAL RELATIVITY   263

Erbin 2021 for a review). However, string theorists are more likely these days to suggest that string theory is an approximation to the unknown 'M-theory', which dualities are supposed to indicate (see §8.5). Either way, we have first a perturbative sum in powers of g_s over worldsheets—the different ways strings can scatter—and then for each such worldsheet, each stringy scattering diagram, the path integral (9.2) to perform. Formally at least, each diagram then corresponds to an amplitude for the fields X^μ and $h^{\alpha\beta}$ to propagate from their given in- and out-states: for a QFT process living on the given worldsheet.

At this point, it is worth leaving the main thread of the discussion to make a few remarks about these diagrams, for they play an important role in understanding the physical content of string theory. First, they play a central role in explaining quantum fields: at low resolution relative to the string length (for instance, those of contemporary high energy physics) sheets in diagrams will be indistinguishable from lines, and stringy diagrams can contribute as ordinary Feynman diagrams. That is, to low order, the X^μ and $h^{\alpha\beta}$ path integral over the sheet yields QFT propagators, and the physics of the interaction region is approximated by a point interaction term (as pictured in the lower half of figure 9.1). Since string excitations correspond to point particles at low energy, these propagators and interactions should be those of such point particles; then the diagrams for a given collection of particles and interactions will define a set of scattering amplitudes, and hence (if all goes well) a QFT. The theories so derived need not be renormalizable: they are cut-off at the string scale, where the sheets can no longer be approximated by lines. Thus even gravity, a field whose quanta are gravitons, can be approximated in this way. (Gravitons, we saw, are states of closed, not open, strings, but much the same points hold for closed strings.)[5]

Looking more closely at the string diagrams, figure 9.2 demonstrates how string interactions are described differently in different spacetime frames. O and O' disagree on when, and on where on the string the vertex of the interaction occurs: i.e., which is the latest time slice on which there is only one string? This fact points to an important feature of string theory (Green et al. 1987, 30f). Because the location of the interaction is not an invariant, there is no point interaction in the theory: such a vertex would be frame-dependent, not Lorentz invariant. That's good, because point interactions lead to infinities in QFT, and in particular to non-renormalizability in quantizations of the gravitational field: the absence of a spacetime point interaction is supposed to lead to the finiteness of string theory.[6]

---

[5] Since the parameters of a string theory fix the particle spectrum, not every QFT can be reproduced in this way, limiting which QFTs can be understood in stringy terms. However, the introduction of D-branes allows for a much richer, and more flexible particle spectrum, including realistic versions of the standard model.

[6] For instance, Polchinski (1998, 78). However, some commentators have expressed skepticism: Penrose (2004, §31.13), Smolin (2007, 186ff).

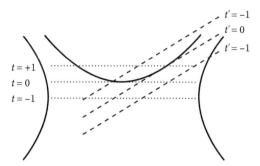

Figure 9.2 A string splitting in two according to two different Minkowski reference frames. In the first frame, O, the splitting occurs at $t = 0$: before that (e.g., at $t = -1$) there is a single string, and after (e.g., at $t = +1$) there are two. In the second (primed) frame, O', the splitting occurs at $t' = 0$, at a different spacetime point. Indeed, the point at which the splitting occurs in either frame is at a time at which there are already two strings in the other!

Moreover, it suggests that, unlike ordinary QFT, there is no constant parameterizing interactions. In QFT, each pointlike interaction vertex is associated with a constant coupling strength (the charge, in a general sense). But in string theory, pick any point at which the split—interaction—occurs in some frame; in other frames (see figure 9.2) it is an endpoint of a free string, with which no such constant is associated. In short, the interaction dynamics should already be included in the free dynamics, (9.1)! That may seem paradoxical, given that we already said that splitting and joining does contribute a string coupling factor, g_s, to amplitudes, but the solution is that on further analysis the value of the string coupling turns out to be determined dynamically (by a field of stringy dilaton quanta).

Returning to the question of computing path integrals for string theory, after accounting for gauge symmetry, the next problem that one faces in calculating the contribution of a diagram is how to perform an integral over such a complex surface to obtain an action: i.e., how is the $\int d^2\sigma$ integral of the action performed? Even in the simplest cases, of 'tree diagrams', with no loops, how is one to integrate over a worldsheet with legs at different positions? Here is another place at which the conformal symmetry of the theory works its magic, to make something complicated tractable. In closed string theory, any tree diagram, with any number of legs, is conformally equivalent to a sphere with a hole for each leg, so one need only perform the integral over such a sphere. (Similarly, a diagram with only open strings is conformally equivalent to a disc, with points missing at the boundary. We focus on closed strings because we will be interested in gravitons, which are closed string states.)

To see how this works, isolate the leg of a diagram, and let it have a coordinate patch $(-T \leq \tau \leq 0, 0 \leq \sigma < \pi)$, where T is the initial time in some Minkowski

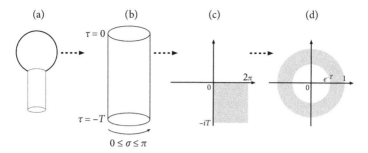

Figure 9.3 (a) Consider a worldsheet with a leg attached; remove it, leaving a hole. (b) Give the removed leg cylindrical coordinates ($-T \leq \tau \leq 0, 0 \leq \sigma < \pi$). (c) Represent the leg in the complex plane, with coordinates $w = 2\sigma + i\tau$, so that it occupies the shaded rectangle. (d) Apply the holomorphic transformation $w \to z = e^{-iw}$, which maps the rectangle into the shaded annulus. Since the transformation is a symmetry of the action, the annulus can be reinserted into the hole left in the sheet.

frame (see figure 9.3). Then introduce a complex coordinate $w = 2\sigma + i\tau$, in order to represent the cylinder as a rectangle in the complex plane. The advantage of the complex representation here is that holomorphic transformations correspond to conformal transformations of real coordinates, which we saw are symmetries of the sigma action. In particular, the holomorphic transformation $w \to z = e^{-iw}$ projects the leg onto a unit annulus in the complex plane: the 'bottom' of the tube is mapped $2\sigma - iT \to e^{-T}(\cos 2\sigma - i \sin 2\sigma)$, the inside of the annulus; while the 'top' is mapped $2\sigma \to (\cos 2\sigma - i \sin 2\sigma)$, the outside. Therefore, because of conformal symmetry, the contribution of any leg can be computed by integrating over an annulus. Applying the same transformation to each leg in tree diagram, replacing each with its equivalent annulus (and a further overall conformal transformation) yields a sphere with a hole for each leg.

But we can go one better, and take $T \to \infty$, as we should for a string scattering from the infinite past to infinite future. In that case $2\sigma - iT \to e^{-T}(\cos 2\sigma - i \sin 2\sigma)$ holomorphically maps the infinite past to the origin. Again, the conformal invariance of the action means that the state of a string in the infinite past is equivalent to an object at a *point* on a spherical world sheet in computations of scattering.

Recall here that there are two manifolds and two sets of coordinates in play, and this transformation concerns only one of them. With respect to Minkowski spacetime, the diagram is unchanged: we still have a string propagating in from the infinite past. Initially we had worldsheet coordinates that pictured things the same way: τ agrees with time in the Minkowski frame, and so $\tau \to \infty$ as $T \to \infty$. The transformation to a point of course is a Weyl rescaling of worldsheet metric, which has no effect on the Minkowski metric: the string at infinite Minkowski past

266 OUT OF NOWHERE

is now represented by a point at the origin of the worldsheet coordinates. But that's what we want, since the Lagrangian is to be integrated over the worldsheet in the action.

An object at a point represents an operator in QFT, so what we have just seen is a 'state-operator correspondence', a standard feature of conformally symmetric QFTs, or 'conformal field theories' (see Tong 2006, 96ff). Thus progressing on the question of scattering amplitudes requires identifying the appropriate 'vertex operators' for different scattering particles, or rather for different string excitations. The simplest case is of course the (unphysical) ground state tachyon with 4-momentum hk, for which $e^{ik\cdot X}$ is the vertex operator.[7] This can be shown, but is intuitively the correct answer. Particle species are distinguished by the representations of symmetries by their states: the tachyon is a scalar, so the states carry no spacetime indices—like the vertex operator. The only non-trivial transformations are translations, and since both the tachyon wavefunction and vertex operator have the same form, e^{ikx}, they transform in the same way. More generally, to have the right transformations, a vertex operator will have the same indices as the corresponding excitation, and will contain the particle wavefunction.

In this case, the tree level term of the path integral for, say, scattering from two tachyons to two tachyons will (for the closed string) have the general form:

$$\int DXDh \; e^{iS_\eta[X^\mu, h]} e^{ik_1\cdot X} e^{ik_2\cdot X} e^{ik_3\cdot X} e^{ik_4\cdot X}, \tag{9.3}$$

where the integral is now over the surface of a unit sphere—summed over conformally inequivalent placements of the four vertex operators. What seemed intractably complicated again turns out to be tractable after all: once again we see that string theory works by formulating physics as a conformal field theory in two dimensions, bringing the powerful and rigid structure of complex analysis to bear.

9.2 Building a curved background

With the basic picture of the path integral formulation of string theory in hand, we can now turn to the promised results of this chapter: the relation between string theory and spacetime geometry.[8] Specifically, to learn about gravity we will study its influence on string scattering. Following our working hypothesis

---

[7] More precisely, the vertex operator is $\int d^2\sigma\sqrt{-h}e^{ik\cdot X}$, since an integral over the worldsheet is needed for the operator to be diffeomorphism invariant. Note that to compute their expectation values, field operators (on the Hilbert space) appear as the corresponding classical fields in a path integral. That is why vertex 'operators' are given as functions. See Green et al. (1987, §1.4.2 and 2.2.3) or Polchinski (1998, §3.6) for a more complete derivation of the vertex operators.

[8] The earliest occurrence of the following argument of which we are aware is Green et al. (1987, 165f), though it is common in the subsequent literature.

THE STRING THEORETIC ACCOUNT OF GENERAL RELATIVITY 267

from §7.3 that the massless traceless-symmetric tensor state of the closed string is the graviton, we therefore need to study scattering diagrams involving such states; which we just learned means including graviton vertex operators in path integrals.

As we noted in (7.60), a graviton has a traceless-symmetric wavefunction $\epsilon_{\mu\nu}e^{ikX}$ (now written in suitable non-lightcone coordinates). Since the action has no indices, the indices must be contracted, requiring X derivatives, so the vertex operator is:[9]

$$V_G = \epsilon_{\mu\nu}e^{ikX} \cdot \partial_\alpha X^\mu \partial^\alpha X^\nu \qquad (9.4)$$

(as in footnote 7, V_G is always to be taken inside a worldsheet path integral).

However, we are interested in *classical* gravity in string theory, so V_G, which represents the effect of a single graviton, is not quite what we want. Instead we need the operator for a quantum state that best describes a classical gravitational field—a 'coherent state'. To progress we need to sketch the basic idea in enough detail to draw the main conclusion of this section. A more thorough treatment is given in appendix 9.A.

Consider a pair of operators with commutation relation, $[\hat{a}, \hat{a}^\dagger] = 1$: as is familiar, this relation has a representation in a basis of states $\{|n\rangle : n = 0, 1, 2, \dots\}$, where $\hat{a}^\dagger|n\rangle = \sqrt{n+1}|n+1\rangle$ and $\hat{a}|n\rangle = \sqrt{n}|n-1\rangle$. That is, $\hat{a}$ and $\hat{a}^\dagger$ are lowering and raising operators, respectively. Then define the following (unnormalized) state (with complex α):

$$|\alpha\rangle \equiv e^{\alpha\hat{a}^\dagger}|0\rangle = \left(1 + \frac{\alpha\hat{a}^\dagger}{1!} + \frac{(\alpha\hat{a}^\dagger)^2}{2!} + \dots + \frac{(\alpha\hat{a}^\dagger)^n}{n!} + \dots\right)|0\rangle$$

$$= |0\rangle + \frac{\alpha}{\sqrt{1!}}|1\rangle + \frac{\alpha^2}{\sqrt{2!}}|2\rangle + \dots + \frac{\alpha^n}{\sqrt{n!}}|n\rangle + \dots. \qquad (9.5)$$

Exponentiating the creation operator produces a superposition of every excitation level (and for that reason cannot be reached by finite excitations). It's straightforward to verify that (9.5) is an eigenstate of the lowering operator (though $\hat{a}$ is not Hermitian): $\hat{a}|\alpha\rangle = \alpha|\alpha\rangle$. Indeed we can take this property to be definitional of a 'coherent state' (though significant generalizations of the concept can be found in the substantial literature on the subject). One can easily see that the expectation value of the number operator $\hat{n} = \hat{a}^\dagger\hat{a}$ is $\langle\alpha|\hat{n}|\alpha\rangle = |\alpha|^2$.

This story generalizes to the field $\hat{\phi}(x)$, understood as an operator annihilating quanta at x: by definition, a coherent state of the field satisfies $\hat{\phi}(x)|\psi\rangle = \psi(x)|\psi\rangle$, for some scalar field $\psi(x)$. $\hat{\phi}(x)$ can be decomposed into momentum modes

---

[9] See Polchinski (1998, §3.6) for more details. His derivation follows from the diffeomorphism invariance of the vertex operator and avoiding negative norm states, showing another route to the existence of graviton states in the string spectrum.

268 OUT OF NOWHERE

$\hat{a}_k$, which annihilate quanta of momentum $p = hk$: $\hat{\phi}(x) = \sum_k \hat{a}_k e^{ikx}$. Using our previous definition of coherent state for each $\hat{a}_k$, (and the fact that all operators commute for $k \neq k'$) it can then be simply verified that

$$|\psi\rangle = \prod_k e^{\alpha_k \hat{a}_k^\dagger}|0\rangle \tag{9.6}$$

is a coherent state for the field, with $\psi(x) = \sum_k \alpha_k e^{ikx}$. Conversely then, any classical field with a Fourier decomposition corresponds to a coherent state: the classical Fourier coefficients define the state through (9.6). It is also easily seen that the expectation value for quanta density, a measure of the squared field strength in a coherent state, $\langle\psi|\hat{\phi}^\dagger(x)\hat{\phi}(x)|\psi\rangle = |\psi(x)|^2$, the classical value. Not only that, coherent states minimize canonical uncertainty relations, and $\psi(x)$ will satisfy classical field equations. In short, in QFT, coherent states of a quantum field are often taken to underwrite the appearance of classical field phenomena, a crucial idea in the following derivation of GR. (However, in appendix 9.A we will discuss limitations to the coherent state account of classical fields.)[10]

Returning to string theory, before considering gravity, let's first apply these ideas to the notationally tidier case of strings scattering in a scalar field background. We want this background to be fundamentally quantum (and stringy), so composed of scalar quanta (ultimately understood as string excitations), but corresponding to a classical background field $\psi(X)$—because we are considering a string background, we use target space coordinates X^μ. You may recall from §7.3 that the scalar particle corresponds to the ground state of the string, and is a tachyon since it has a negative mass. Although this example is then ultimately unphysical (since energy is unbounded from below), it still helps to illustrate how background states can be constructed.

First we note that (9.6) tells us that if $\hat{a}_k^\dagger$ creates a scalar quantum, then a coherent state of the field is:

$$\prod_k e^{\alpha_k \hat{a}_k^\dagger}|0\rangle = \prod_k \left(1 + \frac{\alpha_k \hat{a}_k^\dagger}{1!} + \frac{(\alpha_k \hat{a}_k^\dagger)^2}{2!} + \cdots + \frac{(\alpha_k \hat{a}_k^\dagger)^n}{n!} + \cdots \right)|0\rangle. \tag{9.7}$$

Then we recognize that these quanta are in fact strings in the tachyon mode: stringy tachyons. In §9.1 we learned that because of conformal symmetry, the effect of a stringy particle on scattering can be reduced to the insertion of a vertex operator in the path integral. So the scalar background will contribute an operator obtained by replacing every $\hat{a}_k^\dagger|0\rangle$ in (9.7)—each tachyon—with a tachyon vertex operator $S_T^k = e^{ikX}$:

---

[10] We have used the scalar field as our example for ease of presentation, but identical considerations apply to bosonic spin and tensor fields (see Rosaler 2013, ch. 4 for a discussion of the electromagnetic field). Fermionic and supersymmetric fields are more complex, but similar ideas apply (Combescure and Robert, 2012).

$$\prod_k e^{\alpha_k \hat{a}_k^\dagger}|0\rangle \leftrightarrow \prod_k \left(1 + \frac{\alpha_k S_T^k}{1!} + \frac{(\alpha_k S_T^k)^2}{2!} + \cdots + \frac{(\alpha_k S_T^k)^n}{n!} + \cdots\right)$$

$$= \prod_k \left(1 + \frac{\alpha_k e^{ikX}}{1!} + \frac{(\alpha_k e^{ikX})^2}{2!} + \cdots + \frac{(\alpha_k e^{ikX})^n}{n!} + \cdots\right)$$

$$= \prod_k e^{\alpha_k e^{ikX}} = e^{\sum_k \alpha_k e^{ikX}} = e^{\psi(X)}. \tag{9.8}$$

Thus, to calculate the effect of a stringy tachyon background field $\psi(X)$ on string scattering we need to include a factor $e^{\psi(X)}$ in our path integral (9.2)[11]

$$\int DXDh \, e^{S_\eta} \to \int DXDh \, e^{S_\eta} \, e^{\int d^2\sigma\sqrt{-h}\,\psi(X)}. \tag{9.9}$$

But that is to say that the effect of a classical scalar background field—conceived as a coherent state of stringy tachyons—is simply to add the field to the classical world sheet Lagrangian:

$$S_\eta \to S_\eta + \int d^2\sigma\sqrt{-h}\,\psi(X) = \int d^2\sigma\sqrt{-h}\,(\mathcal{L}_\sigma + \psi(X)). \tag{9.10}$$

The same reasoning holds, mutatis mutandis, for other background fields, including a graviton background field. As we saw in §7.3, gravitons correspond to traceless-symmetric plane wave solutions to the weak field equation (7.61): $h_{\mu\nu} = \epsilon_{\mu\nu}e^{ikX}$ (in suitable coordinates). General—*not only weak field*—symmetric tensors can be expanded in such waves as $s_{\mu\nu} = \sum_k \epsilon_{\mu\nu}^{(k)}\alpha_k e^{ikX}$: these are classical 'graviton background fields'. We reason just as we did for the tachyon background field, except we use this expansion of the field, and the vertex operator (9.4) for the gravitons, which are nothing but suitably excited strings. Then, in the step corresponding to (9.8) we obtain the factor $e^{s_{\mu\nu}\partial_\alpha X^\mu \partial^\alpha X^\nu}$. When we insert this in the path integral (9.2) to represent a background field composed of stringy gravitons we find

$$\int DXDh \, e^{S_\eta} \to \int DXDh \, e^{S_\eta} \, e^{\int d^2\sigma\sqrt{-h}\,s_{\mu\nu}\partial_\alpha X^\mu \partial^\alpha X^\nu}$$

$$= \int DXDh \, e^{\int d^2\sigma\sqrt{-h}\,\eta_{\mu\nu}\partial_\alpha X^\mu \partial^\alpha X^\nu} \, e^{\int d^2\sigma\sqrt{-h}\,s_{\mu\nu}\partial_\alpha X^\mu \partial^\alpha X^\nu}$$

$$= \int DXDh \, e^{\int d^2\sigma\sqrt{-h}\,(\eta_{\mu\nu}+s_{\mu\nu})\partial_\alpha X^\mu \partial^\alpha X^\nu}. \tag{9.11}$$

---

[11] With the necessary worldsheet integral for worldsheet diffeomorphism symmetry.

270 OUT OF NOWHERE

Therefore the effect of a graviton background, $s_{\mu\nu}$, on the string path integral is to shift the metric appearing in the action $\eta_{\mu\nu} \to g_{\mu\nu} = \eta_{\mu\nu} + s_{\mu\nu}$. Put the other way around (and this is the pay-off of this section) in string theory *a curved spacetime background should be understood as a stringy graviton field in a flat target space*—and not, for instance, as a preordained classical curved background spacetime. That is, curved spacetimes correspond—according to this analysis—to different string states.

It is important to emphasize once again that the field coherent states that we have invoked do not exist in the state space of a single string. Increasing excitations of a string do not correspond to additional particles, but to different particles, with increasing masses: from the spacetime point of view the total particle number is always one. So (9.7) and (9.11) suppose that there are multiple strings, and indeed a superposition of different numbers of strings. In short, it is the kind of state that one would expect in a string field theory, in which strings can be created and annihilated, in analogy with the Fock representation of QFT. As discussed in the previous section, such a formulation is not well-understood, so the existence of such states—or more cautiously that such states well approximate fundamental states of some 'M-theory'—is something of a leap of faith at this point in the argument, yet one that we shall take.

In string theory then, observed spacetime curvature is explained in terms of interactions with gravitons, which is just what one expects in the QFT approach to gravity. The explanation is of a different kind to that of curvature in GR, in which 'matter tells spacetime how to curve' through the EFE. Rather it is reductive in some sense: since they constitute the metric field, gravitons *are* curvature. We will discuss how to understand this explanation in more detail, its scope and limitations, and what it tells us about fundamental and classical spacetime physics, in chapter 10. We have presented the string theory account of spacetime geometry largely uncritically in this chapter, but among the important questions to be raised there is the status of the Minkowski geometry, $\eta_{\mu\nu}$, apparently still assumed by the theory: for instance by its explicit appearance in (9.1).

For now, however, focus on the three ingredients of the result. First, there is the fact that the second excited state of the closed string is a massless traceless-symmetric tensor. This already led us to tentatively identify it as the graviton in chapter 7, essentially because that mathematical form allows it to act as a perturbation to a flat background. So this element of the result was to be expected. Second, it is crucial that the graviton couples to derivatives ∂X^{μ}, so that the $s_{\mu\nu}$ term can be added to the $\eta_{\mu\nu}$ term in the path integral. As we discussed, this coupling follows directly from the form of the graviton. Finally, the necessary exponentiation of the graviton vertex operator implies a coherent state of the field: that the result is equivalent to introducing a classical background is another way to see how coherent states indeed describe (approximately) classical fields in QFT.

9.3 The Einstein field equation

Showing that the second excited state of a closed string contributes to the space-time metric is required by its provisional interpretation as the graviton. However, since GR is our empirical theory of gravity, to complete the identification we need to see that the classical field produced by coherent graviton states satisfies (approximately) the EFE. That they do is the foundation of string theory's claim to be a quantum theory of gravity.[12]

The discussion so far shows that in a curved spacetime the string theory path integral has the form (9.11). As we have seen, this describes a collection of interacting strings (which are not in a coherent state), interacting with a background of stringy gravitons, which are in a coherent state. We have seen that the full path integral involves a perturbative sum over (topologically distinct) worldsheets, but clearly even for each particular worldsheet in the series (and each worldsheet metric h), an integral $\int DX$ remains, corresponding to different embeddings of the worldsheet in spacetime. To see what this integral means, focus on the lowest order, tree worldsheet, one with no holes: with respect to spacetime, a cylinder, with multiple legs to past and future infinity.

The action in (9.11), with appropriate constants, is

$$S_\sigma = -\frac{T}{2} \int d^2\sigma \sqrt{-h}\, g_{\mu\nu}(X) \cdot \partial_\alpha X^\mu \partial^\alpha X^\nu \tag{9.12}$$

(where T is the string tension constant). As we saw, conformal symmetry with respect to the worldsheet means that we can take the σ-integral over the unit sphere (with additional vertex operators as necessary to describe any scattering with background particles)—despite the target space geometry of the string. Naively, (9.12) describes an interacting field, $X^\mu(\sigma)$, over such a worldsheet, with a self-coupling $g_{\mu\nu}$; while $\int DX$ is a functional integral over all such fields, familiar territory from QFT. However, from this point of view, the coupling is a function of the 'field strength' X^μ: $g_{\mu\nu} = g_{\mu\nu}(X(\sigma))$, unlike the more familiar constant couplings. This kind of field is known as a 'non-linear sigma model', and therefore (9.11) is formally the path integral for a quantum non-linear sigma model.

Despite its non-linearity, this is a well-understood QFT, and many of the usual ideas apply. In particular, it can be expanded perturbatively, effectively in terms of the X^μ-field 'coupling', $g_{\mu\nu}$—but in practice, as Green et al. (1987, 167ff) explain, in powers of the 'Regge slope', $\alpha' \equiv 1/2\pi T$. (Once again, the path integral for S_σ is not inherently perturbative. Perturbation theory is adopted for the practical purpose of deriving its consequences.)

---

[12] The ideas of this section were presented in Huggett and Vistarini (2015); see also Vistarini (2019, ch. 3). They are well known in the physics literature: e.g., Green et al. (1987, §3.4).

272   OUT OF NOWHERE

As usual, the theory must be regularized and renormalized, and this is the rub, for regularization can break the symmetries of a QFT. For example, one has to be careful to regularize in a way that preserves spacetime Poincaré invariance. However, such regularization generally breaks scale invariance, for example in the form of a cut-off, or mass parameter (in Pauli-Villars regularization). On renormalization, this scale dependence shows up in the renormalized coupling, leading to 'running' couplings, that vary in the energy, or length scale of scattering. In short, a renormalized theory is generally not scale invariant.

The non-linear sigma model is a QFT on the worldsheet, so renormalization of the non-linear sigma model accordingly produces a scale dependence *with respect to the worldsheet* in string theory. But as we have seen repeatedly, worldsheet scale invariance—Weyl invariance—is a crucial feature of string theory, and hence there is a potential conflict. We will discuss in §9.4 why this conflict must be avoided, why Weyl symmetry must not be broken in this way; in the remainder of this section we will take that constraint for granted, and study its consequences. To do so, we need to know how the renormalized coupling runs, something described in renormalization group theory by the β-function, the derivative of a parameter g with respect to changes in scale Λ.[13] Schematically:

$$\beta_g \equiv \frac{dg}{d \ln \Lambda}. \tag{9.13}$$

Clearly, scale invariance of a parameter requires that the corresponding β-function vanish: $\beta_g = 0$. In the case of our sigma model, with action (9.12), Friedan (1980) (also Callan et al. 1985) showed perturbatively that the β-function associated with $g_{\mu\nu}$ is given by

$$\beta_g = R_{\mu\nu} + O(\alpha'^2), \tag{9.14}$$

where $R_{\mu\nu}$ is the Ricci tensor for the background spacetime, $g_{\mu\nu}$. Therefore Weyl symmetry requires that (to first order) $R_{\mu\nu} = 0$.[14] But that's to say that $g_{\mu\nu}$ is 'Ricci flat', which is simply the vacuum EFE, appropriate to (9.12) since it contains no matter fields. That is, Weyl invariance for the sigma model QFT entails that the spacetime metric be an allowed solution of GR (at lowest order in α', for tree diagrams). Note well that assuming $\alpha' \ll 1$ is not at all the same as assuming $|g| \ll 1$, so this result is not limited to a weak, linear gravitational field. We will discuss this point further in §10.5.

So the final step in the identification of string theory as a theory of gravity is complete: a background field composed of massless spin-2 excitations behaves

[13] See Huggett (2002) for a philosophical introduction to these ideas.
[14] Strictly, scale invariance only requires that the perturbative series sums to zero, and setting each term to zero is a stronger assumption. But since (9.14) will have quantum corrections anyway, it is enough for our purposes that it approximately vanish here.

THE STRING THEORETIC ACCOUNT OF GENERAL RELATIVITY   273

as the gravitational field of GR. But a number of points are worth unpacking: Are the appropriate field equations obtained when there are background matter fields? What about quantum corrections to the metric? Why can't classical scale invariance be broken by the quantized theory? Before addressing these, we should emphasize a few points about the theory under consideration.

To start, a quick reiteration: string theory is scale invariant with respect to the worldsheet metric $h_{\alpha\beta}$, not the spacetime metric $g_{\mu\nu}$. In the derivation, the latter is treated as a renormalized, (potentially) scale dependent, non-linear coupling for an X^μ field on the worldsheet; then invariance with respect to rescaling $h_{\alpha\beta}$, entails that $g_{\mu\nu}$ satisfies the vacuum equation of GR. Now, *classically* a string only occupies a 2-dimensional subspace of spacetime, and moreover, as a mathematical fact, the classical action (9.12) simply is scale invariant with respect to $h_{\alpha\beta}$, whatever value $g_{\mu\nu}$ takes: so classically nothing at all follows for the spacetime metric. The result discussed holds when one *quantizes* and performs the path integral $\int DX$ over all possible embeddings in spacetime, weighted by e^{S_σ}. The path integral makes the resulting amplitude dependent on the values of $g_{\mu\nu}$ at *all* points of spacetime, since every point is occupied by a point of the string in some path. Moreover, as we saw, regularization means that β_g is a non-trivial function, which must vanish for scale invariance, and so quantum string theory places a substantive constraint on the metric of spacetime.

It is also important to bear in mind the understanding of $g_{\mu\nu}$ developed in the previous section. According to that picture, $R_{\mu\nu} = 0$ is to be understood as a dynamical constraint on the (coherent) states of stringy gravitons, since that is what $g_{\mu\nu}$ represents. Thus in string theory, the field equations are nothing but approximate descriptions of strings themselves. In particular, the background metric does not represent a new degree of freedom in addition to those of string theory, instead it is a special string state: it is thus not a distinct primitive entity, and hence not a 'background' at all in the sense of an ontologically prior given.

To develop this important point, consider non-vacuum solutions, in which a second background field is posited. As before, these will also represent the presence of many-string coherent states, observed by their contributions to scattering amplitudes. We saw in §7.3 that in addition to the graviton, the first (massless) excited states of the closed string also include a scalar dilaton field Φ, and another tensor field, the Kalb-Raymond field $B_{\mu\nu}$. When coherent states of these stringy quanta are included as background fields, the same story can be told: the form of the vertex operator determines the form of the background term in the sigma action, and requiring that β-functions vanish leads to the appropriate GR equations of motion. For example, if one includes just a dilaton background the action becomes:

$$-\frac{1}{4\pi\alpha'}\int d^2\sigma\sqrt{-h}\left(g_{\mu\nu}(X)\partial_\alpha X^\mu\partial^\alpha X^\nu + \alpha' R^{(\sigma)}\Phi(X)\right), \tag{9.15}$$

274   OUT OF NOWHERE

where $R^{(\sigma)}$ denotes the worldsheet Ricci scalar. The graviton and dilaton β-functions are (in 26 dimensions)[15]

$$\beta_g = \alpha' R_{\mu\nu} + 2\alpha' \nabla_\mu \nabla_\nu \Phi + O(\alpha'^2)$$

$$\beta_\Phi = -\frac{\alpha'}{2} \nabla^2 \Phi + \alpha' \nabla_\mu \Phi \nabla^\mu \Phi + O(\alpha'^2). \tag{9.16}$$

In turn, the conditions of Weyl symmetry, $\beta_g = \beta_\Phi = 0$, are (to first order in α') the Euler-Lagrange equations for the spacetime effective action

$$S_{\text{Eff}} = \frac{1}{2\kappa} \int d^{26}x \sqrt{-g}\, e^{-2\Phi} (R - 4\nabla_\mu \Phi \nabla^\mu \Phi). \tag{9.17}$$

This action is almost that for a scalar field in GR, except for the extra dilaton factor $e^{-2\Phi}$. However, this factor can be absorbed into the spacetime metric: $g_{\mu\nu}(X) \to g'_{\mu\nu}(X) = e^{-\Phi(X)} g_{\mu\nu}(X)$. Then

$$S_{\text{Eff}} = \frac{1}{2\kappa} \int d^{26}x \sqrt{-g'}(R' - 4\nabla_\mu \Phi \nabla^\mu \Phi); \tag{9.18}$$

once again, the correct action, this time for a scalar field, with κ as the gravitational coupling constant $\sqrt{8\pi G_N}$. As before, Weyl symmetry is the crucial condition.

The rescaling of the metric to achieve this result is worth noting. It is of course not justified by spacetime conformal symmetry, since neither string theory, nor GR has this symmetry. Instead, the passage from (9.17) to (9.18) indicates that from the point of view of an effective spacetime dynamics, the effect of a dilaton field is to rescale the metric. Conversely, one can rescale a metric field, at the price of introducing a dilaton field. Thus the two descriptions are equivalent from the point of view of scattering amplitudes. Of course, it is unmotivated to introduce a dilaton, as a Poincaré-like universal force, simply for this purpose in GR. But dilatons are part of the string spectrum, so unavoidable in this case. The descriptions (9.17) and (9.18) are called the 'matter frame' and 'Einstein frame', respectively, but 'frame' does not denote a coordinate system, rather the way that degrees of freedom are arbitrarily divided between matter and geometry in such a theory. This kind of equivalence of matter and geometry calls for philosophical study.

Note that along similar lines, a constant part of $\Phi(X)$ could be pulled out of the integral, and used to rescale κ, or equivalently G_N. In other words, we see that the dilaton dynamically controls the strength of gravity! (And indeed, on further analysis, the strength of the string coupling, g_s. See Green et al. 1987, §3.4.6.)

---

[15] The derivation is not particularly useful for our purposes, so the reader is referred to Green et al. (1987, §3.4.5) and Polchinski (1998, §3.4), from which this example is taken.

THE STRING THEORETIC ACCOUNT OF GENERAL RELATIVITY   275

A more realistic case involves the 'heterotic' string. The heterotic string has fermionic in addition to bosonic degrees of freedom, as discussed in §7.4. The points of a heterotic string have fermionic coordinates, θ, in addition to the familiar bosonic ones σ: in effect the string has some fermionic 'thickness', expressed by Grassman numbers (these anti-commute even classically, so that $\gamma^2 = 1/2\,(\gamma \cdot \gamma + \gamma \cdot \gamma) = 0$). Here, the crucial point is that the heterotic string has quanta of Yang-Mills fields among its excitations, and hence models matter fields. Again, one can include the effect of coherent states by including background fields in the action:

$$S_\sigma \sim T \int \mathrm{d}^2\sigma \mathrm{d}\theta \,\left(g_{\mu\nu}\partial X^\mu \partial X^\nu + A_{\mu\alpha}(X)\partial X^\mu j^\alpha + \theta \psi^i \partial \psi^i\right). \qquad (9.19)$$

Since the heterotic string has both bosonic and fermionic coordinates, both must be integrated over. $A_{\mu\alpha}$ is a background gauge field, and ψ the fermions to which it couples (j^α is their worldsheet current): both are excitations of the heterotic string. Callan et al. (1985) investigated this action, showing that to lowest order $g_{\mu\nu}$ is Ricci flat, and that the standard free Yang-Mills equations must be satisfied. Moreover, they also showed that at first order in α' the β-function for the target metric has a term corresponding to the stress-energy of the Yang-Mills field.

$$\beta_G = R_{\mu\nu} - \frac{\alpha'}{2}\mathrm{tr}(F^2_{\mu\nu}) + O(\alpha'^2) \text{ terms for the coupling to } g_{\mu\nu}. \qquad (9.20)$$

Thus, once again, worldsheet conformal symmetry, $\beta_G = 0$, means that to order α', the EFE holds, even when matter is present.

These two examples indicate the generality of the result that background fields (approximately) satisfy the EFE, but also emphasize the ontological picture behind it. The EFE is in fact a consistency constraint between modes of the string, arising in quantum string theory because of Weyl symmetry. (In the last two examples, we obtained such a relation between different coherent states, including the graviton.) The situation is like that in GR, insofar as there are many solutions, many consistent pairings of matter and metric. However, in string theory it is absolutely clear that these two things—matter and metric—have a common nature, as string excitations: so the field equations do not relate two different kinds of things, for there is no fundamental distinction between the two. In this sense, string theory provides a significant ontological unification.[16]

Some final points should be emphasized. First, the results that we have discussed are only to lowest order in the sigma model expansion (so order α', that is), and tree

---

[16] Of course, there are those (e.g., Rovelli 1997, 193f) who interpret the metric as of essentially the same kind of thing as other physical fields, even at the classical level. But in string theory there is no choice.

diagrams (so lowest order in the string coupling, g_s): our world sheets are topological spheres with legs/vertex operators, not tori. At this level, backgrounds behave effectively as classical fields in a classical spacetime, according to standard classical equations. But of course, higher level contributions will entail quantum corrections to these classical equations, including the EFE. String theory only entails GR approximately—in generality, it predicts observable deviations, though perhaps unobservable by foreseeable technology.[17] (Of course, away from the classical limit that we have been assuming, effective classical behavior completely breaks down, and there will only be a quantum description of strings.)

Second, string theorists use classical equations to describe background spacetimes in a very wide variety of cases: see Peet (2001) for an indication of the range of black hole models in (super)string theory, or Gasperini (2007) for string models of big bang cosmologies. These models involve quantum corrections, but to an effective classical background, presumably on the assumption that (away from their singularities) the classical big bang, and classical black holes can be modeled as stringy coherent states in the way we have considered. The legitimacy of this assumption will be addressed in chapter 10.

Finally, while the route to relativistic spacetime described here is part of string theory lore, it is not thought to be the full, or most satisfying story, by most string theorists. Instead, they look to a more fundamental formulation, to which the formalism discussed here is supposed a perturbative approximation. As we have mentioned, it is not currently known what that formulation is: at one time string field theory was a popular candidate, then various forms of 'M-theory', none of which has (yet) succeeded as a fundamental formulation. However, a common, long-standing hope is that such a theory will not involve spacetime degrees of freedom in its formulation—that spacetime will be fully 'emergent'.[18] That such is the case in string field theory, and that dualities physically identify different spacetimes are offered as evidence that it will also be the case in M-theory. However, our goal here has been to pursue the question in the perturbative theory that we have at present, and we have indeed seen that even at this level spacetime can be derived in a sense: specifically, the metric is a dynamical feature of string states, and indeed has the same explanation in terms of string modes as all other fields.

9.4 Anomalies

The derivation of the EFE is the ground for string theory's claim to be a theory of gravity. And clearly, a key component of that derivation is the stipulation that quantum string theory be Weyl invariant. It is a straightforward mathematical fact

---

[17] See Kiefer (2004, §2.2.3) for general considerations about gravity as an effective theory.
[18] See Rickles (2014, ch. 9) for some of the history of this desideratum for string theory.

THE STRING THEORETIC ACCOUNT OF GENERAL RELATIVITY 277

that the classical theory is, but in our presentation it was assumed that the same is true at the quantum level: the β-functions were not automatically zero, but rather we set them to zero, with important physical consequences. For this and other reasons, it is not an exaggeration to say that Weyl invariance of quantum string theory is one of the most important pillars of the theory; so we need to understand its status. The issue is especially pressing given that the derivation involves perturbation theory: one might wonder whether the non-trivial β-functions are artifacts of a faulty perturbative approach, or in the necessary regularization, and that an 'exact' treatment of the sigma action might not permit the derivation to proceed.

There is in fact a straightforward answer to the question of 'why Weyl invariance?'. In gauge theory (including the non-linear sigma model), classical local gauge symmetries that are broken in quantum theory lead to pathologies, especially violations of unitarity: quantum theories, on pain of inconsistency, must avoid local 'anomalies'. String theory rests on this general principle. Despite its central role in a range of gauge field theories, the significance of the principle has been almost entirely neglected by philosophers of physics.[19] Because Weyl symmetry is such a central premise in the interpretation of string theory, we therefore need to devote a little space here to explicating anomalies in general, and in string theory in particular.

The important background to this discussion is Noether's (first) theorem, in both its classical and quantum forms. Because it is something of a detour to the main thread of this chapter, an intuitive demonstration and explication of the theorem (especially the less familiar quantum version) has been placed in appendix 9.B, rather than here. In short, suppose that $\varphi(x) \to \varphi'(x) = \varphi(x) + \delta\varphi(x)$ is an infinitesimal symmetry of the action: $S[\varphi(x)] = S[\varphi'(x)]$, for any $\varphi(x)$ (not just the physical ones, which minimize S). Then the classical Noether theorem delivers a conserved current, $\partial j(x) = 0$ (where ∂ is the spacetime divergence). The action has the same form in the corresponding quantum theory, but it *alone* is not the central quantity. Rather, the path integral is, which involves *both* the action and the measure of functional integration, $D\varphi$: the path integral is given by $\int D\varphi \, e^{iS[\varphi]}$.

If $\varphi(x) \to \varphi'(x) = \varphi(x) + \delta\varphi(x)$ is a symmetry of *both* action and measure, then the quantum Noether theorem delivers a current with a conserved expectation value: $\partial\langle j_\varphi(x)\rangle = 0$. But a symmetry of the action need not also be a symmetry of the measure, and so not all classical conserved currents are also conserved by the corresponding quantum theory. When this occurs, the quantum theory is said to possess an 'anomaly'. As physicists came to understand in the 1960s and '70s (see Fine and Fine 1997), anomalies have some very important implications for

---

[19] The exceptions that we have been able to find are a longer, but largely historical, treatment by Fine and Fine (1997), and brief discussions by Cao (1998, 218ff), Healey (2007, 182f), and Huggett and Vistarini (2015). See also the excellent discussion by Duncan (2012, §15.5) in the physics literature.

278 OUT OF NOWHERE

QFT, allowing more physical possibilities in some cases, and ruling them out in others. To emphasize how central these field-theoretic ideas are to string theory (and because they are not at all well-known to philosophers of physics) it is worth briefly sketching this history.

Anomalies were first investigated in the case of 'chiral' symmetries.[20] Consider flat spacetime with a *classical* massless Dirac field $\psi(x)$: a 4-component vector field that is acted on by the Dirac matrices γ^μ ($\mu = 0, 1, 2, 3$). Suppose that this field is coupled to a background SU(N) gauge field A_μ: defining $\bar{\psi} \equiv \psi^\dagger \gamma^0$, the action is given by

$$S = \int d^4x \, i\bar{\psi}\gamma^\mu(\partial_\mu + A_\mu)\psi \equiv \int d^4x \, i\bar{\psi} \, \nabla \, \psi, \qquad (9.21)$$

where ∇ is the Dirac operator.[21] Define $\gamma^5 \equiv -\gamma^0\gamma^1\gamma^2\gamma^3$. It is straightforward to check that for constant α, $\psi \to e^{i\alpha\gamma^5}\psi$ is a symmetry of the action; and, using the methods described in appendix 9.B, to show that the corresponding classically conserved 'chiral' (or 'axial') current is $f_5^\mu(x) = \bar{\psi}\gamma^\mu\gamma^5\psi$. The name comes from the inclusion of γ^5, which has eigenvalues ± 1, distinguishing positive from negative chirality states.

In the quantum theory the path integral is $\int D\psi D\bar{\psi} \, e^{-S}$. We already know that the exponent is symmetric under $\psi \to e^{i\alpha\gamma^5}\psi$, but is the measure? As Adler (1969), and Bell and Jackiw (1969) discovered in the late 1960s, the answer is 'no'. As explained in the appendix, calculating quantum currents involves calculating the transformed measure: the Jacobian of the transformation. Quoting from Nakahara (2003, §13.2.1) the result is that

$$\partial\langle j_\varphi(x)\rangle = \frac{1}{4\pi^2}\text{tr}\left[\epsilon^{\kappa\lambda\mu\nu}\partial_\kappa(A_\lambda\partial_\mu A_\nu + \frac{2}{3}A_\lambda A_\mu A_\nu)\right], \qquad (9.22)$$

the abelian 'chiral anomaly'—the (generally non-zero) amount by which the current fails to be conserved.

This anomaly arises from a quantum breaking of a classical global symmetry, not of a non-abelian local gauge symmetry. It has important physical consequences (e.g., Aitchison 2007, §8.3): if the classical symmetry held in quantum mechanics, then the conserved current would in fact prohibit the observed decay of pions. (It was the discovery that the anomaly permits pion decay which really motivated interest in anomalies.) In the context of our concern that the Weyl anomaly might be caused by an incorrect approach to regularization or perturbation theory, it is worth noting that the chiral anomaly was also first discovered in

---

[20] This discussion follows Nakahara (2003, §13.2), who in turn follows Fujikawa's approach.

[21] (9.21) is the action when time has been Wick rotated into the complex plane, and an equivalent theory in 4-dimensional Euclidean space is given. In the path integral below, the path weight is thus e^{-S} rather than e^{iS}.

THE STRING THEORETIC ACCOUNT OF GENERAL RELATIVITY   279

perturbation theory, and so similar questions can be asked. But it is in fact independent of the form regularization takes (e.g., Aitchison 2007, 154), and by the methods described in appendix 9.B, it can be understood non-perturbatively. The same points hold for the sigma model of string theory.

The reason that the abelian anomaly causes no problems is that it does not couple to the gauge field; there is no j_5^μ-A^μ interaction. That's not surprising, given that the symmetry in no way involves A^μ. However, there is a related *local* chiral symmetry with a classically conserved current that does couple to the gauge field.[22] It too is broken in the quantum theory, but now a coupling to A^μ means that the associated current shows up in Feynman diagrams, and its non-conservation means that it does not conserve probability—the theory is non-unitary. Far from having beneficial effects, the non-abelian chiral anomaly implies that the theory cannot be quantized at all. Moreover, the reasons, just given, that a violation of local gauge symmetry leads to non-unitarity is general. Once this point was understood, that local gauge symmetries must hold in quantum mechanics became an important guiding principle in QFT research, methodologically comparable to—though logically stronger than—the gauge principle, or a demand for renormalizability. (For completeness, we should note that the standard model also violates local chiral symmetry: but in this case other anomalies in the theory exactly cancel out, leaving the theory safe overall.)

That brings us back to strings, for Weyl symmetry $h_{\alpha\beta} \to e^{\phi(\tau,\sigma)} h_{\alpha\beta}$ is a local gauge transformation, and so all the foregoing applies to quantizing string theory, as we shall now see.

In fact there is a very direct way to derive a conservation law for the classical string (bypassing the method described in appendix 9.B). Vary the string action (9.1) by an infinitesimal Weyl transformation, $h^{\alpha\beta} \to e^{\phi(\tau,\sigma)} h^{\alpha\beta} = (1 + \phi(\tau,\sigma)) \cdot h^{\alpha\beta}$: thus $\Delta h^{\alpha\beta}(\tau,\sigma) = \phi(\tau,\sigma) h^{\alpha\beta}(\tau,\sigma)$. Then according to the calculus of variations:

$$\Delta S_\sigma = \int d\tau d\sigma \frac{\delta S_\sigma}{\delta h^{\alpha\beta}} \cdot \Delta h^{\alpha\beta} = \int d\tau d\sigma \frac{\delta S_\sigma}{\delta h^{\alpha\beta}} \cdot \phi(\tau,\sigma) h^{\alpha\beta} = 0, \qquad (9.23)$$

since the transformation is a symmetry. But $\phi(\tau,\sigma)$ is an arbitrary function, and so (from the fundamental theorem of the calculus of variations),

$$\frac{\delta S_\sigma}{\delta h^{\alpha\beta}} \cdot h^{\alpha\beta} = 0. \qquad (9.24)$$

Interpreting this quantity as the divergence of a 'Weyl current' entails that the current is conserved. In fact, we have already seen part of this expression in (7.22)

---

[22] First investigated by Bell and Jackiw (1969); but see Nakahara (2003, §13.3) for a treatment along the lines we have followed.

280 OUT OF NOWHERE

of §7.2, where we identified it as the worldsheet stress-energy tensor. That is, we have

$$T_\alpha^\alpha(\tau, \sigma) \equiv T_{\alpha\beta} \cdot h^{\alpha\beta} = -\frac{2}{T}\frac{1}{\sqrt{-h}}\frac{\delta S_\sigma}{\delta h^{\alpha\beta}} \cdot h^{\alpha\beta} = 0. \tag{9.25}$$

Thus interpreting (9.24) as the conservation law associated with Weyl symmetry, amounts to the demand that the *trace* of the stress-energy tensor vanish. This result can readily be checked in the case of (7.28): since $T_{\tau\tau} = T_{\sigma\sigma}$ the diagonal terms cancel when the tensor is contracted with the Minkowski worldsheet metric $\eta_{\alpha\beta}$.

But for the reason discussed—possible non-symmetry of the path integral measure—it is not guaranteed that the symmetry will be preserved in a quantum mechanical path integral. The trace of the stress-energy may not vanish, so the quantized Weyl current may not be conserved, with a consequent failure of unitarity. Of course, we saw in the previous section that just such a possibility occurs if the spacetime background fails to satisfy (at lowest order) the field equations (or indeed, if spacetime has the wrong dimensionality). There we took for granted that Weyl symmetry must hold in quantized string theory: now we have seen that quantum theories require that such local gauge symmetries must hold, and that in particular quantum string theory *must* possess Weyl symmetry.

For completeness, let us determine the form of the anomaly (Polchinski 1998, 43, 92). First, the (expectation value of the) stress-energy tensor can be computed simply by the method just described, in the case of a flat worldsheet metric, $h_{\alpha\beta} = \eta_{\alpha\beta}$: the result again is that $T_\alpha^\alpha(\tau, \sigma) = 0$. Since Weyl symmetry is in question, one cannot conclude that $T_\alpha^\alpha(\tau, \sigma) = 0$ in all conformally equivalent worldsheet metrics, and so this special case does not demonstrate quantum Weyl symmetry. However, one can now reason as follows: any Weyl anomaly appears because the theory must be regularized to make sense (this applies in the exact theory, as well as the perturbative theory), but regularization can be chosen to preserve Poincaré invariance and diffeomorphism symmetry (of worldsheet and spacetime), so that they remain quantum symmetries. Thus $T_\alpha^\alpha(\tau, \sigma)$ must be symmetric under those transformations, and so can only contain quantities symmetric under them. But we just saw that $T_\alpha^\alpha(\tau, \sigma)$ vanishes for a flat worldsheet, from which it follows that only the worldsheet scalar curvature can contribute: hence we must have

$$T_\alpha^\alpha(\tau, \sigma) = a_1 R^{(\sigma)}, \tag{9.26}$$

the 'Weyl anomaly'.

This result is illuminating for two reasons. First, some more calculation (e.g., Polchinski 1998, §3.4) shows that a_1 does not vanish unless $D = 26$, which it must in general for the trace to vanish and the Weyl current to be conserved.

THE STRING THEORETIC ACCOUNT OF GENERAL RELATIVITY 281

Thus this result shows more directly how the dimensionality of spacetime is a consequence of Weyl symmetry. Second, it is worth noting that anomalies typically have their origins in the topology of the fields. For example (see Nakahara 2003, 440f), the chiral anomaly (9.22) is the second Chern character of the fiber bundle of the gauge field, A_μ; and its integral over spacetime is a topological invariant of the bundle, the second Chern class.[23] In the case of string theory, the integral of the Weyl anomaly (9.26)—i.e., the worldsheet Ricci scalar $R^{(\sigma)}$— over the worldsheet is the Euler number; again a topological invariant, this time of the worldsheet. Broadly speaking, $\mathcal{D}\varphi$ can 'see' the whole of space, and is sensitive to its topology, because it is the measure in the space of *complete* field configurations.

The aim of this section was to provide an elementary introduction to anomalies, and to their role in QFT, through a general theoretical framework and some concrete examples (further background, concerning Noether's theorem, and the derivation of conserved currents and anomalies, can be found in appendix 9.B). The first important lesson is that these considerations are not intrinsically perturbative, but rather concern the path integral taken as an exact object, in this case the path integral of the sigma action over a worldsheet. Of course, one may need to apply perturbation theory to calculate quantities as a practical matter: the β-functions were given this way, for instance. But one should not think that anomalies occur only as a result of a particular perturbation scheme, and might be absent (or different) in another. When they occur, they are features of the exact QFT, and show up in any perturbative scheme.

The second lesson is that local gauge anomalies—of which the abelian chiral anomaly is not an example, but the Weyl anomaly is—cannot be tolerated, because they lead to non-unitarity. Hence we were fully justified in demanding scale invariance for the β-functions: the requirement is one of mathematical necessity, not an external desideratum. (And bear in mind that the presence of an anomaly is equally a mathematical fact for any given theory: it is a purely formal matter whether the classical action has a conserved current, and a purely formal matter whether it is conserved in the quantum theory.) In short, as soon as we wrote down the classical sigma action, it was logically necessary that the background fields of the quantized theory satisfy the EFE.[24]

---

[23] In fact, it is equal to $\nu_+ - \nu_-$, where $\nu_\pm$ is the number of positive (or negative) chirality eigenstates of $\nabla\!\!\!/$ with eigenvalue 0 (i.e., configurations $\psi(x)$ for which $\gamma^5\psi = \pm\psi$ and $\nabla\!\!\!/\psi = 0$.) This identity between objects characterizing topological and differentiable structure is an instance of the Atiyah-Singer 'index theorem'.

[24] Not quite. As discussed in Polchinski (1998, §9.9) and Vistarini (2019, §3.3.1), if one breaks Weyl symmetry, unitarity entails that new terms must appear in the action with an 'equal and opposite' anomaly, to cancel the Weyl anomaly. The argument of this paragraph assumes that no such terms are added.

282 OUT OF NOWHERE

9.5 Summary

We will review the derivation of the field equations at the start of the next chapter, before considering their implications, so here we will only recap the main steps and ideas. In this presentation, we first argued that if one wants to describe scattering in the presence of a coherent state of gravitons, then one should replace the flat target space metric $\eta_{\mu\nu}$ with an arbitrary curved one $g_{\mu\nu}$ in the action. That's because the coherent excitation is formally equivalent to an exponentiated rank-2 tensor coupled to the string X^μ fields; so it appears in the path integral as a contribution to the metric (and similarly for the coherent states of other string excitations). This formal story suggests that spacetime curvature should be understood as in a sense *apparent*, the effects of gravitons in an *actually* flat spacetime. We will question this understanding in the next chapter. The important ideas for now are those of vertex operators, which arise from the conformal symmetry of the theory, and of coherent states, based on the QFT explanation of classical fields.

The second step relies on the fact that string theory has an anomaly which becomes apparent when the path integral is regularized. Since the anomaly is associated with Weyl symmetry, which is local, varying in spacetime, it is a pathology, leading to a failure of unitarity, and would make string theory inconsistent. However, (amongst other necessary conditions, including $D = 26$) the anomaly vanishes if, to lowest order, the target space metric satisfies the EFE with any other 'background' fields (as coherent states are known when they are incorporated into the action). Hence the laws of GR are entailed for the metric of target space in quantum string theory. To better understand this point we discussed anomalies at some length.

So formally speaking, we now understand why, according to string theory, matter couples to the metric as required by GR. In the next chapter we will continue to ask how this result amounts to a derivation of GR, and especially whether such a derivation of the metric amounts to a derivation of spacetime.

9.A Appendix: coherent states of quantum fields

It is clear from the preceding that the idea that the appearance of classical fields is to be understood in terms of coherent states of quantum fields is crucial to the claim of string theory to recover GR: the metric field is taken to be a coherent state of stringy gravitons. This idea is a common one in QFT, and clearly of the greatest importance for the interpretation of QFT, yet it has been almost entirely neglected in the philosophy of physics literature (two significant exceptions being Wallace 2012 and Rosaler 2013). Absent a satisfactory treatment

THE STRING THEORETIC ACCOUNT OF GENERAL RELATIVITY 283

elsewhere, we have to include our own here to investigate the general claim; because it contributes to the interpretation of QFT more generally, rather than quantum gravity specifically, we have placed it outside the main discussion, in this appendix.[25]

We will develop the essential ideas of coherent states in the context of the real scalar field for simplicity, but the same general considerations apply to the gravitational/metric field. For our purposes, two points in particular are important: that coherent quantum states best represent classical fields (so they represent classical background fields in the action), and that they correspond to an infinite superposition of excitation levels (so that they produce the correct contribution to the actions (9.9) and (9.11)). Both of these points will be explained in this appendix. Our strategy will generally be to work with the simplest possible case, in order to illuminate the general points that hold more widely.

9.A.1 Massless scalar field

First we review the basic ideas of QFT. So consider a massless (for simplicity) classical scalar field, satisfying the Klein-Gordon equation, with a background source $\rho(t)$:

$$\partial_t^2 \varphi(x, t) - \nabla^2 \varphi(x, t) = \rho(x, t), \tag{9.27}$$

for which the Lagrangian (density) is

$$\mathcal{L} = 1/2[(\partial_t \varphi)^2 - (\nabla \varphi)^2] - \rho \varphi. \tag{9.28}$$

This section presents some standard material, but in addition to serving as a brief introduction or reminder for those who would like it, it also differs from standard presentations[26] by allowing for a source term. We are ultimately interested in GR, and in part in the linearized theory, in which the equations of motion take the form $\partial_t^2 \bar{h}^{\mu\nu} - \nabla^2 \bar{h}^{\mu\nu} = 16\pi T^{\mu\nu}$: the matter energy-momentum $T^{\mu\nu}$ acts as a source term for the gravitational field $\bar{h}^{\mu\nu}$. So we should similarly allow for a source in our analogy.[27] Let us emphasize that there is no back-reaction on the source in (9.27); it has no dynamics even though it is time-dependent. Moreover, the source will remain classical when the field is quantized.

[25] We want to thank Joshua Rosaler, both for discussion of these matters and for his excellent thesis on classical limits of QFT just cited, which was very helpful in preparing this treatment. We have also benefited from discussion with David Wallace.

[26] We draw on the nice treatments found in Lancaster and Blundell (2014), especially chapter 11, or Tong (2006).

[27] Even though we will ultimately show that it makes very little difference to the main conclusions; indeed, we will drop ρ terms at certain points for simplicity of expression.

284   OUT OF NOWHERE

According to the standard framework of Lagrangian mechanics, the canonical momentum is thus given by the functional derivative

$$\pi(x,t) = \delta\mathcal{L}/\delta(\partial_t\varphi) = \partial_t\varphi. \tag{9.29}$$

While the Hamiltonian (density) is

$$\mathcal{H} = \pi\partial_t\varphi - \mathcal{L} = 1/2[(\partial_t\varphi)^2 + (\nabla\varphi)^2] + \rho\varphi = 1/2[\pi^2 + (\nabla\varphi)^2] + \rho\varphi. \tag{9.30}$$

To keep things as simple as possible, we will work in 1+1-dimensional Minkowski spacetime, and in a particular frame rather than in a manifestly covariant way; we also suppose 2π-periodic boundary conditions (to avoid cluttering expressions with factors keeping track of the measure over phase space). These assumptions are readily relaxed, without changing the fundamental lessons.

We first Fourier-expand the field and source in momentum components:

$$\varphi(x,t) = \sum_{k=-\infty}^{+\infty} \alpha_k(t)e^{-ikx}$$

$$\rho(x,t) = \sum_{k=-\infty}^{+\infty} \rho_k(t)e^{-ikx}. \tag{9.31}$$

Inserting into the Klein-Gordon equation (9.27), we find that $\ddot{\alpha}_k + k^2\alpha_k = \rho_k$, so that

$$\alpha_k(t) = \alpha_k(0)e^{-ikt} + C_k(t), \tag{9.32}$$

for some source-dependent function $C_k(t)$, which vanishes for $\rho = 0$; which we can suppose is the case at $t = 0$ for notational convenience. The form of $C_k(t)$ does not matter for our purposes: what is important is that because it depends only on the source, it will be a c-number rather than an operator on quantization.[28]

It is a property of the Fourier expansion that in order for φ to be real, $f_k(t)e^{ikx} + f_{-k}(t)e^{i(-k)x}$ must be real for each value of k; hence $f_{-k}(t) = f_k^*(t)$. Thus (ignoring an overall constant)

---

[28] Its physical significance is clear: we have a forced oscillator, whose motion decomposes into an oscillatory part and a part describing the motion of its 'center of mass'. It is the oscillations that produce the salient features in our discussion; the source terms are included only to show that they do not make a relevant difference.

$$\varphi(x,t) = \sum_{k=1}^{\infty} \left(\alpha_k(t)e^{-ikx} + \alpha_k^*(t)e^{ikx}\right) \equiv \varphi^{(+)}(x,t) + \varphi^{(-)}(x,t), \qquad (9.33)$$

where the fields are identified with corresponding sums over the given terms. Similarly, we can write the canonical momentum (9.29) as

$$\pi(x,t) = \partial_t \cdot \sum_{k=1}^{\infty} \left(\alpha_k(t)e^{-ikx} + \alpha_k^*(t)e^{ikx}\right)$$

$$= \sum_{k=1}^{\infty} -ik \cdot \left(\alpha_k(t)e^{-ikx} - \alpha_k^*(t)e^{ikx} + D_k(t)\right), \qquad (9.34)$$

taking the time derivative of (9.32). Again, the form of the source-dependent $D_k(t)$ terms is unimportant for the general points to be made: what is important is that they are c-number-valued on quantization.

In units in which $\hbar = 1$ canonical quantization follows from the equal time commutation relations,

$$[\hat{\varphi}(x,t), \hat{\pi}(x',t)] = i\delta(x - x'). \qquad (9.35)$$

Note that we will work in the Heisenberg picture, so that operators are time-dependent. Rewrite (9.33) and (9.34) as operator equations:

$$\hat{\varphi}(x,t) = \sum_{k=1}^{\infty} \left(\hat{\alpha}_k(t)e^{-ikx} + \hat{\alpha}_k^\dagger(t)e^{ikx}\right) \equiv \hat{\varphi}^{(+)}(x,t) + \hat{\varphi}^{(-)}(x,t), \qquad (9.36)$$

and

$$\hat{\pi}(x,t) = \sum_{k=1}^{\infty} -ik \cdot \left(\hat{\alpha}_k(t)e^{-ikx} - \hat{\alpha}_k^\dagger(t)e^{ikx} + D_k(t)\right). \qquad (9.37)$$

Then (suppressing t-dependence) the canonical commutation relations entail that for any k,[29]

$$[\hat{\alpha}_k, \hat{\alpha}_{k'}^\dagger] = \frac{\delta_{k,k'}}{2\pi k}. \qquad (9.38)$$

Or defining $\hat{a}_k \equiv \sqrt{2\pi k} \cdot \hat{\alpha}_k$

$$[\hat{a}_k, \hat{a}_{k'}^\dagger] = \delta_{k,k'} \qquad (9.39)$$

---

[29] Using $\delta(x) = \frac{1}{\pi}\sum_{n=1}^{\infty} \cos nx$ for the 2π-periodic Dirac delta (up to the constant term that we are ignoring for brevity).

286 OUT OF NOWHERE

so that $\hat{a}_k^\dagger$ and $\hat{a}_k$ are raising and lowering operators, respectively, in a number basis for the kth mode,

$$\hat{a}_k^\dagger|n_k\rangle = \sqrt{n_k + 1}|n_k + 1\rangle, \qquad \hat{a}_k|n_k\rangle = \sqrt{n_k}|n_k - 1\rangle, \qquad (9.40)$$

in the way familiar from the simple harmonic oscillator (SHO). From (9.36) we have that $\hat{\varphi}^{(+)}(x,t) = \sum_{k=1}^{\infty} \hat{a}_k(t)e^{-ikx}$, which means that the action of the operator is to *annihilate* a single quantum of the field at (x,t). Conversely, $\hat{\varphi}^{(-)}(x,t)$ *creates* a quantum—the indices on the operators indicate whether they come from the $\pm k$ terms of the Fourier expansion, rather than their effect on the field.

Finally, we can ask about the Hamiltonian. Write the Hamiltonian density (9.30) as an operator equation and integrate over space, x:

$$\hat{H} = \int 1/2[\hat{\pi}^2 + (\partial_x\hat{\varphi})^2] + \rho\hat{\varphi} \, dx. \qquad (9.41)$$

Inserting $\hat{\pi}(x,t)$ and $\partial_x\hat{\varphi}(x,t)$ from above yields, after some manipulation,[30] and application of (9.39)

$$\hat{H} = \sum_{k=1}^{\infty} 2k \cdot (\hat{a}_k^\dagger\hat{a}_k + 1/2) + \sqrt{\frac{2\pi}{k}}(\rho_k\hat{a}_k^\dagger + \rho_k^*\hat{a}_k). \qquad (9.42)$$

The source-free ($\rho = 0$) field thus decomposes into modes that *independently* evolve according to the Hamiltonian for a SHO. (And of course $\hat{a}_k^\dagger\hat{a}_k$ is the number operator, $N_k|n_k\rangle = n_k|n_k\rangle$, so the mode-number eigenstates form the energy basis.) When a source is present ($\rho \neq 0$), then each mode behaves as an independent SHO with a driving term, which we will discuss presently.

9.A.2 Coherent states of a simple harmonic oscillator

At this point, it will be useful to review some important ideas concerning raising and lowering operators, and the states on which they act.[31] So note (temporarily suppressing the k index) that even though $\hat{a}$ is not Hermitian, it still has eigenstates, though with complex eigenvalues: $\hat{a}|\alpha\rangle = \alpha|\alpha\rangle$. Such 'coherent states' take the (unnormalized) form

---

[30] Using $\delta_{kk'} = \frac{1}{2\pi}\int_0^{2\pi} e^{i(k-k')x}dx$.

[31] Zwiebach (2013) provides a useful discussion.

$$|\alpha\rangle = e^{\alpha\hat{a}^\dagger}|0\rangle \tag{9.43}$$

$$\equiv \sum_{n=0}^{\infty} \frac{(\alpha\hat{a}^\dagger)^n}{n!}|0\rangle \tag{9.44}$$

$$= \sum_{n=0}^{\infty} \frac{\alpha^n}{\sqrt{n!}}|n\rangle, \tag{9.45}$$

as the reader can easily verify.[32] It is of course important for our purposes that such a state is a superposition of every excitation level: of every number of momentum k modes. This is exactly the property that lead us to exponentiate the vertex operators in (9.11), in the derivation of the EFE.

Although these SHOs arose as modes of a field, of course the same formalism applies to particles. That is, the canonical commutation relation $[\hat{x}, \hat{p}] = i$ entails that

$$\left[\frac{1}{\sqrt{2}}(\hat{x} + i\hat{p}), \frac{i}{\sqrt{2}}(\hat{x} - i\hat{p})\right] \equiv [\hat{a}, \hat{a}^\dagger] = 1, \tag{9.46}$$

defining raising and lowering operators for particle states. The SHO Hamiltonian $\hat{p}^2/2m + \lambda\hat{x}^2$ can be rewritten as $\omega(\hat{a}^\dagger\hat{a} + 1/2)$ so that number eigenstates are also energy eigenstates. But the number basis exists for any Hamiltonian, for instance for the free particle Hamiltonian $\hat{p}^2/2m = (\hat{a}\hat{a}^\dagger - \hat{a}^2 - \hat{a}^{2\dagger} - 1)/4m$, even though the number states are not now energy eigenstates.

The particle position representation brings out the first indication of the classicality of coherent states, which we shall pursue for the field in §9.A.4. That is, the coherent state property entails

$$\hat{a}\psi(x) = \sqrt{2}(\hat{x} + i\hat{p})\psi(x) = \sqrt{2}\left(x + \frac{d}{dx}\right)\psi(x) = \alpha\psi(x), \tag{9.47}$$

whose solutions are Gaussians $\sim e^{-(x-A)^2/2}$, which naturally represent localized particles. Moreover, their Fourier transforms are also Gaussians, so such states are also localized in momentum space. Indeed, once normalized, one can compute $\Delta x = \Delta p = 1/\sqrt{2}$ (where $(\Delta o)^2 \equiv \langle\hat{o}^2\rangle - \langle\hat{o}\rangle^2$); so that the Heisenberg uncertainty relation, $\Delta x\Delta p \geq 1/2$, is both saturated and simultaneously minimized by position and momentum. Hence coherent states are the most classical states possible given quantum uncertainty.[33]

Returning to the number representation, a significant property of coherent states is that they remain coherent when the Hamiltonian is $\hat{H} = \omega(\hat{a}^\dagger\hat{a} + 1/2)$,

---

[32] The identity $\hat{a}(\hat{a}^\dagger)^n = n(\hat{a}^\dagger)^{n-1} + (\hat{a}^\dagger)^n\hat{a}$ (resulting from the commutation relations for $\hat{a}$ and $\hat{a}^\dagger$) is useful in so doing. The normalization factor is $e^{-|\alpha|^2/2}$.

[33] For a field, coherent states can be represented as Gaussian wavefunctionals over classical field configurations: see Rosaler (2013, §3.1.2).

288   OUT OF NOWHERE

namely that for the SHO—*and that for any given mode in the source-free* ($\rho = 0$) *scalar field* (9.42). This is easily shown: the Heisenberg equations of motion give

$$\dot{\hat{a}} = i[\hat{H}, \hat{a}] = -i\omega\hat{a}, \quad \text{so that} \quad \hat{a}(t) = e^{-i\omega t}\hat{a}. \tag{9.48}$$

Thus (i) $\hat{a}^\dagger(t)$ and $\hat{a}(t)$ satisfy the commutation relations of raising and lowering operators (since $\hat{a}$ and $\hat{a}^\dagger$ do), and (ii) $\hat{a}(t)|\alpha\rangle = e^{-i\omega t}\alpha|\alpha\rangle$ (since $\hat{a}|\alpha\rangle = \alpha|\alpha\rangle$)—an eigenstate of the annihilation operator remains an eigenstate.[34] In other words, the state remains coherent, while its eigenvalue rotates about a circle radius $|\alpha|$ in the complex plane.

Important for our discussion is that coherence is also preserved when the oscillator is driven by an external (classical) force, so that the Hamiltonian becomes $\hat{H} = \omega(\hat{a}^\dagger\hat{a} + 1/2 + f(t)(\hat{a}^\dagger + \hat{a}))$, where $f(t)$ is a (rescaled) forcing term (recall that $\hat{x} \sim \hat{a}^\dagger + \hat{a}$). By the same reasoning as before, we then have

$$\dot{\hat{a}} = i[\hat{H}, \hat{a}] = i\omega[\hat{a}^\dagger\hat{a} + 1/2 + f(t)(\hat{a}^\dagger + \hat{a}), \hat{a}] = -i\omega\hat{a} - i\omega f(t) \tag{9.49}$$

so that

$$\hat{a}(t) = e^{-i\omega t} \cdot \left(\hat{a} - i\omega \int_0^t dt' f(t') e^{i\omega t'}\right). \tag{9.50}$$

Then we again have (i) that $\hat{a}^\dagger(t)$ and $\hat{a}(t)$ satisfy the commutation relations of raising and lowering operators. Moreover,

$$\text{(ii)} \quad \hat{a}(t)|\alpha\rangle = e^{-i\omega t} \cdot \left(\alpha - i\omega \int_0^t dt' f(t') e^{i\omega t'}\right)|\alpha\rangle, \tag{9.51}$$

so that an eigenstate of the annihilation operator remains an eigenstate. The state again remains coherent, but the trajectory of the eigenstate has two components: one ($e^{-i\omega t}\alpha$) rotating constantly in the complex plane, and one ($-i\omega \int_0^t dt' f(t') e^{-i\omega(t-t')}$) depending on the external force.

Now, the modes of a scalar field in the presence of a source have the Hamiltonian (9.42), and not exactly that of the forced SHO just considered. However, it is easy to see that the same reasoning will apply if we replace $f(t)(\hat{a}^\dagger + \hat{a})$ with

---

[34] To emphasize this point, it is perhaps helpful to make it also in the Schrödinger picture. Then we are interested in the time-dependent state $|\alpha\rangle(t) = \hat{U}(t)|\alpha\rangle$. We see that $\hat{a} \cdot |\alpha\rangle(t) = \hat{U}(t)\hat{a}(t)\hat{U}^\dagger(t) \cdot \hat{U}(t)|\alpha\rangle = \hat{U}(t)\hat{a}(t)|\alpha\rangle = e^{-i\omega t}\alpha|\alpha\rangle(t)$ (using point (ii) in the final step). Thus $|\alpha\rangle$ remains an eigenstate of $\hat{a}$ in a SHO potential: indeed, up to a possible time-dependent phase, $|\alpha\rangle(t) \propto |e^{-i\omega t}\alpha\rangle \equiv |\alpha(t)\rangle$. (It is straightforward to find the explicit form of the state by acting on (9.43) with $e^{-i\hat{H}t}$.)

$p(t)\hat{a}^\dagger + p^*(t)\hat{a}$ in (9.49), so that any mode of the field which starts in a coherent state will also remain coherent, *even when there is a classical source for the field*.

The property of remaining coherent does not, however, hold of general Hamiltonians: our result depended on $[\hat{H}, \hat{a}]$ having only c-number terms and terms linear in both $\hat{a}$ in (9.49). For instance, in contrast a free particle Hamiltonian has (inter alia) terms proportional to $(\hat{a}^\dagger)^2$, and $[(\hat{a}^\dagger)^2, \hat{a}] = -2\hat{a}^\dagger$; but the Heisenberg equation of motion $\dot{\hat{a}} = -2\hat{a}^\dagger$ does not have the nice solution upon which our result depended. Indeed, it can easily be shown that the uncertainty in position of a free particle which is initially in a coherent state increases as $\Delta x(t) = \sqrt{(\sigma + t^2\hbar^2/2m^2)}$, which is the sign of dispersion, the loss of coherence. Thus it is important to bear in mind that while coherent states play an important role in understanding classical systems in quantum mechanics, they cannot be the whole story. For instance, a freely moving speck of dust does not remain localized, but slowly spreads (unlike a speck in a SHO potential). However, since the mass suppresses the rate of spread, freely moving macroscopic objects will generally remain coherent for time intervals of interest. Suppose our speck of dust weighs 10^{-13}kg, has a radius of 10^{-6} m, and an uncertainty of 10^{-8}m (an order of magnitude below the resolution of optical microscopes): then it would take over 100,000 years for its uncertainty to merely double.

Because of their classicality, and persistence in such cases as this, coherent states seem promising candidates for representing observed classical systems, such as suitably large bodies or classical fields, in quantum mechanics. But we will see below limitations on this interpretational postulate. First, we need to develop coherent states of the field itself, not only its SHO modes.

9.A.3 Coherent states of the scalar field

We have seen two things so far: that the scalar field decomposes into dynamically independent oscillator modes (§9.A.1), and that coherent states of oscillators remain coherent (§9.A.2). (And noted the classicality of such states.) Now we want to apply these ideas to the scalar field; we will do so by first considering (normalized) eigenstates, not of the mode annihilation operator $\hat{a}_k$, but of the field annihilation operators, $\hat{\varphi}^{(+)}$, defined in (9.36):

$$\hat{\varphi}^{(+)}(x, t)|\psi\rangle = \psi(x, t)|\psi\rangle \tag{9.52}$$

(since $\hat{\varphi}^{(+)}$ is not Hermitian, $\psi(x, t)$ may be complex).

290   OUT OF NOWHERE

We can readily see that such states exist by working in the mode expansion. If $|\alpha_k\rangle$ is the α_k-eigenvalued coherent state of the kth mode, $\hat{a}_k|\alpha_k\rangle = \alpha_k|\alpha_k\rangle$, then let[35]

$$|\varphi\rangle \equiv |\alpha_1\rangle \otimes |\alpha_2\rangle \otimes \ldots |\alpha_k\rangle \otimes \ldots . \tag{9.53}$$

Making the tensor product explicit, the field annihilation operator (9.36) is written (at a fixed time)

$$\hat{\varphi}^{(+)}(x) = \sum_{k=1}^{\infty} e^{-ikx} I \otimes I \otimes \cdots \otimes \hat{a}_k \otimes I \ldots, \tag{9.54}$$

where $\hat{a}_k$ appears in the kth slot of the product. Thus

$$\hat{\varphi}^{(+)}(x)|\varphi\rangle = \sum_{k=1}^{\infty} e^{-ikx}\alpha_k \cdot |\varphi\rangle \equiv \varphi^{(+)}(x)|\varphi\rangle. \tag{9.55}$$

So states of the form (9.53) are coherent states of the field. And indeed, since each mode in the state involves a superposition of every number of quanta, so do the field coherent states, as we assumed in the construction of the corresponding vertex operator (9.8).

Therefore we can define our state label φ as in (9.33) to be (up to an overall constant):

$$\varphi(x) = \sum_{k=1}^{\infty} \left(\alpha_k e^{-ikx} + \alpha_k^* e^{ikx}\right) \equiv \varphi^{(+)}(x) + \varphi^{(-)}(x). \tag{9.56}$$

So we see immediately in (9.56) that a necessary and sufficient condition for $|\varphi\rangle$ to be a coherent states is that $\varphi(x)$ have a Fourier decomposition—or in the case without periodic boundary conditions, that it have a Fourier transform. Thus the αs are *both* the coherent state labels of a quantum field mode decomposition (9.53), *and* the coefficients of a classical field mode decomposition (9.56). And thus coherent states are in formal correspondence with well-behaved classical scalar fields.

---

[35] Since the modes Fourier decompose the field, (i) they take independent states, and (ii) specifying every mode completely determines a field: so tensor products of modes are field states. States of the form (9.53) constitute an overcomplete basis. However, because such states involve the superposition of an infinite number of excitation levels of at least one mode, they are unitarily inequivalent to the representation containing the vacuum (Ruetsche, 2011, §10.4).

THE STRING THEORETIC ACCOUNT OF GENERAL RELATIVITY   291

Moreover, from (9.36), (9.55), and (9.56) $\langle\varphi|\hat{\varphi}^{(-)}(x) = \langle\varphi|\varphi^{(-)}(x)$, so that for a normalized coherent state

$$\langle\varphi|\hat{\varphi}(x)|\varphi\rangle = \langle\varphi|\hat{\varphi}^{(+)}(x) + \hat{\varphi}^{(-)}(x)|\varphi\rangle = \langle\varphi|\varphi^{(+)}(x) + \varphi^{(-)}(x)|\varphi\rangle$$
$$= \varphi(x). \tag{9.57}$$

But the field expectation value is the quantum prediction of the field strength, and so the coherent states describe classical fields in the usual sense in quantum mechanics: they are quantum states of the fields corresponding to their labels. Note that they are not eigenstates of $\hat{\varphi}(x)$ (for which (9.55) also holds). We could alternatively (with reference to (9.57)) have labeled coherent states by their $\hat{\varphi}^{(+)}(x)$ (complex) eigenvalues $\varphi^{(+)}(x)$, but we choose not to in order to emphasize the connection to classical fields.

9.A.4 The classical limit

With the basic framework of coherent states in hand, we will make two points about understanding them as representing classical fields.

(a) First, on quantization, the Heisenberg equation of motion reads

$$\partial_t\hat{O}(x,t) = i[\hat{H}, \hat{O}(x,t)] = i\int dx' \, [\hat{\mathcal{H}}(x',t), \hat{O}(x,t)]$$
$$= i\int dx' \, [1/2(\hat{\pi}^2(x',t) + (\nabla\hat{\varphi}(x',t))^2) + \rho\hat{\varphi}(x',t), \hat{O}(x,t)], \tag{9.58}$$

using the operator version of (9.30). Using the canonical commutation relations (9.35) we have as special cases

$$\partial_t\hat{\varphi}(x,t) = i[\hat{H}, \hat{\varphi}(x,t)] = \hat{\pi}(x,t)$$
$$\partial_t\hat{\pi}(x,t) = i[\hat{H}, \hat{\pi}(x,t)] = \nabla^2\hat{\varphi}(x,t) + \rho(x,t), \tag{9.59}$$

for the field and its conjugate momentum. These immediately entail that

$$\partial_t^2\hat{\varphi}(x,t) - \nabla^2\hat{\varphi}(x,t) = \rho(x,t) \tag{9.60}$$

so that $\hat{\varphi}$ satisfies the Klein-Gordon equation with source, the same equation (9.27) as the classical field, φ. This result is just a special case of the more general correspondence between the equations of motion for field operators and their classical

292   OUT OF NOWHERE

counterparts, resulting from the parallel Hamilton and Heisenberg equations of motion.[36]

Hence for *any* state $|\psi\rangle$

$$\langle\psi|\partial_t^2\hat{\varphi}(x,t)|\psi\rangle - \langle\psi|\nabla^2\hat{\varphi}(x,t)|\psi\rangle = \langle\psi|\rho(x,t)|\psi\rangle, \qquad (9.61)$$

---

[36] Since it's an important aspect of the classical limit for quantum fields in general, it is worth just sketching why the correspondence holds. In the first case, suppose the Hamiltonian contains a term polynomial in the canonical variables, $\varphi^m\pi^n$. Then the Poisson bracket form of Hamilton's equation for φ will contain a contribution

$$\{\varphi, \varphi^m\pi^n\} = \frac{\delta\varphi}{\delta\varphi} \cdot \frac{\delta\varphi^m\pi^n}{\delta\pi} - \frac{\delta\varphi}{\delta\pi} \cdot \frac{\delta\varphi^m\pi^n}{\delta\varphi} = n\varphi^m\pi^{n-1}.$$

While on the quantum side, now treating the quantities as operators, satisfying $[\hat{\varphi},\hat{\pi}] = i$, we have $\hat{\pi}^n\hat{\varphi} = \hat{\varphi}\hat{\pi}^n - in\hat{\pi}^{n-1}$, which we can use to show that

$$[\hat{\varphi}, \hat{\varphi}^m\hat{\pi}^n] = \hat{\varphi} \cdot \hat{\varphi}^m\hat{\pi}^n - \hat{\varphi}^m\hat{\pi}^n \cdot \hat{\varphi} = in\hat{\varphi}^m\hat{\pi}^{n-1}.$$

Thus $\{\varphi, \hat{\varphi}^m\hat{\pi}^n\}$ ' $=$ ' $\frac{1}{i}[\hat{\varphi}, \hat{\varphi}^m\hat{\pi}^n]$, and the term's contribution to Hamilton equations (in Poisson bracket form) and Heisenberg equations of motion will be the same. The story continues to hold for a power of the spatial derivative, $(\partial^m\varphi)^n$, appearing in the Hamiltonian. φ will commute with such a term, so it will not contribute to the field equation of motion; the effect on the dynamics will only be through the canonical momentum. We have to be a little more careful to pay attention to spatial dependence, and use the field commutation relations (9.35) and explicitly integrate the term over space (which was done implicitly previously):

$$\int dx\, \{\pi, (\partial^m\varphi)^n\} = n\int dx\, \Big(\frac{\delta\pi}{\delta\varphi} \cdot \frac{\delta\partial^m\varphi}{\delta\pi} - \frac{\delta\pi}{\delta\pi} \cdot \frac{\delta\partial^m\varphi}{\delta\varphi}\Big) \cdot (\partial^m\varphi)^{n-1}$$

$$= -n\int dx\, \partial^m\frac{\delta\varphi}{\delta\varphi} \cdot (\partial^m\varphi)^{n-1}$$

$$= -n\int dx\, \partial^m\delta(x-y) \cdot (\partial^m\varphi)^{n-1}.$$

(For $m = 1$ this expression can be integrated by parts, to obtain $n(n-1)\partial^2\varphi$.) In the first step we used the identity $\{A, B^n\} = n\{A, B\}B^{n-1}$, which also holds, mutatis mutandis, for a commutator (providing $[A, B]$ is a c-number). Thus the parallel calculation goes through in the quantum case:

$$\int dx\, [\hat{\pi}, (\partial^m\hat{\varphi})^n] = n\int dx\, [\hat{\pi}, \partial^m\hat{\varphi}] \cdot (\partial^m\hat{\varphi})^{n-1}$$

$$= n\int dx\, \partial^m[\hat{\pi}, \hat{\varphi}] \cdot (\partial^m\hat{\varphi})^{n-1}$$

$$= -in\int dx\, \partial^m\delta(x-y) \cdot (\partial^m\hat{\varphi})^{n-1}.$$

And so on for terms of other forms. Because the Poisson brackets and commutators take the same form for the same Hamiltonian, and because Hamilton's equations (in Poisson bracket form) and Heisenberg's equations (up to a factor of i) take the same form, canonical operators will obey the same equations as the canonical variables.

THE STRING THEORETIC ACCOUNT OF GENERAL RELATIVITY   293

from which it follows that

$$\partial_t^2 \langle \psi | \hat{\varphi}(x,t) | \psi \rangle - \nabla^2 \langle \psi | \hat{\varphi}(x,t) | \psi \rangle = \rho(x,t) \langle \psi | \psi \rangle = \rho(x,t), \qquad (9.62)$$

if the state is normalized. The first term follows because in the Heisenberg picture the state has no time dependence, and the third term because $\rho(x,t)$ is a multiple of the identity. That $\langle \psi | \nabla^2 \hat{\varphi}(x,t) | \psi \rangle = \nabla^2 \langle \psi | \hat{\varphi}(x,t) | \psi \rangle$ can be readily shown using the path integral expression of the expectation value; crucially, this result requires that the term is *linear* in the field.[37]

Thus the field expectation value satisfies the appropriate classical equation of motion *in any state*. Of course, as Rosaler (2013, 217) points out, more is required for a state to be approximately classical; it should also be peaked around the classical value as well.

This result is a version of the Ehrenfest theorem in the context of field theory. It used only two ingredients: the Hamilton-Heisenberg correspondence for (9.60), and the linearity of the field equation for (9.27) (hence quadratic Hamiltonian). The necessity of the latter becomes clear if we add an interaction term $\lambda \varphi^n$ ($n \geq 2$) to the equations of motion.[38] The analogs of (9.60) and (9.61) will hold, with a $\hat{\varphi}^n$ term, but the analog of (9.62) will not follow, because expectation values do not in general factorize.

To see that this is so, first recall that physically meaningful expectation values 'normal order' the field operators, placing any annihilation operators to the right of all creation operators. (In short, we exploit the ambiguity in operator order that arises in quantization to cancel off the infinite contribution arising from the finite ground state energy of each of the infinity of modes, in (9.42).) Then in general,

$$\langle \psi | : \hat{\varphi}^n(x,t) : | \psi \rangle \neq \langle \psi | : \hat{\varphi}(x,t) : | \psi \rangle^n, \qquad (9.63)$$

where colons indicate normal ordering. Indeed, for $n = 2$ their difference is the (squared) quantum uncertainty. Hence if a $\lambda \varphi^n$ (or other non-linear) term appears in the equations of motion, then the analog of (9.61) will not imply the analog of (9.62). (The same holds for the particle Ehrenfest theorem when the potential is more than quadratic.) The extent to which it holds *approximately* is the extent to which expectation values approximately factorize: the extent to which the uncertainty vanishes.

---

[37] It is worth pointing out explicitly that since a mass term—$m^2 \varphi$—is also linear, the same conclusions hold for the massive scalar field.

[38] Note that in what follows we are effectively perturbing around the free vacuum, so that the creation and annihilation operators are those for the free scalar field.

294 OUT OF NOWHERE

There is, however, an important exception to non-factorizability. First use (9.36) to write:

$$\langle\psi| : \hat{\varphi}^n(x,t) : |\psi\rangle = \langle\psi| : \left(\hat{\varphi}^{(+)}(x,t) + \hat{\varphi}^{(-)}(x,t)\right)^n : |\psi\rangle$$

$$= \sum_{i=1}^{n} \binom{n}{i} \langle\psi|(\hat{\varphi}^{(-)}(x,t))^i \cdot (\hat{\varphi}^{(+)}(x,t))^{n-i}|\psi\rangle,$$

where the normal ordering has been carried out explicitly in the final step. Now we see that a coherent state $|\varphi\rangle$, for which $\hat{\varphi}^{(+)}(x,t)|\varphi\rangle = \varphi^{(+)}(x,t)|\varphi\rangle$ and $\langle\varphi|\hat{\varphi}^{(-)}(x,t) = \langle\varphi|\varphi^{(-)}(x,t)$, are a special case. For if we have $|\psi\rangle = |\varphi\rangle$, then we can use (9.57) to continue

$$= \sum_{i=1}^{n} \binom{n}{i} \langle\varphi|(\varphi^{(-)}(x,t))^i \cdot (\varphi^{(+)}(x,t))^{n-i}|\varphi\rangle$$

$$= \langle\varphi|\left(\varphi^{(+)}(x,t) + \varphi^{(-)}(x,t)\right)^n|\varphi\rangle$$

$$= \langle\varphi|\varphi^n(x,t)|\varphi\rangle = \langle\varphi|\varphi(x,t)|\varphi\rangle^n, \tag{9.64}$$

assuming that the state is normalized. In other words, because of normal ordering, field expectation values factorize exactly for coherent states.[39]

We saw at the end of §9.A.3 that any physically reasonable classical field defines a coherent state. Working in the Schrödinger picture, since this can be done at one time it can be done over time by selecting an appropriate series of coherent states $|\varphi(t)\rangle$, in such a way that the resulting time-dependent field $\langle\varphi(t)|\hat{\varphi}(x)|\varphi(t)\rangle$ satisfies the *classical* equations of motion that one desires. Or put another way, in the equivalent Heisenberg picture there exists a state and a series of field operators $\hat{\varphi}(x,t)$ such that $\langle\varphi|\hat{\varphi}(x,t)|\varphi\rangle$ takes the specified field values. Such a series of operators would then correspond to the evolution of a classical field. However, nothing entails that such a series is a solution of the *quantum* dynamics: that the operators satisfy the correct Heisenberg equations of motion. Perhaps the quantum dynamics is such that the system doesn't even remain in a coherent state.

---

[39] Indeed, Glauber defined the concept of 'coherent' state in terms of factorization in his classic (1963). We just saw that terms of the form

$$\langle\varphi|\hat{\varphi}^{(-)}(x_1,t)\hat{\varphi}^{(-)}(x_2,t)\ldots\hat{\varphi}^{(-)}(x_n,t)\hat{\varphi}^{(+)}(x_{n+1},t)\ldots\hat{\varphi}^{(+)}(x_{2n},t)|\varphi\rangle$$

factorize. But this quantity measures the correlation between field quanta at $x_1, x_2, \ldots, x_n$ with those at $x_{n+1}, x_{n+2}, \ldots, x_{2n}$, so the fact that it factorizes in such states means that there are no such correlations. The terminology is a little confusing, given that it denotes the *absence* of correlations. But it is important to understand that a coherent state lacks correlations between the 'particle' content of the field, not between the field strength at different locations: broadly, it lacks quantum but not classical correlations. Historically the term 'coherent' comes from classical optics, in which monochromatic light is the paradigm example—in which the field at different points is highly correlated! (The notion of coherent state has evolved further since Glauber's treatment. As the introduction to Klauder and Skagerstam 1985 reviews, the most general notion is of any set of continuously parameterized states $|\lambda\rangle$ that (i) are strongly continuous in the norm, and (ii) span the state space. An orthonormal basis of course satisfies (ii), but not (i). The 'canonical coherent states' that we are discussing do satisfy these properties.)

THE STRING THEORETIC ACCOUNT OF GENERAL RELATIVITY   295

However, for a moment suppose that the quantum dynamics *is* such that if a system is in a coherent state at one time, then it is at all times. Let the field operators satisfy some Heisenberg equation of motion $\triangle\hat{\varphi}(x,t) = 0$. By the Hamilton-Heisenberg correspondence, the corresponding classical equation of motion is $\triangle\varphi(x,t) = 0$. Equally, by factorization (e.g., 9.64) and supposition, if the state is initially coherent then we have at all times

$$\triangle \langle\varphi|\hat{\varphi}(x,t)|\varphi\rangle = \langle\varphi| : \triangle\hat{\varphi}(x,t) : |\varphi\rangle = 0, \tag{9.65}$$

according to the Heisenberg equation of motion. Thus the field expectation value would satisfy the appropriate classical equation of motion.

To recap, if the classical equations of motion are linear then the field expectation value *in any state* will satisfy the classical equations of motion. If the equations of motion are non-linear then in general they will not. However, even in this case any time-dependent classical field can be constructed from a series of coherent states, though in general it will not satisfy the quantum dynamics. However, if the system is such that coherence is preserved then the field will satisfy the correct classical dynamics. Obviously the next question is 'when is coherence of the field preserved?'

(b) The question is not hard to answer: we saw in §9.A.2 that an individual mode k will be remain in a coherent state if and only if the commutator $[\hat{H}_k, \hat{a}_k]$ equals a term linear in $\hat{a}_k$ (plus a c-number): then we have (9.49) and (9.50). If that is the case for every mode then the same reasoning that leads to (9.48) or (9.55) will, in the source-free case, lead to

$$\hat{\varphi}^{(+)}(x,t)|\varphi\rangle = \sum_{k=1}^{\infty} \frac{e^{-ik(x+2t)}}{\sqrt{2\pi k}} \alpha_k \cdot |\varphi\rangle. \tag{9.66}$$

(Comparing (9.42) with the SHO Hamiltonian, and recalling that $\hat{a}_k \equiv \sqrt{2\pi k} \cdot \hat{\alpha}_k$.) And similarly in the presence of a source. That is, $|\varphi\rangle$ will remain an eigenstate of $\hat{\varphi}^{(+)}(x,t)$—a coherent state of the field.

But if the field equations are non-linear—as they are for interacting fields—containing powers of the field or its derivatives, then the Hamiltonian will contain greater than quadratic terms. (This is most easily seen by noting that the Lagrangian must contain greater than quadratic powers, so that the Euler-Lagrange equation $\partial\mathcal{L}/\partial\varphi = \partial_\mu \cdot \partial\mathcal{L}/\partial(\partial_\mu\varphi)$ contains non-linear terms.) In that case, when the Hamiltonian is expanded in modes, as in (9.42), one will obtain terms that are non-linear in either $\hat{\alpha}$ or $\hat{\alpha}^\dagger$, and the Heisenberg equation of motion for $\hat{\alpha}(t)$ will be non-linear. And in that case, as we noted in §9.A.2, individual modes will not remain in coherent states. In short, *coherence will in general only be preserved in linear—non-interacting—fields.*

In other words, although non-linear theories left a loophole for coherent state expectation values to satisfy classical field equations (9.65), that very non-linearity then closes off the loop: the quantum dynamics implies that the system will not remain in a coherent state, and the interpretation as a classical field eventually will be lost.

As we explained in §9.3, the derivation of the EFE for a background metric does not depend on the validity of linearized gravity. However, we now see that linearity *is* relevant to the string theoretic account of gravity in another way: the identification of the classical background field with a coherent state of stringy gravitons. For linearity is required for the preservation of coherence, which in turn is required so that field expectation values satisfy classical equations of motion. But of course, gravity—and any other interesting field—is non-linear. In such cases then, the coherent state approach to the classical limit will remain valid if the non-linearity is weak, so that states remain approximately coherent for suitable time scales. So the question now becomes that of quantifying closeness to coherence, and the rate at which a state loses coherence. This question will be addressed in §10.5.2.

9.B Appendix: Noether's theorem

It is important to have a grasp of Noether's theorem, in classical and quantum field theory, in order to appreciate the above results. But more than that, the presentation here is aimed at preparing the reader to follow the reasoning found in textbooks on QFT and string theory, as they investigate these topics beyond the scope of this book. In particular, the proofs that are given here are constructive: they provide methods for computing conserved currents and anomalies. Thus they are crucial tools for the pursuit of a deeper understanding. Our starting point is Feynman (1967, 103ff), which gives an intuitive demonstration of the connection between symmetries and conservation laws.

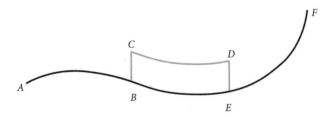

Figure 9.4 A minimum of the action *ABEF*, and its symmetric deformation *ABCDEF*.

Figure 9.4 shows a path *ABEF* in phase space which minimizes the action: $\delta S = 0$ for any infinitesimal variation, and hence the path describes a physically possible trajectory according to the principle of least action. Deform the path so that an arbitrary proper part *BE* is shifted by an infinitesimal symmetry transformation, to obtain a new path *ABCDEF*. (i) Because the transformation is a symmetry, the action is the same when evaluated along *BE* and *CD*; hence the difference

THE STRING THEORETIC ACCOUNT OF GENERAL RELATIVITY   297

between the two paths is $\delta S = S_{BC} + S_{DE}$. But (ii) since the original path is a solution to the equations of motion, it satisfies Hamilton's Principle and $\delta S = 0$ for any infinitesimal transformation, including this one; so $S_{BC} = -S_{DE} = S_{ED} \equiv \epsilon Q$ (ϵ an infinitesimal constant). But the positions of BC and ED along the trajectory are arbitrary, so for any physical trajectory, there must be a conserved quantity, Q, associated with any (continuous) symmetry. 'QED'.

This intuitive 'proof' captures the logic of a more formal derivation of the classical Noether theorem. Consider a theory for a field $\varphi(x)$ (with x the set of spacetime coordinates), and an infinitesimal transformation $\varphi(x) \to \varphi(x) + \delta\varphi(x)$ that is a symmetry of the action. The 'trick' is to apply a transformation

$$\varphi(x) \to \varphi'(x) = \varphi(x) + \epsilon\alpha(x) \cdot \delta\varphi(x), \tag{9.67}$$

where $\alpha(x)$ is an arbitrary function, and ϵ an infinitesimal constant; then calculate the variation in the action, and set it to zero for a physical trajectory (of course, (9.67) may not itself be a symmetry). We can quickly see that the result will be some conserved current. (i) Because we have symmetry of the action, if $\alpha(x) = $ constant then there is no variation, and hence δS must have no $\alpha(x)$ dependence. So it must depend on derivatives $\partial\alpha(x)$.[40] So schematically, with dx and ∂ representing spacetime integration and divergence, respectively:

$$\delta S = \epsilon \int_M dx\, j(x) \cdot \partial\alpha(x). \tag{9.68}$$

Since $\alpha(x)$ is a field, the unknown weight $j(x)$ must be too; since S is not a function of x, we must integrate over spacetime manifold M. (ii) Suppose $\alpha(x)$ vanishes outside of some finite spacetime region, so again only a proper part of the path is displaced. We can integrate (9.68) by parts,

$$\delta S = \epsilon[j(x)\alpha(x)]_M - \epsilon \int dx\, \alpha(x)\partial j(x). \tag{9.69}$$

The first term vanishes because it is evaluated outside the support of $\alpha(x)$ (just as AB and EF did not contribute). Then again, because an infinitesimal variation vanishes for a physical trajectory,

$$\int dx\, \alpha(x)\partial j(x) = 0, \tag{9.70}$$

and because $\alpha(x)$ is arbitrary, the fundamental theorem of the calculus of variations tells us that $\partial j(x) = 0$. So $j(x)$ is a conserved current associated with the symmetry. QED.

---

[40] We will ignore higher derivatives since the reasoning is the same if they are involved.

298 OUT OF NOWHERE

This derivation of Noether's theorem is nice because by closely following Feynman's argument it makes conceptually clear the connection between symmetry and conservation. Moreover, the 'trick' involved is a method for computation of the conserved current for a given symmetry: calculate the variation δS for the transformation (9.67), and insert it into (9.69) to give an equation for $j(x)$. For example, see Nakahara (2003, §13.2.1) for the derivation of the chiral current for (9.21). Our proof shows that such currents will be conserved along physical trajectories, those satisfying Hamilton's principle.

In addition, both Feynman's intuitive derivation, and this more formal one, are readily adapted to quantum mechanics. In Feynman's path integral approach we are interested in $Z = \sum_\lambda e^{iS_\lambda}$, so now we sum over all paths, indexed λ, each weighted by a phase, e^{iS_λ}, depending on the corresponding action. Referring to figure 9.4, $S_\lambda = S_{A_\lambda B_\lambda} + S_{B_\lambda E_\lambda} + S_{E_\lambda F_\lambda}$, so

$$\exp iS_\lambda = \exp iS_{A_\lambda B_\lambda} \cdot \exp iS_{B_\lambda E_\lambda} \cdot \exp iS_{E_\lambda F_\lambda}. \tag{9.71}$$

Classically we deformed a section of the physical path by an infinitesimal symmetry, but in quantum mechanics we apply it to every path in the sum: $A_\lambda B_\lambda E_\lambda F_\lambda \rightarrow A_\lambda B_\lambda C_\lambda D_\lambda E_\lambda F_\lambda$. Hence

$$\begin{aligned}
Z \rightarrow Z' &= \sum_\lambda \exp iS_{A_\lambda B_\lambda} \cdot \exp iS_{B_\lambda C_\lambda} \cdot \exp iS_{C_\lambda D_\lambda} \cdot \exp iS_{D_\lambda E_\lambda} \cdot \exp iS_{E_\lambda F_\lambda} \\
&= \sum_\lambda \exp iS_{A_\lambda B_\lambda} \cdot \exp iS_{B_\lambda C_\lambda} \cdot \exp iS_{B_\lambda E_\lambda} \cdot \exp iS_{D_\lambda E_\lambda} \cdot \exp iS_{E_\lambda F_\lambda} \\
&= \sum_\lambda \exp iS_{B_\lambda C_\lambda} \cdot \exp iS_{D_\lambda E_\lambda} \cdot \exp iS_\lambda, \tag{9.72}
\end{aligned}$$

where we have used the fact that the transformation is a symmetry, $S_{C_\lambda D_\lambda} = S_{B_\lambda E_\lambda}$, and (9.71). But $B_\lambda C_\lambda$ and $D_\lambda E_\lambda$ are infinitesimal path segments, so

$$\begin{aligned}
Z' &\approx \sum_\lambda (1 + iS_{B_\lambda C_\lambda}) \cdot (1 + iS_{D_\lambda E_\lambda}) \cdot \exp iS_\lambda \\
&\approx Z + \sum_\lambda (iS_{B_\lambda C_\lambda} + iS_{D_\lambda E_\lambda}) \cdot \exp iS_\lambda \\
&= Z + \sum_\lambda (iS_{B_\lambda C_\lambda} - iS_{E_\lambda D_\lambda}) \cdot \exp iS_\lambda. \tag{9.73}
\end{aligned}$$

In the corresponding point in the classical proof we used Hamilton's minimum principle to set the variation in the action to zero. In the quantum case we instead appeal to the fact that we are summing over *all* paths, so that the systematic deformation does not produce a sum over different paths, but rather merely permutes the summands: both Z and Z' are appropriately weighted sums over all paths and so $Z = Z'$. Hence from (9.73),

$$\sum_\lambda S_{B_\lambda C_\lambda} \cdot \exp iS_\lambda = \sum_\lambda S_{E_\lambda D_\lambda} \cdot \exp iS_\lambda, \tag{9.74}$$

and therefore this is a conserved quantity (times an infinitesimal constant). What is important to note is that this conserved quantity is itself a weighted sum over paths, and therefore is an expectation value. That is, as one should expect, the quantum Noether theorem shows that for every symmetry there is an observable whose expectation value is conserved. 'QED'.

More formally (following Polchinski 1998, 41f) we need to consider a path integral $\int D\varphi\, e^{iS[\varphi]}$, rather than the action of a single path. Symmetry then is not a matter of invariance of the action, but of the combination of path integral measure and action—this is the crucial difference between the classical and quantum Noether theorems, and explains how one can hold while the other fails, resulting in a quantum 'anomaly'.

So suppose that infinitesimal $\varphi(x) \to \varphi(x) + \delta\varphi(x)$ leaves $D\varphi\, e^{iS[\varphi]}$ invariant, and again apply the transformation (9.67). Again, because $\delta\varphi(x)$ is a symmetry, the variation in $D\varphi\, e^{iS[\varphi]}$ vanishes if $\alpha(x) =$ constant, so we only consider $\partial\alpha(x)$-dependence,

$$D\varphi'\, e^{iS[\varphi']} = D\varphi\, e^{iS[\varphi]} \cdot \left(1 + \epsilon \int dx\, j_\varphi(x)\partial\alpha(x) + O(\epsilon^2)\right). \tag{9.75}$$

The subscript on $j_\varphi(x)$ is to note that this unknown coefficient depends on the field configuration: the variation in $D\varphi\, e^{iS[\varphi]}$ of course depends on $\varphi(x)$, and here that dependence is in $j_\varphi(x)$. Next, again let $\alpha(x)$ vanish outside some spacetime region[41] and integrate by parts exactly as in (9.69), and conclude

$$D\varphi'\, e^{iS[\varphi']} \approx D\varphi\, e^{iS[\varphi]} \cdot \left(1 - \epsilon \int dx\, \alpha(x)\partial j_\varphi(x)\right). \tag{9.76}$$

Now, from a formal point of view, in a functional integral over *all* paths, $\phi \to \phi'$ merely replaces the variable of integration, and so cannot change the path integral:

$$\int D\varphi\, e^{iS[\varphi]} = \int D\varphi'\, e^{iS[\varphi']}. \tag{9.77}$$

As in the intuitive proof, the sum is over all paths in both cases, so the limits of the integration are unchanged. Taking the path integral over (9.76), using (9.77), and rearranging the order of operations yields

---

[41] If one is interested in a path integral for given field operators $\varphi(x_i)$ at fixed locations x_i (i.e., an amplitude for particles created at such positions), then the following argument goes through by letting $\alpha(x)$ having its support away from the x_i. In other words, as in the Feynman-style intuitive approach, the proof applies to path integrals in general, not just Z.

300 OUT OF NOWHERE

$$\int D\varphi \left(e^{iS[\varphi]} \int dx\, \alpha(x)\partial j_\varphi(x) \right) = 0 \tag{9.78}$$

$$= \int dx \left(\alpha(x) \cdot \partial \int D\varphi\, e^{iS[\varphi]} j_\varphi(x) \right).$$

Finally, since $\alpha(x)$ is arbitrary we must have

$$0 = \partial \int D\varphi\, e^{iS[\varphi]} j_\varphi(x) = \partial \langle j_\varphi(x) \rangle, \tag{9.79}$$

where the final step follows because path integrals with 'inserted' functions of the fields, such as $j_\varphi(x)$, are expectation values for the corresponding observables.[42] QED.

The quantum Noether theorem shows that in quantum theory any symmetry (of the combination of action and path integral measure) entails a current whose expectation value is conserved—the same current (as an operator) as in the classical theory. (Though it can be computed using (9.76) and explicit computation of the effect of (9.67).) We have noted two differences in the classical and quantum proofs: first the role of the classical minimum principle, $\delta S = 0$, is played in the quantum case by the fact that because a transformation is merely a change of variables in a path integral, $\delta Z = 0$. More importantly for our purposes, we have seen that classically symmetries of the action lead to conserved currents; while quantum mechanically it is symmetries of the combination of action *and* path integral measure that lead to conserved currents.

However, symmetry of the action does not entail symmetry of the measure, and when they come apart, anomalies result, as discussed in §9.4. In particular, if $\varphi \rightarrow \varphi + \delta\varphi$ is a symmetry of the former but not the latter, then when we transform according to (9.67) we find

$$\int D\varphi'\, e^{iS[\varphi']} = \int D\varphi\, |J_{\varphi,\varphi'}|\, e^{i(S[\varphi]+\delta S)}, \tag{9.80}$$

where $J_{\varphi,\varphi'}$ is the Jacobian for the transformation (now non-trivial, because $\delta\varphi$ is not a symmetry of the measure). Then using (9.77) and (9.69) on the left and right, respectively, we have

$$\int D\varphi\, e^{iS[\varphi]} = \int D\varphi\, |J_{\varphi,\varphi'}| \left\{ e^{iS[\varphi]} \cdot \left(1 - \epsilon \int dx\, \alpha(x)\partial j_\varphi(x) \right) \right\}, \tag{9.81}$$

or

$$\epsilon \int dx\, \alpha(x) \cdot \int D\varphi\, e^{iS[\varphi]} \partial j_\varphi(x) = \int D\varphi\, e^{iS[\varphi]} \left(|J_{\varphi,\varphi'}| - 1 \right). \tag{9.82}$$

---

[42] Any QFT textbook will explain this point. A nice recent text, suitable for philosophers, is Lancaster and Blundell (2014, chs. 21–27)

Comparison with (9.79) shows that the presence of the Jacobian means that the current will now not be conserved. Moreover, the Jacobian can be calculated (it of course has ϵ and $\alpha(x)$ dependence), and the fact that $\alpha(x)$ is arbitrary exploited, to yield an equation for $\partial \langle j_\varphi(x) \rangle$. For a concrete application of this technique, see the computation of the chiral anomaly (9.22) in Nakahara (2003, §13.2.1).

Chapter 10
The emergence of spacetime
in string theory

This chapter builds on the results of the previous two to investigate the extent to which spacetime might be said to 'emerge' in perturbative string theory. Our starting point is the string theoretic derivation of general relativity (GR) explained in depth in the previous chapter, and reviewed in §10.1 below (so that the philosophical conclusions of this chapter can be understood by those who are less concerned with formal detail, and so skipped the previous one). The result is that the consistency of string theory requires that the 'background' spacetime obeys the Einstein Field Equation (EFE)—plus string theoretic corrections. But their derivation, while necessary, is not sufficient for spacetime emergence. So we will next, in §10.2, identify spacetime structures whose derivation would justify saying that a generally relativistic spacetime 'emerges': this section will be important for establishing a fruitful way of approaching the question. In effect, at this stage we will have discharged step SF1 of the functionalist program described in §2.4, and enumerated some of the 'roles' played by spacetime (and the classical formalism by which we describe those roles). The second step, SF2, then requires that an explanation be given of how these roles are realized according to string theory; thus the remainder of the chapter, §10.3–10.5, will investigate how these structures arise as empirical phenomena in string theory: at this point we will also draw on the results concerning T-duality from chapter 8 (again summarizing the essential ideas). Naturally, the derivation of the EFE is a crucial element of this explanation, but as it demonstrates a 'merely formal' correspondence more is needed: the 'physical salience' of the string-spacetime correspondence (§2.2). To understand what assumptions underpin the claim that the derivation is physically salient, explaining how the spacetime roles are played in string theory, we will consider how they might be operationalized, especially in high energy scattering experiments. A critical question in this discussion will be the recurring one of this book: whether these structures are indeed emergent, or just features already present in the more fundamental string theory.

Out of Nowhere. Nick Huggett and Christian Wüthrich, Oxford University Press. © Nick Huggett and Christian Wüthrich (2025). DOI: 10.1093/oso/9780198758501.003.0010

THE EMERGENCE OF SPACETIME IN STRING THEORY   303

10.1 Deriving general relativity

So first, an overview of the derivation; refer to the previous chapter (which itself draws on Green et al. (1987, §3.4) and Polchinski (1998, §3.7)) for more details. A caution: almost all of this section should be read as a discussion of a formal mathematical framework, rather than making metaphysical, ontological, or interpretational claims—that will come in the remainder of the chapter. First, a distinction.

We will use Feynman's 'sum over paths' (or 'path integral') formulation of quantum mechanics. The fundamental idea is that each *classical* path γ between states A and B (in a given time) is assigned a quantum amplitude $e^{iS(\gamma)/\hbar}$, where $S(\gamma)$ is the action along γ. It's important to appreciate that *every* path (for which S is well-defined) connecting A and B is thus assigned an amplitude, not just the one that minimizes the action: every path counts, not only the one allowed by the classical equations of motion. Then the quantum probability amplitude for propagation from A to B is given by the *sum* of path amplitudes: the amplitude for the system to evolve from A to B (in the given time) is $\sum_j e^{iS(\gamma_j)/\hbar}$, where the γ_j are the different paths.[1] (The $A \to B$ transition probability is then obtained by squaring this amplitude.) The formulation can be applied to a particle to calculate the probability to propagate from one position to another: sum the amplitudes of every path between the two positions (and square). Or to a field to find the probability that it will evolve from one configuration to another: sum over all evolutions connecting them.

However, there is another famous sum associated with Feynman (building on the framework developed by Dyson and others), which should be distinguished from that over classical paths: the sum over *Feynman scattering diagrams*. The latter is a recipe for constructing successively smaller quantum corrections—'perturbations'—to a given classical system: the diagrams are a powerful heuristic representation of these corrections, and summing them (if all goes well!) should yield successively better approximations to the probability amplitude. But there is nothing intrinsically perturbative about the path integral approach described above: in particular, *a weighted classical path is not the same as a scattering diagram*—in the case of fields, the former is a classical field configuration over spacetime, while the latter depicts a process involving (virtual) quanta. The sum over paths defines an exact quantum theory (specifically a complete set of amplitudes), while the sum over diagrams converges on the same set of quantum amplitudes. (At least, that situation is the mathematical ideal: reality is typically less compliant.) But no proper part of one sum is the same as any part of the other.[2]

---

[1] Of course, in typical cases the paths are not countable as suggested here, and the sum is replaced by an integral over paths—hence the more common name for the approach. For our discussion the difference is not important.

[2] Carlo Rovelli has pointed out to us in correspondence that the sum over scattering diagrams can be thought of as a sum over *quantum* paths.

304   OUT OF NOWHERE

With these formal tools distinguished, we can review the derivation of GR (bearing in mind the caution that the following merely describes a formal framework, not an interpretation).

(A) Feynman's perturbative approach is assumed to apply to quantum string theory. In fact the usual logic is reversed: instead of giving an exact formulation from which a perturbative theory can be derived by Feynman's methods, the theory is given in perturbative form, and an exact—but currently dimly perceived—theory which it approximates is postulated. (Most naturally one would suppose it to be a string field theory, e.g., Taylor (2009) or Erbin (2021), in which strings are created and annihilated, but most string theorists believe that something more radical—'M-theory'—will be required.) Broadly speaking, the perturbative recipe is as follows. Suppose one is interested in the amplitude for a given group of incoming particles to interact and produce a given group of outgoing particles. Since (as discussed in §7.3.4) particles correspond to different excitations of strings, the given initial and final string states are fixed. Then the amplitude is computed through a perturbative sum over *string scattering diagrams*.

More specifically, each term in the sum corresponds to a different string worldsheet compatible with the given initial and final states: for example, two incoming strings might join to form one string, then split into two again; but they might also join, split, but then rejoin, and resplit; or they might split and rejoin twice; or three times; and so on (see the top row of figure 9.1). Each of these evolutions corresponds to a distinct worldsheet with an increasing number of 'holes'.[3] All need to be included in the sum: and the more splitting, the smaller the contribution from the diagram—just as in particle scattering, in which a diagram contributes less the more interaction vertices it contains. (Because each interaction is associated with a 'small' interaction factor $\lambda \ll 1$, so that n interactions suppress a term by a factor λ^n.)

(B) However, the probability amplitude for any given worldsheet will be the sum of the amplitudes of each of the different trajectories that the string might take through spacetime between given initial and final states. So the first perturbative sum over worldsheets contains a *second* sum over all the different ways that each worldsheet might be embedded in spacetime: a non-perturbative (for now) sum over classical worldsheet paths, which we now explain.

To describe an embedding formally, as in chapter 7 we assign spacetime coordinates X^μ to spacetime ($\mu = 0, 1, \ldots D$, where there are D spatial dimensions); and timelike and spacelike worldsheet coordinates τ and σ, respectively, to the string itself, as a 2-dimensional spacetime object. Then $X^\mu(\tau, \sigma)$ is a function that takes each worldsheet point (τ, σ) to the point of target space at which it is embedded,

---

[3] For the results discussed in this chapter the sum over topologies just described suffices. In fact, for full scattering calculations there are additional parameters—'moduli'—that need to be summed over in the path integral. These describe global properties of the worldsheet; for instance, how skewed a torus is (see Polchinski 1998, ch. 5).

THE EMERGENCE OF SPACETIME IN STRING THEORY   305

X^μ. With flat metric $\eta_{\mu\nu}$, the Minkowski 'sigma action' for a string with embedding $X^\mu(\tau, \sigma)$ is the worldsheet integral[4]

$$S_\sigma[X^\mu] = -\frac{T}{2} \int \eta_{\mu\nu}\left(\partial_\tau X^\mu \partial_\tau X^\nu - \partial_\sigma X^\mu \partial_\sigma X^\nu\right) d\tau d\sigma, \tag{10.1}$$

where T is the constant internal tension. Thus, choosing units in which $\hbar = 1$ (we also set $c = 1$), the contribution to the amplitude from a particular worldsheet path is $e^{iS_\sigma[X^\mu]}$; and the total contribution of a particular worldsheet—itself a *summand* in the perturbative scattering series—is the sum (or rather integral) of $e^{iS_\sigma[X^\mu]}$ over every possible target space embedding of the worldsheet, $X^\mu(\tau, \sigma)$.

Schematically then,

$$\text{total amplitude} \quad \approx \quad \underbrace{\sum_{\text{worldsheets } j} \lambda_j}_{} \quad \cdot \quad \underbrace{\sum_{\substack{\text{embeddings } k \\ \text{of worldsheet } j}} e^{iS_\sigma[X^\mu_{jk}]}}_{}. \tag{10.2}$$

(10.2) represents the scattering dynamics of quantum string theory.[5] Reading from the right, each term (j, k) corresponds to the kth embedding of the jth world-sheet, $X^\mu_{jk}(\tau, \sigma)$; their sum is an exact path integral for that worldsheet. But the first sum, over j, is a perturbative Feynman expansion over worldsheets of increasing number of holes; λ_j decreases with j because the contributions decrease the more string interactions occur. So overall the expression is a perturbative approxima-tion. The first sum, over j, is often said to describe the quantum aspects of quantum strings, because it treats them much like field quanta; the second sum, over k, is correspondingly said to capture the peculiarly stringy nature of strings, because it depends on their extended nature, unlike quantum field theory (QFT). In the next two steps we will focus on this rightmost 'stringy' sum, describing two of its crucial features.

(C) First, string excitations correspond to the quanta of quantum fields, where the mass and type depends on the excitation level of the string: for example, a negative mass tachyon, a massless graviton, and so on. (As discussed in §7.3.4 and below, these identifications make use of Wigner's idea that particle types can be distinguished according to the way their states transform under spacetime sym-metries.) It's worth emphasizing that the difference between different quanta is not one of kind according to string theory; to change a quantum from one 'type' to another the string need only change its state.

---

[4] See (7.16). $\partial_\tau X^\mu$, e.g., is the spacetime vector expressing the rate at which the string's location changes with respect to its τ coordinate; $\eta_{\mu\nu}\partial_\tau X^\mu \partial_\tau X^\nu$ is its length squared. So the action is propor-tional to the difference between the squared rates of change with respect to τ and σ, integrated over the worldsheet.

[5] Amongst other things, a more rigorous treatment would Wick rotate the system into Euclidean space by setting $t \to it$.

306   OUT OF NOWHERE

Consider what difference it would make to its worldsheet amplitude if a string interacted with such a stringy particle during its evolution—if, that is, a second, appropriately excited string interacted with the worldsheet. For reasons discussed in chapter 9, the effect can be captured by including an appropriate factor in the path amplitude: specifically, the worldsheet integral of a 'vertex operator'. For instance, the vertex operator for a tachyon is e^{ikx}, and for a graviton $s_{\mu\nu}(\partial_\tau X^\mu \partial_\tau X^\nu - \partial_\sigma X^\mu \partial_\sigma X^\nu)$, with $s_{\mu\nu}$ a traceless symmetric tensor. That is, if the worldsheet interacts with a graviton, the appropriate path integral is now the sum over all embeddings $X^\mu(\tau, \sigma)$ of

$$ e^{iS_\sigma[X^\mu]} \times \int s_{\mu\nu}(\partial_\tau X^\mu \partial_\tau X^\nu - \partial_\sigma X^\mu \partial_\sigma X^\nu)\, d\tau d\sigma. \qquad (10.3) $$

If scattering occurs in the presence, not of a single quantum of gravity, but of n gravitons, then the amplitude will have n such factors. Note that exciting a single string doesn't produce more particles, but changes the type of particle associated with the string. Thus n quanta means that there are n similarly excited strings, and so a vertex operator for each one.

What, though, if scattering occurs in the presence of a classical gravitational field? In QFT, classical fields can be described in terms of 'coherent states', a superposition of every number of excited quanta: paradigmatically, a state of the form $\sum_n |n\rangle/n!$, where $|n\rangle$ is a state of n quanta (this concept, and some of its limitations, were explained in detail in appendix A to chapter 9). Remembering that different 'types' of quanta are nothing but different states of strings, we thus see that in string theory all classical fields are understood in the same way, in terms of many-string states, whose strings are in different excited states. In particular, gravitational and matter fields are unified in string theory, as simply different manifestations of the same basic 'stuff' (a point emphasized in Read 2019).

Since we no longer have n background gravitons, but a superposition of gravitons for every n, the path integral now involves a sum over powers of the graviton vertex operator

$$ e^{iS_\sigma[X^\mu]} \times \sum_n \frac{1}{n!} \left(\int s_{\mu\nu}(\partial_\tau X^\mu \partial_\tau X^\nu - \partial_\sigma X^\mu \partial_\sigma X^\nu)\, d\tau d\sigma \right)^n $$

$$ = \exp\left(\int \eta_{\mu\nu}(\partial_\tau X^\mu \partial_\tau X^\nu - \partial_\sigma X^\mu \partial_\sigma X^\nu)\, d\tau d\sigma \right) $$

$$ \times \exp\left(\int s_{\mu\nu}(\partial_\tau X^\mu \partial_\tau X^\nu - \partial_\sigma X^\mu \partial_\sigma X^\nu)\, d\tau d\sigma \right) $$

$$ = \exp\left(\int (\eta_{\mu\nu} + s_{\mu\nu})(\partial_\tau X^\mu \partial_\tau X^\nu - \partial_\sigma X^\mu \partial_\sigma X^\nu)\, d\tau d\sigma \right), \qquad (10.4) $$

THE EMERGENCE OF SPACETIME IN STRING THEORY   307

where we substituted for S_σ using (10.1), then used the Taylor expansion for e^x, and that $e^x e^y = e^{x+y}$.

But (10.4) is exactly the amplitude we would have found if the Minkowski sigma action (10.1) had been modified by $\eta_{\mu\nu} \to g_{\mu\nu} = \eta_{\mu\nu} + s_{\mu\nu}$: the sigma action appropriate to a curved target space metric $g_{\mu\nu}$, rather than the flat one we supposed originally! In short, *as far as the path integral is concerned, there is no difference between a curved spacetime, and a flat one with a coherent excitation of gravitons.* All this follows because the string action and the graviton vertex both involve a tensor coupled to X^μ derivatives (the latter because the tensor gravitational field $s_{\mu\nu}$ can be expanded in graviton plane waves, as discussed in §9.2); and because the quantum mechanical path integral exponentiates the action, while a coherent state exponentiates the graviton vertex operator. (There is nothing special about gravitons and the gravitational field in this analysis: similar points hold in the presence of coherent states of any of the fields composed of string quanta.)

To flag the crucial interpretational point that will arise from this mathematical fact: since all physical quantities can be derived from the path integral (according to the usual understanding), there simply is *no* physical difference between the graviton and curved spacetime descriptions, and curvature *is* a coherent state of gravitons. This idea is found in Green et al. (1987, 165ff) and Polchinski (1998, 108) (amongst others), and with more conceptual detail in Motl (2012). That the metric is *constituted* by gravitons in the way just described (not caused by them, say), and so of a different nature than in a classical spacetime theory, leads us to speak of its 'emergence' (in our extended sense).

(D) Second, now we change perspective in a significant way. A worldsheet has space and time coordinates, σ and τ, so formally is itself a (2-dimensional) spacetime.[6] From this point of view, the value of $X^\mu(\tau, \sigma)$ at a point is formally the value of a $(D + 1)$-component vector[7] at that point, rather than its spacetime location. From this perspective then, $X^\mu(\tau, \sigma)$ is not the spacetime embedding of the worldsheet, but a *field* on the worldsheet, described by the action (10.4); but these perspectives are formally equivalent, so the amplitude for a worldsheet embedding in spacetime is just the amplitude for the field $X^\mu(\tau, \sigma)$, which describes a classical evolution on the worldsheet. But now we realize that Dyson-Feynman perturbation theory can be applied to calculate an expansion for the sum over these classical field amplitudes: this is exactly the sum over embedding amplitudes in (10.2), which gives the contribution of each worldsheet.

---

[6] As discussed in §7.1.2 following (7.23), these coordinates represent space and time in the sense that the worldsheet possesses a Lorentzian 'auxiliary' metric h_{ab}, chosen to be flat in (10.1). This should be distinguished from the 'induced' metric inherited from the embedding in spacetime. In particular, because of the 'Weyl symmetry' (under $h_{\alpha\beta}(\tau, \sigma) \to \Omega^2(\tau, \sigma) h_{\alpha\beta}(\tau, \sigma)$) of string theory, $h_{\alpha\beta}$ does not encode metrical information beyond causal structure on the worldsheet (on which it agrees with the induced metric in classical solutions).

[7] Note that X^μ does not live in the tangent space to the worldsheet; from the point of view of the worldsheet, X^μ lives in an 'internal' vector space, on which $g_{\mu\nu}$ is the inner product.

308   OUT OF NOWHERE

Hence we can schematically rewrite (10.2) as

$$\text{total amplitude} \approx \sum_{\text{worldsheets } j} \lambda_j \cdot \sum_{\substack{X^\mu \text{field scattering} \\ \text{diagrams } k \text{ on} \\ \text{worldsheet } j}} (\alpha')^{n_k} G_{jk}. \qquad (10.5)$$

G_{jk} is the contribution of the kth scattering diagram for the X^μ field (picturing virtual processes for its quanta), evaluated on the jth worldsheet. $\alpha' \sim 1/T$ is the 'Regge slope', a small parameter in which the expansion can be made; the relation $\alpha' \sim 1/T$ to the string tension constant gives another sense in which this sum is uniquely 'stringy'. The perturbative sum is organized into powers of α', so into diminishing contributions. Compared to (10.2) we have replaced the exact path integral for X^μ with a perturbative expansion, so we now have a doubly perturbative expansion. It is this second sum over X^μ scattering diagrams that is relevant to the derivation of the field equations of GR.[8]

Thought of as describing a classical field theory on the worldsheet, $S_\sigma[X^\mu]$ is significantly different from familiar field dynamics in two ways. First, most fields depend on the metric of the spacetime in which they live, but $S_\sigma[X^\mu]$ does not depend on any worldsheet metric: it possesses 'Weyl' symmetry. (A little more precisely, it is only dependent on the worldsheet lightcone structure, inherited from target space: the distinction between spacelike, timelike, and lightlike curves restricted to the worldsheet. See footnote 6.) As a matter of mathematical necessity, such a symmetry must also be possessed by the corresponding quantum theory; otherwise the resulting 'anomaly' would render the theory inconsistent (§9.4).

Second, familiar fields have a constant interaction strength: for instance, electromagnetic interactions depend on the constant electrical charges of the quanta. But from (C) we know that in the presence of a gravitational field, or equivalently in a curved background spacetime, the action for the field on the worldsheet is given by (10.1) with $\eta_{\mu\nu} \to g_{\mu\nu}$. From the form of the action, it follows that when quanta of the X^μ-field interact on the worldsheet, the strength of the interaction depends on $g_{\mu\nu}$, which is a function of position on the worldsheet, not constant. Such quantum fields—known as 'sigma models'—have been studied, and the critical result for us is that they are *Weyl symmetric only if, at first order in* α', $g_{\mu\nu}$ *satisfies the EFE*, which defines GR. Higher order terms produce corrections to the generally relativistic equations, but we emphasize that an expansion in α' is very different from a weak gravitational field approximation: it is the *full* EFE that is derived, rather than its linear approximation.[9] In this sense, gravity is a logical consequence—one could say 'prediction'—of string theory.

---

[8] Specifically, for the tree level string worldsheet.

[9] We will say little more about the technical details of the derivation here, since they were described at length in the previous chapter (this chapter concerns the conceptual issues); but it is worth mentioning a couple of important points from that discussion. (i) If there are no other 'background fields' (similarly

THE EMERGENCE OF SPACETIME IN STRING THEORY   309

To run (C)–(D) backward, we first learn (from Weyl symmetry) that quantum strings can only live in a curved spacetime that satisfies (to first order in α') the laws of GR. Then we find out that curved spacetime is nothing but a state of string graviton excitations, of strings themselves, in Minkowski spacetime. Together, those show that GR is an effective, phenomenal theory describing the collective dynamics of strings (in certain quantum states) in target space. (Or at least, they show it for the class of models of GR lying in the scope of perturbative string theory assumed in the derivation.) The job of this chapter is to unpack this explanation, and try to understand better what is achieved, and how.

Before we proceed, it is worth pointing out the technical doubts that have been raised about the derivation of GR. If there are serious questions, is there any point considering what it tells us about the emergence of spacetime? In *The Trouble with Physics* (2007, 184ff), Smolin critically discusses the claim that the 'derivation' described above amounts to a string theory prediction or explanation of gravity. His main points are that, first, the bosonic string theory which we have discussed is not a fully coherent physical theory because it contains a faster than light tachyon (which means that the theory is energetically unstable). Second, although this fact motivated string theorists to develop supersymmetric string theory to avoid the tachyon problem, such models are only known to exist in stationary spacetime backgrounds: for instance, black holes. They have not been shown to exist in evolving cosmological solutions that might describe our universe: for instance, in a Friedmann-Lemaître-Robertson-Walker spacetime. These points are valid restrictions on the argument we just reviewed: it doesn't *prove* that a part of GR sufficient to describe our universe can be derived from supersymmetric string theory. Rather, the derivation within bosonic string theory, and the (as yet) partial success of supersymmetric string theory, are *evidence* that such a derivation is possible.

After his largely negative assessment of this and other string theory 'accomplishments', Smolin's conclusion is not that string theory should be abandoned, but rather that it should be one option pursued amongst others: as his work (and this book) exemplifies. In fact, string theory "succeeds at enough things so that it is reasonable to hope that parts of it, or perhaps something like it, might comprise some future theory" (198). (One possibility, of course, is that this future theory will be the 'M-Theory' that string dualities putatively indicate. Then the derivation of the EFE, and the supersymmetric spacetimes are evidence that that theory can explain GR.) One could argue with his assessment, but because of the considerations of

---

understood as coherent states of string excitations), then the result is that $g_{\mu\nu}$ is 'Ricci flat', the vacuum EFE. If there are other background fields, coherent states of other string quanta, then $g_{\mu\nu}$ will couple to them according to the corresponding EFE. (ii) The full analysis uses the stronger condition that the theory be Weyl symmetric at each order of perturbation theory. This condition provides an identity for $g_{\mu\nu}$ in powers of α'. To first order in α', this identity is the EFE, while higher order terms produce small, stringy corrections to the classical solution.

310   OUT OF NOWHERE

our introductory chapter, even Smolin's conclusion is sufficient to motivate our investigation of the 'emergence' of curved spacetime in string theory. We explicitly expect to be operating in a field of incomplete models, to identify and understand the ways in which the spatiotemporal might be an effective form of the non- (or not fully) spatiotemporal, in a future more complete theory. Thus, while recognizing Smolin's important caveat, we will continue to analyze the question of spacetime emergence in the context of the bosonic string that we have described.

10.2 Whence spacetime?

While the 'emergence' of the metric from the graviton field is clearly crucial for the recovery of classical, relativistic spacetime (and of course for calculating quantum string corrections to GR), the discussion so far does not address how other aspects of classical spacetime arise in string theory. To what extent are they fundamental features of string theory? How do those aspects that are not emerge? As we discussed in chapter 2, we follow a 'spacetime functionalist' approach to make progress on these questions: we start with a list of relevant structures, which may or may not be exhaustive. Then we can attempt to trace out their origins in the theory.

First a conceptual and terminological clarification. As we have stressed in earlier chapters, one should distinguish, at least conceptually, the classical spacetime of a theory like GR or QFT from the background, 'target' spacetime of string theory. String theory stands in the relations of being more fundamental than, and (putatively) explanatory of, GR or QFT etc; conversely GR and QFT are (putatively) effective descriptions of string theory in an appropriate limit. Although both fundamental and effective theories may involve formally similar structures—in this case Lorentzian spacetimes—one should not immediately conclude that these represent one and the same physical object; indeed chapter 8 argued that they do not. The general point is that formal models are representations of natural systems, and simply because two share a common formal structure, it does not follow that it represents a common natural object—even if the models stand in the relation of fundamental and effective.

To mark the (at least conceptual) distinction between the spacetime of string theory and that of GR or QFT, it is useful to have a clear terminology. Thus we speak of the background spacetime assumed in string theory as 'target space(time)', and the spacetime of GR or QFT as 'classical spacetime', or 'relativistic spacetime', but most often simply as 'space(time)'. So, target space is (relatively) fundamental, while space may be effective or emergent instead. Or from a more epistemic point of view, spacetime is well-confirmed by the empirical success of QFT and GR, while target space is to be inferred from the success of string theory in recovering QFT and GR, and especially from successful novel predictions (were any to exist!).

Returning to the question of derived spacetime structures, suppose that one has a model of classical spacetime, compatible with available measurements: a smooth manifold, with a metric and matter fields, satisfying the relevant dynamical equations—of GR, or of QFT, or of semi-classical gravity, depending on the situation described. Such a model has a topology, local and global, as well as a metric; we will enquire after the origin of each within string theory, focusing on the derivation of the EFE.

Of particular salience, the local open set structure and the metric give meaning to locality in a general sense: the size of open regions, whether they overlap, how far apart they are, whether or not they are spacelike separated, and so on. Of course, the full open set structure of spacetime constitutes its global topology, but by a 'local open set structure' we mean a proper subset of the open sets with their relations of overlap (subject only to the constraint that their union is a connected region). Informally, such a structure is the minimum required to make sense of 'where' questions, such as 'where did such-and-such an event occur?': a meaningful answer indicates an open set or sets (and excludes others). For our analysis, the paradigmatic events are *scattering events*, so the paradigmatic answer will be that 'the particles interacted inside *that* region': the collision chamber of the CMS (compact muon solenoid) detector at CERN, say, a region of order $1m^3$. Therefore, to understand the origin of the phenomenal local open set structure of spacetime—the location of particle scattering events—we want to know how string theory represents such events; the answer could be the simple one that they are represented by the target space locations of stringy processes, but it might not be.

We want to make two points about the methodology just adopted. First, we are investigating the emergence of more than the EFE: by enquiring into the local open set structure, we are enquiring into the origin of the structure represented by the smooth manifold in a classical spacetime model. If you like, the derivation of the EFE just shows how classical gravity emerges in string theory, not how spacetime itself emerges. Second, by raising this question we are impinging on debates that go by the names 'absolute-relative' or 'substantivalism-antisubstantivalism'— what does the manifold itself represent? On these questions readers may have strongly divergent presuppositions: manifold substantivalists (if any exist—the term is from Earman and Norton 1987) think that spacetime points are on an ontological par with physical entities, taking spacetime geometry quite literally; anti-substantivalists think not; others will simply not feel the pull of these debates at all. However, the question we have framed is neutral on such questions, because we are not asking about the origin of 'the manifold itself', but about the derivation of the local open set structure of spacetime, which is something all should be able to interpret according to their preferred stance on classical spacetimes. We view this neutrality as an important virtue of the approach adopted here. (Of course, it may be that the analysis of the derivation itself bears on the issue of substantivalism, but we will not pursue that topic here.)

312   OUT OF NOWHERE

To return to the question of emergence, thus framed, when one asks how a phenomenal theory—one accommodating all current observations—can be found as an 'effective' description of a (more) fundamental theory in some regime, it is not a reasonable requirement that the phenomenal and fundamental theories be in perfect agreement. All that can be demanded is that their predictions agree in that domain of possible experiments, and to that degree of precision, for which the phenomenal theory is believed to hold. After all, what we expect is that the fundamental theory will be a better match with experiment at some higher resolution: it will get things right that the phenomenal theory gets wrong.[10]

In particular, because the available energy places a practical restriction on the shortest measurable length, we should study spacetime emergence on the supposition of *smallest measurable open sets* in the derived theory: in QFT or GR, especially. For instance, CERN's 14TeV Large Hadron Collider (LHC) probes at length scales around 10^{-19} m. Any variations in quantities, including the metric, are indistinguishable from constant values across such a set: a measurement of variation would amount, contrary to supposition, to a measurement within the set. We emphasize that such constraints on measurability and distinguishability are not a priori philosophical restrictions, but reflect the contingent limitations of the technology used to test the empirical consequences of the phenomenal theory. But the upshot is that (i) the local open set structure of the smallest measurable open sets, plus (ii) a spacetime metric compatible with them to experimental accuracy, plus (iii) the observable global topology, constitute empirical elements of classical spacetime which need to be understood within string theory. It is through these structures that spacetime performs the functions of making well-defined 'where', 'how big', and global facts. Our question thus is how these structures arise; whether they are in fact derived, or instead occur as fundamental components of string theory. To answer that, we need to examine the empirical basis of the structures.

In the derivation of GR from string theory, the phenomenal theory involves the scattering of quantum particles—so QFT—in a classical spacetime with metric $g_{\mu\nu}$, while string theory describes the scattering of corresponding strings in a target space of metric $g_{\mu\nu}$ (since quantum corrections to the metric are by assumption unobservable by the relevant measurements, we ignore them). (i) The open sets of interest are then those in which particle scattering events (real or possible) are located. (ii) In both string and field theory, $g_{\mu\nu}$ enters as an undetermined parameter in the action, and so appears as a variable in scattering cross-sections: thus with all other parameters fixed, $g_{\mu\nu}$ can be determined at the location of the scattering event, by the observed value of the cross-section.[11]

---

[10] Though it is also logically possible that the phenomenal theory gets some things more right than the more fundamental. Just because an account is better overall, it does not follow that it is perfect, or even better in every regard. That is, scientific progress need not be strictly cumulative.

[11] A couple of important points should be made here, which will be addressed later. First, note that instead of a more familiar appeal to rods and clocks, we are giving the metric 'chronogeometric

THE EMERGENCE OF SPACETIME IN STRING THEORY   313

At first glance these two spacetime structures seem to be directly given by corresponding structures of target space: it appears that (i) the measured open sets just are the locations of string scattering in target space, and (ii) the observed metric just is that which determines the string scattering cross-section. That is, target space just is the space of the phenomenal QFT, and our observations simply fail to resolve the spatial structure of strings, so that they appear as quanta. In the following sections we will argue that this correspondence is misleading (as of course should be expected given (C) above and chapter 8).

We do not have a great deal to say about question (iii), of global topology. For as seen in §8.5, there are dualities which show that spacetime topology is not an invariant of string theory. Given our interpretation of dualities, it follows immediately that the observed topology is emergent not fundamental.

A final methodological remark: we do not claim that the simple picture just sketched is a full account of how a stringy effective theory could be brought into practical, specific correspondence with a theory of classical spacetime. For example, Wilson (2008) investigates some of the many ways in which classical theories of different levels mesh, and convincingly shows that they never conform *in detail* to the kind of story just sketched. And there is no reason to suppose that the situation won't be equally messy in the current case. But no matter. What is proposed is an idealized scheme to allow some points to be made about the relations between classical spacetime and string theory. To the extent that the idealization is accurate, the points are correct: such is always the situation, in physics or philosophy, when we wish to say something manageably succinct about a complicated reality. The only problem (which Wilson calls 'theory-T syndrome') arises if one mistakes such idealizations for claims about absolute metaphysical reality. But we won't do that! In particular, we don't claim that this construction is the most accurate possible description of our empirical knowledge of the topological and metrical properties of spacetime. The analysis only requires that it is good enough for our points to stand. To dispute that, it is not enough to show that our picture is incomplete: one would also need to show that our conclusions fail in a better picture. If that can be done clearly, we will indeed have a better understanding of these matters than that presented here.

10.3 Whence *where*?

We first turn our attention to the origin of the observable open set structure of spacetime. As noted, things at first appear quite simple: naively, the open sets

---

significance' through scattering cross-sections. Our choice has the advantage that we do not introduce primitive objects external to the theory. Second, as we will discuss later, T-duality means that scattering does not in fact uniquely determine the background metric, as supposed here. This fact will not affect the arguments until it is addressed.

314   OUT OF NOWHERE

of space in which scattering events occur are nothing but the open sets of target space—spatiotemporal open set structure is given at the fundamental level in string theory, and is in no sense 'emergent'. But this appearance is deceptive, and we shall see that there are obstacles to this conclusion (§10.3.2). Ultimately we will argue that localization is emergent, with the implication that strings are not literally spatial objects at all (§10.3.3). Before giving that argument we will explore an alternative to the target space view that seems to make the open set structure emergent, and show why it does not.

10.3.1 The worldsheet interpretation

One possible route to emergence was mentioned in §7.2: Witten (1996) argues for a view that is popular, at least as a heuristic, amongst string theorists—that the string *worldsheet* rather than target space is the fundamental spacetime structure. We saw in (D) above that formally one can view the contribution of a given worldsheet as a scattering process for a vector field, X^μ living on the worldsheet: a picture in which the worldsheet is viewed as a 2-dimensional spacetime. Moreover, as discussed in §7.2.1, the 2-dimensionality of the string is important for the mathematics of the theory (its conformal symmetries, and the applicability of complex analysis). Witten's 'worldsheet interpretation' proposes that this formal picture is in some sense the correct way to view perturbative string theory (until a more complete, non-perturbative formulation is available).

From this point of view, what is the origin of the open set structure of spacetime in GR or QFT? If the string worldsheet is the fundamental spacetime, then a first suggestion is that the origin lies in the open set structure of the worldsheet. This proposal (which Witten does not make) may not seem terribly plausible, since the worldsheet is only 2-dimensional. However, as described in §8.5, and discussed below, there are dualities that make spaces of different dimensions (at least) empirically indistinguishable. Moreover, the second series (over k) in (10.5) formally describes a quantum field on the worldsheet, so there is something to the idea, and it is worth disposing of it with a little care. The problem is that this proposal misinterprets the double sum over paths, for it is the first sum, over worldsheets j, that corresponds to *particle* scattering. As we explained in (B), the particles in a QFT scattering process are represented in string theory by strings in given excited states: so the terms in a Feynman expansion for a quantum field correspond to the different string worldsheets. And we are taking the phenomenal open sets to be the locations of, paradigmatically, such particle scattering events, say some particular event occurring in the CMS detector at the LHC; hence to the worldsheets j of the first sum (see figure 9.1). Thus localization of observed scattering depends on the space in which the strings are localized, namely target space. The second series quantum mechanically describes the localization of a given world sheet in

THE EMERGENCE OF SPACETIME IN STRING THEORY   315

target space; and so presupposes the open set structure of target space. It is thus localization in target space that determines the 'where' of the scattering event, and so it is to the open set structure of target space, not the worldsheet that we should look. Or rather, from the worldsheet perspective, we should look to the open set structure of the space of *possible values* of the field, X^μ, since that is how target space is interpreted.

So we seem to have returned to the original naive idea, which the worldsheet interpretation was attempting to replace: that strings are localized in classical, observed spacetime. But what about the fact that X^μ is treated as a *quantum field*, rather than coordinates of a classical target space? Does that imply a sense in which relativistic spacetime arises from something other than a presupposed spacetime? No, because it is still the structure of the space of possible classical values of the X^μ field that do the work. From the target space perspective, the quantum string does not follow a single classical path in target space, but 'explores' an extended region; the amplitude involves a sum over target space paths (10.2). From the worldsheet perspective, the quantum X^μ field does not have a single classical configuration over the string, but a range of configurations are explored; reflected formally in the same way in (10.2). But this sum over paths simply assumes the standard open set structure of target space/the manifold of possible values of X^μ, and it is this structure that corresponds to the structure of the locations of phenomenal scattering events. In short, the worldsheet interpretation doesn't provide a different account of the origin of event locations from that in the straightforward, naive view with which we started. Either way it is taken to be the open set structure of target space, whether this is interpreted as a spacetime or as a space of possible classical field values: either way, the phenomenal open set structure is simply given in string theory, not emergent.

Therefore, the only way to argue for the emergence of a spatiotemporal open set structure from the worldsheet would be to take a hard ontological line on Witten's interpretation: target space *really* represents field values and not a spacetime. Then one could claim that spacetime emerges from something that is in fact a field, not itself a spacetime. It is unlikely that Witten himself intended such a strong claim, rather than intending to demonstrate the heuristic value and conceptual naturalness of the worldsheet point of view (his more forceful points regarding 'the fate of spacetime' concern dualities). Nor do we think it especially fruitful to pursue arguments on the question of whether X^μ is 'really' this or that, unless its relevance to the development of string theory can be established (for instance, perhaps one of the interpretations aligns better with whatever the degrees of freedom of the exact theory turn out to be). Indeed, we view it as a virtue of the way we have posed the question of the emergence of spacetime—in terms of local open set structure— that it has already led to a substantive question with some traction on the physics involved; while avoiding debating whether there 'really' is a field or spacetime, a cousin of the substantivalism debate that we put to one side earlier. We have seen

316   OUT OF NOWHERE

quite clearly that both the target space and worldsheet views say that the structure of phenomenal event locations is not emergent, but directly built into string theory.

10.3.2 T-duality and scattering

But not so fast! What we learned earlier of dualities should make us doubt such conclusions. We argued in chapter 8 that dual theories are physically equivalent, so that only features on which they agree are physical. But so far we have only considered how the observed open set structure arises in one theory; what about its duals? If it is derived from a different open set structure in different duals, then those structures are unphysical (beyond their 'common core'), and so not assumed in string theory after all. They are in some sense mere formal representations, and the open set structure of spacetime must arise from some other string theoretic structure that *is* invariant under dualities. Indeed, given the conclusion that target spacetime is not spacetime, it seems something like that must be the case.

To develop this line of thought, for simplicity let's focus on T-duality, and the case of flat background spacetime.[12] Recall, we claimed on the basis of a formal isomorphism that string theory on a space with a closed dimension of (large) radius R is physically equivalent to a string theory in a space with (small) radius $1/R$ (in units in which the small 'string length' constant $\ell_s = 1$). We argued that as a result only those quantities on which the duals agree are physical. So for example, there is no determinate physical radius to target space; and since there is a determinate (effective) physical radius to the space of GR and QFT, it therefore cannot be target space.

Of course, since T-dual target spaces have the same cylindrical topology, even with their reciprocal radii, they have exactly the same topological open set structures: that much is physical (with respect to T-duality). But this observation alone does not entail that the open set structure of space simply arises from a single shared topology. For the open sets of target space are mapped to those of space by identifying (possible) scattering events in each: again, as we discussed in the previous section, open sets are understood as possible answers to 'where' questions about phenomenal processes, so understanding their origin requires identifying the corresponding fundamental stringy processes, and whatever stringy structure 'localizes' them. But that cannot be achieved merely by considering the topology of target space; we need to identify the stringy process in each dual, to see what they have in common.

---

[12] We shall see later that T-duality can be extended to general background metrics, and of course we have discussed dualities that do more violence to the topology of spacetime. But the same basic considerations apply in such more complicated cases.

THE EMERGENCE OF SPACETIME IN STRING THEORY 317

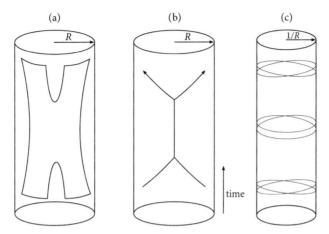

Figure 10.1 Scattering in a space with a closed dimension (other spatial dimensions are not shown). (b) is the observed QFT process, in a dimension of radius R, in which two particles interact to produce a third, which then decays to two outgoing particles. (a) is the string description in a target space of radius R: two strings join into one, which then splits into two. (c) is the dual—physically equivalent—description in a target space of radius $1/R$: two once-wound strings join to form a single twice-wound string, which then splits back into two once-wound strings.

So consider a phenomenal particle scattering event: say two incoming quanta interacting at the LHC to produce two outgoing quanta. The radius, R, of the space in which we observe this event is of course large—much larger than the CMS detector in which it occurs, say. This particle scattering event then has a string theoretic description in a large target space, also of radius R, in which the quanta are replaced by appropriately excited strings, in the way that we have discussed. The situation is pictured in figure 10.1 (a) and (b), at the lowest 'tree' level of perturbation theory, for an interaction allowing the production of an intermediate particle from the incoming pair. The key point, shown clearly in the figure, is that scattering occurs in a region of the cylinder homeomorphic to the plane, in space and in large radius target space.

But the kinetic energy of a string in a large radius theory is dual to the winding energy of the dual string: the potential energy it has due to being stretched around the closed dimension a number of times under its internal tension. So the T-dual scattering process is quite different: where before there were strings carrying kinetic energy, now there are wound strings. A process that involves strings joining and splitting in one dual corresponds to strings changing their winding numbers in the other dual. Thus the T-dual process is also pictured in figure 10.1 (c), as two once-wound strings joining to form an intermediate twice-wound string, which splits into two again. Of course, these are dual—physically equivalent— representations of the single observed particle scattering process. Therefore, the

318   OUT OF NOWHERE

T-dual region of target space in which scattering takes place is drastically changed; because the strings are wound, the whole of space is involved, a region with the topology of a cylinder. (No surprise: because target space is much smaller than ordinary space, one would hardly expect their open sets to be in simple correspondence!) Compare this with the first dual, according to which scattering took place in a proper subregion, homeomorphic to the plane.[13]

Although the dual target spaces are homeomorphic cylinders then, the physical question of locality depends on how the locations of events in one space correspond to those in the other—and T-duality does not preserve the topology of these locations, as this example shows. So the duals disagree over where in target space a scattering event occurs, and hence the representations of where scattering occurs in the duals are not fully physical (and only physical insofar as the duals agree): specifically, there is no physical fact about whether an event occurred in a planar open region, or in a cylindrical region of target space. Therefore, although the spatial location of a scattering event will formally agree with that of the large radius dual, since the latter isn't physical, phenomenal scattering regions are not directly built into string theory. On the other hand, just as for the phenomenal radius, there is a fact of the matter about the phenomenal scattering location: and so the locality structure of space—which one can say *is* space in a pre-metrical sense—is also derived or emergent in string theory, because of T-duality.

Before we unpack this argument, we want to address the puzzle of how a process localized around an entire closed dimension of target space could possibly appear as if localized in an open region of space. In §8.1 we discussed the similar question of how a target space of small radius appears as a space with an observably large radius, and how Brandenberger and Vafa (1989) answer. To recap: T-duality works because quantized strings require two spaces to represent states, target space, and a 'winding space' of reciprocal radius; spatial wavefunctions in target space represent string momentum, while a 'winding wavefunction' in winding space represents the quantum state of winding; under T-duality, spatial wavefunctions in target space are mapped to winding wavefunctions and vice versa; hence processes in a target space of radius R, are dual to processes in a winding space of equal radius, $1/(1/R) = R$; so if a phenomenal process that measures the radius of space (timing a photon around the closed dimension is their example) corresponds to a process in a target space of radius R in one theory, in its dual it corresponds to a process in a winding space of radius R—either way, the measurement will be R, an observed large radius.[14]

---

[13] For now we will let the figure carry weight of the argument for this conclusion; below we will consider objections.

[14] The other assumption that is required to guarantee a *large* radius is that the measurement is a low energy process: a long wavelength spatial wavefunction for the radius R dual, and a low winding state for the the radius $1/R$ dual.

THE EMERGENCE OF SPACETIME IN STRING THEORY   319

The path signposted by Brandenberger and Vafa can be followed in this case too, to explain how a scattering event localized around an entire cylindrical dimension of a small radius target space appears as a scattering event localized in a planar open set of space. The winding pictured in target space corresponds to a quantum wavefunction living in winding space: that's how the theory in fact represents winding. So the quantum description of scattering in the radius $1/R$ dual is as a process in winding space—indeed, formally the very same processes found in target space in the radius R dual. (So the leftmost, radius R cylinder in figure 10.1 can be interpreted as picturing a tree level winding space transition amplitude for its dual.) That is to say, according to the small radius dual, the open set structure of space comes from that of winding space instead of target space. And that's how the closed region of target space appears as an open region of space.[15]

10.3.3 Scattering and local topology

With that puzzle resolved, we can turn to assessing the duality argument for the emergence of open set structure more carefully. We claimed that because the duals disagree on the localization of scattering in target space, such target space localization must be indeterminate; while the observed localization in classical space is determinate (setting aside ordinary quantum uncertainty). There were two steps to this argument. First, that the physical content of T-duals is restricted to their common part: whatever they disagree on is indeterminate. The argument for this claim was given at length in chapter 8, so we will not rehearse it again here—we claim that the move from metrical issues (the radius of space) to topological ones (the open set structure) does not change matters.

The second step of our argument notes that a scattering process occurs in an open region of target space in one dual, and in a closed cylindrical region in the other. Then the physical indeterminacy of target space localization follows immediately (as the indeterminacy of the radius did previously). And then because open set structure *is* determinate for space, it cannot be identified with target space open set structure, but is emergent. However, so far the second step was supported simply by figure 10.1, picturing the dual scattering regions, which is rather quick.

That figure shows a tree diagram for a scattering process, rather than a full scattering event: only the first term of the leftmost sum in (10.2). What should we say of the full process, summed over all diagrams? Is that still localized in an open region? And even a single scattering diagram represents an integral over all of space, not just a proper subset of it: the rightmost sum in (10.2) embeds the worldsheet in

---

[15] And recall from §8.4 that we should indeed read the duals this way, as describing different target space processes.

320 OUT OF NOWHERE

every possible way in space, not just a subset of it. So is the diagram misleading about the localization of scattering?

To answer these questions, consider that they are not specific to string theory, but apply equally to scattering in QFT, in which we certainly take concrete scattering events to be effectively localized, in the CMS detector, say. What holds in that case should also hold in the case of string scattering, which we take it to approximate. Indeed, (as spelled out in §9.1) the connection between string and QFT scattering diagrams is extremely tight. In the terms of the large radius dual, QFT scattering diagrams are taken to directly approximate string diagrams, at long length scales and low energies (relative to the string length and energy): see figure 9.1.[16] So if QFT scattering diagrams represent a localized process—as indeed they do—then so do the corresponding string diagrams.

Although a QFT S-matrix encodes amplitudes for particles coming from infinity to interact and produce new particles at infinity, integrated over the whole of space,[17] it is understood that this is a good approximation to the physical situation being modeled, in which the whole process is contained and observed in a finite region, say the CMS detector. Particle wavefunctions can never be localized for a finite time, so it is true that interactions of tails do happen over all space. But interactions are effectively localized, and of course for the identifications of phenomenal local regions, it is the effective localization that matters. Formally, that a finite volume interaction can be approximated by an infinite volume one is justified because in the limit that the finite volume goes to infinity, they agree (see Fraser 2006, especially §3.2.1, for a clear discussion). If the finite volume is 'big enough', the differences will be unobservable. Then, since the particle diagrams are approximations to stringy diagrams, the latter can be taken in the same way, as approximating the same localized process, despite the sum over all target space.

As for the small radius dual, the argument that the scattering region includes the entire circumference of the closed dimension is straightforward: the in- and out-states are for winding eigenstates, and hence for strings that are wound around the dimension. There simply is no way for them to interact locally, unless the interaction is also around the dimension, for at least part of its duration. (And again, because T-duality exchanges space and winding space, represented in terms of the winding wavefunction, we have a process effectively localized in winding space: formally the same process that occurs in target space in the large radius dual.)

---

[16] As previously noted, this identification has to be taken with a grain of salt: realistic, standard model particles do not arise simply as excitations of strings, but require specific tuning of background topology and a system of D-branes. The string diagrams that we are discussing do correspond to particle-scattering diagrams, just (for the most part) not of particles observed in our universe. However, the argument pursued in this section still should go through in a more realistic scenario.

[17] Note that the infinite volume approximation is not entirely apt: integrating over arbitrarily large distances can produce unphysical infrared divergences.

And so indeed, the naive reading of figure 10.1 is supported by more careful analysis: the duals disagree on how to represent the target space localization of a scattering event, and so target space is indeterminate with respect to that question. Moreover, the determinate localization of phenomenal scattering cannot simply be that of target space—localization in classical spacetime is in that sense derived or emergent, not fundamental.

Conversely, strings do not scatter in spacetime, taken as a literal fundamental statement: they scatter in target space. As one describes nature at smaller and smaller spatial scales, it is not the case that increasing spatial detail is included until particles are understood to be objects extended in one spatial dimension; instead, around the string scale the emergent spatial picture breaks down altogether and is replaced by the distinct target+winding spaces one. In short, *strings are not literally tiny objects*, because such a statement implies a comparison with spatial objects, which strings are not fundamentally. This is another aspect of the claim of string theorists that there is a shortest length scale, below which the concept of space breaks down.

That event localization is emergent in string theory is the main conclusion of this section, but it serves another important purpose, independent of whether that result is accepted. What we have aimed to do with this example is illustrate the fruitfulness of the approach we have described for addressing the question of spacetime emergence; we specifically want to urge it as more fruitful, and better defined than attempting to import the substantivalism debate into quantum gravity. To summarize, identifying the locations of phenomenal events—scattering events in this case—has allowed us to ask a well-formed question about the origin of that location in the more fundamental theory: *what is it about the fundamental description of the process that corresponds to that localization?* The answer might have been straightforward: perturbative string theory is formulated in target space, so particle-event localization in space is simply understood as string-event localization in target space. But the conclusion of chapter 8 that target space is not space should already make that answer seem questionable. And indeed, the T-duality argument shows that there is no physical fact about the localization of string scattering in target space. Yet in either dual the string interaction is localized in some space (winding if not target), so the very notion of localization in an open set of *some* manifold is still valid—things are still 'spatial' to that extent.

10.4 Whence the metric?

The previous chapter, and the summary above, explained the standard account of how a metric satisfying the EFE can be derived in string theory as the excitations of graviton modes of strings: how string theory has GR, the classical theory of gravity, as a low energy limit. The key step in this derivation occurs in (10.4). On the

322　OUT OF NOWHERE

one hand, graviton coherent states (seen explicitly on the LHS of the equation) contribute to scattering exactly as a metric (seen on the RHS): they *are* the gravitational field according to the theory. (And analogously for fields of other stringy quanta.) On the other hand, the observed, classical metric at a location (a smallest observable open set) is operationalized by the outcome of scattering in the location: in scattering cross-sections, $g_{\mu\nu}$ appears as a parameter, and so for fixed values of other parameters can be measured by the observed cross-section. Specifically, a cross-section calculated using (10.4) will measure $s_{\mu\nu}$, the tensor characterizing the graviton states, or equivalently $g_{\mu\nu} = \eta_{\mu\nu} + s_{\mu\nu}$, the observed metric.[18] Weyl symmetry then entails that to first order the measured field satisfies the EFE; more sensitive measurements will, according to string theory, reveal stringy corrections. Or, from the graviton point of view, Weyl symmetry constrains the possible coherent states: $s_{\mu\nu}$ must be such that $\eta_{\mu\nu} + s_{\mu\nu}$ satisfies the EFE (plus higher order corrections).

This account is central to the understanding of string theory as a theory of gravity, but it raises many questions, both to its conceptual and technical coherence, and to its claim to derive gravity as opposed to presupposing it. In this section and the next we will take up these issues, clarifying and defending the account as a derivation of gravity. This section deals with some of the broader conceptual issues; its sequel with some important technical limitations of the account.

10.4.1 'Background independence'

First, in a critical analysis, Roger Penrose complains that "string theory does not really properly come to terms with the problem of describing the dynamical degrees of freedom in the spacetime metric. The spacetime simply provides a fixed background, constrained in certain ways so as to allow the strings themselves to have full freedom." (Penrose 2004, 897).[19] This quotation alone may be a little ambiguous, but in the context of the chapter, the most reasonable reading is that Penrose is thinking of only one half of the derivation of GR: that avoiding the Weyl anomaly requires the 'background' metric to satisfy the EFE. His discussion of gravitons (896) appears to be restricted to the point that as excitations of

---

[18] As noted above, T-duality means that the fields are not uniquely determined. We will discuss this complication at the end of the section; for now we will ignore it for simplicity, as it will not affect the following arguments.

[19] Chapter 31 of Penrose's *Road to Reality* is highly recommended and insightful. It contains serious technical challenges to string theory as an exact, complete story, in opposition to the strongest claims for the theory. He does not deny that string theory is a promising partial story of quantum gravity, and so these points need not be addressed here. Once again, we assume only that string theory is at least a partial theory for the emergence of spacetime, not that it is the final truth: that is enough for the kind of work that we defended in the introduction.

THE EMERGENCE OF SPACETIME IN STRING THEORY   323

the string they will contribute perturbatively to scattering processes. He does not mention the essential point (C) above, that in coherent states they contribute to the 'background': that they *are* the gravitational field. Only by considering the partial picture that he presents does it appear that there is no more than a restriction on the kind of 'container' in which strings can live. The full picture changes the situation completely.

To better understand this point, let us clarify an ambiguity in the term 'background' (hence the scare quotes in the previous paragraph). Suppose we use (10.4) to write the string action as $\int g_{\mu\nu}(\partial_\tau X^\mu \partial_\tau X^\nu - \partial_\sigma X^\mu \partial_\sigma X^\nu)\, d\tau d\sigma$. *Formally*, this action can be understood to describe a 2-dimensional hypersurface in a manifold with a given Lorentzian metric $g_{\mu\nu}$. The conventional terms for this metric and manifold are the 'background metric field' and 'background spacetime'. This terminology reflects the mathematical fact that formally one is studying quantum perturbations around a classical 'background' solution, itself assumed to approximate some quantum state: a standard approach to quantum mechanics.[20] String theorists use these terms in this formal sense, which has no interpretational implication: its use is completely consistent with a physical understanding that fundamentally $g_{\mu\nu}$ describes a coherent state of strings, and only effectively a classical spacetime.

However, in the context of gravitational physics, and in Penrose's comment, 'background' has a second meaning. GR is generally understood to make spacetime a dynamical object: in contrast with the given, fixed 'background' of Newtonian or Minkowski spacetimes. This feature of GR is thus termed 'background independence', and often urged as an important lesson of the theory, one that should be observed in a quantum theory of gravity, as we discussed in §5.2. Thus, one might think that the 'background' metric of string theory is a wrong turn![21]

But that of course would be to equivocate on the term: in the standard terminology of string theory the phrase 'background metric' simply designates a part of the mathematical machinery, without any connotation that it is essentially non-dynamical. In particular, this use is not intended to express the claim that the theory is background dependent in the sense relevant to GR: that the metric is non-dynamical. Hence the question of whether 'background spacetime' (in the language of string theory) violates 'background independent' (in the language of GR) remains.

---

[20] In quantum gravity (and general gauge theory) it was developed by DeWitt in the 1960s (DeWitt 1967; see also Kiefer 2004, §2.2). Concretely, his approach relies on the fact that S-matrix elements depend on asymptotic in and out-states, which can be taken to be Minkowski to define particle states, while the interior, interaction region has an arbitrary classical geometry.

[21] Since the work of Kuhn (1962) philosophers and historians of science are more aware that such alleged deep principles are liable to change with major changes in theory. However, it seems at least reasonable to attempt a theory of quantum gravity that respects the principle: some restrictions on the possible avenues of exploration are needed.

324 OUT OF NOWHERE

As we noted, the main part of Penrose's objection appears to be that string theory draws a distinction between space and its metric on the one hand (the 'fixed background' to which he refers), and string dynamics, including graviton excitations, on the other. But if our discussion of coherent states is correct, then Penrose's view is incorrect: far from spacetime curvature being distinct from the string state, it is *constituted* by it, and so could not be less distinct! As we have said before, the consistency requirement is on graviton states themselves, not on a distinct 'container' in which the string dwells. In that sense the metric *is* dynamical in string theory, because it captures the dynamics of the string, specifically of the graviton modes: a point made in Huggett and Vistarini (2015) and Vistarini (2019, §2.4— her chapter 6 gives a different, novel argument for the background independence of string theory).

This response to Penrose illustrates further the difficulty in applying background independence as a desideratum in quantum theories of gravity. For example, in §5.2 we discussed general covariance under active diffeomorphisms as a formulation of the condition. Now, formally string theory violates it, if we consider general diffeomorphisms acting on strings but not target space geometry, so that $g_{\mu\nu}$ is fixed. Does it follow that string theory is background *dependent*? In the first place, we saw that general covariance is only a sufficient condition for background independence; so as a point of logic the answer is 'no'. But more than this, string theory shows that as a formal principle, general covariance can beg the question on the central issue, namely just what objects should stay fixed, and which should be transformed. But the issue of background independence *is* the issue of whether all objects are dynamical, and so should be transformed; thus one already has to have a handle on whether there are fixed structures in order to adjudicate background independence! In the present case, target space formally acts as a 'container' of string scattering processes, so naively its geometry should be held fixed, producing a violation of general covariance; but of course this reasoning already assumes the non-dynamic nature of $g_{\mu\nu}$, the very issue in question. Moreover, we have argued that $g_{\mu\nu}$ is a dynamical object, representing (to lowest order) a coherent state of gravitons; so less naively it would be wrongheaded to keep it fixed if one wanted to evaluate the general covariance of string theory. In short, except in a formalistic, question-begging sense, it arguably does not violate general covariance; but even if it does, it can readily be understood as another counter-example to the view that general covariance is necessary for background independence in the sense that we described in §5.2.[22]

That said, if string theory really is background independent, one could certainly ask for a formulation that demonstrates that fact more perspicuously. Of course

---

[22] The background independence of string theory is studied at much greater length in Read (2016); he reviews many different proposals, both classical and quantum, and gives a generally positive answer to the question of whether string theory is background independent.

THE EMERGENCE OF SPACETIME IN STRING THEORY   325

we have other reasons to desire such a reformulation: as discussed in chapter 8 we would like a formulation that captures only the 'common core' of dual theories, namely M-theory. The nature of the dualities suggests that in such a formulation spatiotemporal structures would not be fundamental; but *non*-spatiotemporal structures trivially cannot be a spatiotemporal background! We emphasize that in our view the desirability of such a formulation lies not in moving beyond a background dependent theory, but in its greater perspicuity in making explicit the implicit background independence of current string theory (and of course in its validity beyond perturbation theory).

At this point, having raised and acknowledged the question of the background independence of string theory, we will move on (although the following discussion bears on the issue). For the issue is secondary in the approach to emergence that we have outlined, which is to trace and analyze the *derivation* of classical spacetime elements from the structures of quantum theories of gravity. Such an account is preliminary to determining to what extent string theory does or does not respect the 'lessons' of GR; and to judging how various formal notions might or might not capture those lessons. (And arguments about whether or not one *should* respect those lessons are even further downstream, beyond the concerns of this book.) It is to the extent that it helps us understand the explanation of classical spacetime that the issue of background independence is relevant in this chapter: especially to address the question of to what extent are spacetime structures built in to string theory, and to what extent emergent, within the formal derivation we have described?

10.4.2 Is there a Minkowski background?

With the question of background independence acknowledged, and the somewhat different focus of our investigation into emergence emphasized, we should address an issue regarding metric emergence that will likely have occurred to the reader. Namely, even granted the graviton interpretation, in the action (10.4), $g_{\mu\nu}$ is composed of two parts: $s_{\mu\nu}$ representing a tensor field contribution from the coherent graviton state, *and* $\eta_{\mu\nu}$ representing a Minkowski metric. Doesn't that mean that a particular—Minkowski—metric structure is baked into string theory, rather than emergent? We will resist that conclusion, which will involve conceptually unpacking the derivation of the EFE more carefully. (Thereby also resisting a significant sense of background dependence: one invoked by Earman 2006b, 21.)

First, we note and put to one side three possible replies. (i) Read (2016, 101f) asks whether the Minkowski part of the background metric can also be interpreted as a coherent string state. He points out that $\eta_{\mu\nu}$ could never be a coherent state of gravitons, since it is not traceless ($\text{tr}[\eta_{\mu\nu}] = D - 1$), while the graviton states are. But, he notes, there is a massless spin-2 string state with non-vanishing trace,

326 OUT OF NOWHERE

composed of the graviton and a scalar excitation, which could form appropriate coherent states. (These states form a reducible representation of the spacetime
symmetries, so correspond to composite particles in the usual understanding.) So
could both parts of $g_{\mu\nu}$ be built as coherent states of stringy quanta in this way? He
correctly replies that this idea won't work, because the very framework for talking
about symmetries of string states, and the corresponding quanta—not to mention the framework of QFT on which the notion of coherent state rests—assumes
Minkowski spacetime. (ii) If one followed Witten's worldsheet interpretation, then
one would understand $\eta_{\mu\nu}$ to be the inner product on a vector field X^μ, rather than
a metric on 'real' spacetime. But this response brings us back to the question of
whether target space or the worldsheet is 'really' the fundamental spacetime, which
we have shelved. (iii) We note finally that string dualities mean that the background
spacetime is not unique: T-duality means that the metric is not determinate, and
duals with different topologies cannot be isometric. We will return to consider the
implications of T-duality below, but our response to the issue of the Minkowski
part of the metric does not appeal to duality. Instead, consider the following.

In arguing for the coherent state picture, we appealed to the fact that in
(10.4) $g_{\mu\nu}$ could be decomposed as $\eta_{\mu\nu} + s_{\mu\nu}$ without making any difference to
the action, hence path integral, hence amplitudes. But the same holds trivially for
an *arbitrary* division of $g_{\mu\nu}$ into a metric and tensor field: $g_{\mu\nu} = \gamma_{\mu\nu} + s'_{\mu\nu}$. (With
$s'_{\mu\nu} = s_{\mu\nu} + \eta_{\mu\nu} - \gamma_{\mu\nu}$.) The split is irrelevant to the action, which only depends on
the total, $g_{\mu\nu}$. As Motl (2012) says, "there's only one perturbative superstring theory
in this sense—whose spacetime fields may be divided to 'background' and 'excitations' in various ways."[23] If this is the case, the resultant metric is not just 'emergent'
in the sense that it involves a component from the string itself, but in the further sense that only the total has physical significance: there is no significance to
any particular split into metric and coherent state contributions. In particular, the
metric part is conventional, without physical meaning. This argument, though we
generally endorse it, requires some unpacking and qualifying.

First some clarification of its terms: up to now, we have, in line with standard
use, referred to $g_{\mu\nu}$ as the 'background' metric, while in the passage quoted, Motl
refers to its metric *part* as the background! (Of course, because he is addressing the
question of whether string theory is independent of this particular 'background'
structure.) For the purposes of our discussion of this proposal it will help keep
things clear to adopt a different terminology: we will call $g_{\mu\nu}$ the 'full' metric, and
its components the 'partial' metric and graviton/tensor field. Then we have argued
that the full metric is emergent because of its graviton field part; but the question
currently addressed is whether the partial metric is a non-emergent spacetime

---

[23] We have found many useful insights into string theory on his blog, although his posts are regret-
tably often marred by personal attacks. The piece cited (which addresses background independence)
is free from these.

structure. The answer we take from Motl is that the partial metric is no structure at all, just an arbitrary component in a split of the only physical quantity, the emergent full metric. This split is necessitated, not by nature, but by the perturbative formalism currently used to describe string theory. (To make the existence of a formalism which does not require such a split a little more than a promissory note, he points out that string field theory, mentioned above, can be given without it.)

Then we reason as follows. The perturbative approach to quantum mechanics means selecting a classical solution with full metric $g_{\mu\nu}$, understood to approximately represent some quantum state, and then expand in quantum mechanical corrections. In just this way, perturbative string theory takes a solution to GR, and expands around it: along the way one finds that the background actually *had* to be a model of GR. Moreover, the basic physical quantities that are defined within the theory—scattering amplitudes (collectively, the 'S-matrix')—are identical to that of a model of involving stringy graviton excitations in Minkowski spacetime. That is, one could have taken a different classical solution as the starting point, and found the very same string physics, but within a different perturbative framework. But similarly, one could in principle have started with any (of some range of) classical solutions, with partial metric $\gamma_{\mu\nu}$, and studied quantum string perturbations around that: the excitations in that framework would also form collective states, including that represented by $s'_{\mu\nu}$. But since the path integral only cares about the full metric $g_{\mu\nu}$, once again the same S-matrix will be found. Although the formalism of perturbation theory requires a classical background, which one is an arbitrary choice, without direct physical significance.[24] All that can be said is that the *choice* of a Minkowski metric produces a simple and clear formal structure.

That is, as discussed in chapter 7, in relativistic quantum mechanics—including QFT—elementary particle species are defined in terms of their (irreducible, unitary) representations of the symmetries of spacetime: how the particle states transform in Hilbert space under the action of the corresponding spacetime transformations.[25] Spacetimes with different symmetries will have different (perhaps overlapping) representations, and hence may allow the existence of different kinds

---

[24] It is worth pointing out that the S-matrix contains only asymptotic amplitudes, not the full set of correlation functions required to define an exact QFT via the Wightman axioms. Hence this equivalence is indeed firmly at the level of perturbative string theory—which of course is more or less all that exists at present.

[25] This conception is due to Wigner (1939), and is a staple of texts on relativistic QM and QFT: a particularly good treatment is found in Weinberg (1995, 2.5). Note that talk of 'particles' in this context is loose (for instance, Malament 1996), and we do not intend to beg the question against such arguments concerning the particle concept in relativistic QFT. 'Quanta' would be a better term, though see Ruetsche (2011) for an extended analysis of the tenability of even the standard quanta concept, in contexts relevant to curved spacetime. At base, our 'particle' concept here just is Wigner's, of irreducible, unitary representation.

328   OUT OF NOWHERE

of particles, or no differentiation into particle species at all, if there are insufficient symmetries. However, the relatively small regions in which QFT is typically explored on Earth are suitably flat, so that the states of observable particles do form representations of the Poincaré symmetries. Thus, the identification of string states with particles that we have followed is based on their common representations of the Poincaré symmetries, and so on the assumption of Minkowski spacetime. That is, the existence of stringy quanta—the scalars, tensors, etc. that make up the string spectrum—has, so far, been predicated on Minkowski geometry.

Thus, one cannot strictly describe the $s'_{\mu\nu}$ tensor field in an arbitrary metric+field decomposition as a '*graviton* coherent state', for gravitons will not be defined in an arbitrary spacetime geometry. However, that is not to say that a similar conception is impossible in curved spacetimes: one asks, (i) what form elementary particle states take, and (ii) what the string excitation spectrum is, and see whether they can be identified. Away from flat spacetime, the formal situation is much less tractable, but in the next simplest cases the answers are, (i) there is a massless spin-2 graviton representation of the de Sitter and anti-de Sitter spacetime symmetries.[26] While (ii), Larsen and Sánchez (1995) find that there is indeed a massless spin-2 representation in the (closed) string spectrum which can be identified as the graviton, just as in Minkowski spacetime. So, indeed a split between partial metric and graviton field is not fixed spacetime structure, but can be chosen at least between Minkowski, de Sitter, and anti-de Sitter metrics. All the same, these particle identifications depend on a rich set of symmetries: in a general spacetime, Wigner's concept of particle will only hold approximately, in local patches with approximately symmetric metrics. In other backgrounds, the perturbative stringy excitations need not be 'particles' in Wigner's sense at all; except approximately in small enough regions, or 'at infinity' if spacetime is asymptotically Minkowski (say, if one considers scattering through a finite curved region). This fact does not change the arbitrary nature of the metric-tensor field split, but only shows that the string-quanta identifications are themselves an artifact of a particular approximation scheme. (This point will be addressed again, from a different perspective, in §10.5.) But this itself is to be expected in a theory of quantum gravity in which the metric is not fundamental, for there is then no particular metric to which particle classification could be referred.

With these qualifications, however, we agree with Motl's proposal: string theory does not postulate a Minkowski background spacetime, in the sense that the choice of the Minkowski background in the equations is conventional. Other choices are equivalent, and the choice is made for mathematical convenience. In that case, no part of the full metric can be identified as 'given', as it could equally (ontologically

---

[26] The issue of how to define mass is subtle: see Garidi (2003) for a recent treatment. The earliest classifications are found in Thomas (1941) and Newton (1950).

speaking) be redefined as a dynamical string state; and so we conclude that the metric is indeed derived, not assumed, in string theory.

10.4.3 T-duality

Read (2019) raises the question of T-duality in the emergence of the metric; an issue discussed further in Read and Menon (2021). A basic question is whether the duality is with respect to the partial metric ($\eta_{\mu\nu}$, say) or full background metric $g_{\mu\nu}$? The former we have already argued to be conventional. But if T-duality applies to the latter, then from the arguments of chapter 8 it will *not* be identified with the observed, classical metric: that will have to be understood in terms of invariant measurements, along the lines described by Brandenberger and Vafa. To answer this question we turn to the 'Buscher Rules', which describe how to implement T-duality in certain curved spacetime backgrounds.[27] In particular, they describe the duality that occurs in the *full* background $g_{\mu\nu}$, in the case of closed dimensions, with suitable symmetries of the metric: so in the terminology of chapter 8, the full background metric is the 'target space' metric. In those cases then, observations cannot uniquely determine the background metric: and, as in the simpler Minkowski case, low energy observations will agree on the metric observed, in both duals. In a Minkowski background we saw, for instance, that an experiment timing a photon around the universe would have the same result whether target space were large or small: a large observed radius. Read and Menon propose that the result would be the same in the more (but not completely) general case of the curved background T-duality: the large spacetime that we in fact do observe. We concur. Therefore, we need to revise our earlier understanding of how the metric (and other fields) are operationalized through scattering observations: scattering cross-sections are only observable up to differences that would allow discrimination of dual metrics.[28]

In this section we have defended the claim that the metric field should be understood as derived in string theory against the major objections to this view. In so doing we have aimed to clarify the content of the claim, and especially its relation to the perturbative nature of string theory as it exists. However, while we endorse this position, we also argue that it is circumscribed, by its perturbative nature, and by the assumption that coherent states provide a classical limit of QFT, as the next section explains.

---

[27] Blumenhagen et al. (2012, §14.2) is a good introduction.

[28] Note that scattering assumes in- and out-states at infinity, and so is not defined in a compact dimension. Local scattering measurements can measure the local metric, but scattering is not an appropriate way to measure global metrical quantities, such as the circumference. To do that, one needs an experiment like Brandenberger and Vafa's photon circumnavigation.

330 OUT OF NOWHERE

10.5 Quantum field theoretic considerations

The string theoretic interpretation of the metric works on the understanding (i) that suitable string states correspond to quanta of the gravitational field, and (ii) that such stringy gravitons form collective—i.e., 'coherent'—states; and on the assumption (iii) that a classical metric field corresponds to such coherent quantum states, in the classical limit. We have discussed the reasons for holding (i) and (ii), but clearly (iii) is equally important, and relatively neglected. In this section we will therefore investigate this question, and argue that while generally justified, its scope is limited, and indeed importantly unknown in the case of gravity. The following discussion is necessarily more technical than the foregoing, and could be skipped by those wanting to understand the overall argument for the derivation of the metric, and read by those wanting a firmer grasp of its theoretical underpinnings.

It is worth stressing that much of this section is not specific to string theory, but concerns general approach to quantum gravity, in which gravity is treated as a quantum field, whose quanta are gravitons. As this chapter (and the last) have explained, such a theory is obtained in an appropriate sector of string theory, but the same is expected of other approaches to quantum gravity, such as loop quantum gravity (and causal set theory if it can be successfully quantized). If gravity is quantum at all, then there should be a regime in which it behaves approximately as a quantum field, a regime recovered by a more fundamental theory of quantum gravity of the kind discussed in this book. On the other hand, QFTs generally develop pathologies at high energies, and are generally not themselves thought to be good candidates for fundamental theories.[29] But because of its place between (fundamental) quantum and classical theories, the question of how the classical gravitational field can be understood in terms of the graviton field is an important one for any approach to quantum gravity, not only string theory. Indeed, since many of the issues arise for the classical limit of *any* quantum field, we believe the following sketch to be of very general importance for the interpretation of QFT—we very much hope that it stimulates further investigation of these questions.

10.5.1 The graviton concept

First, we should recognize that the concept of a 'graviton' is dependent, not only on spacetime symmetries as discussed above, but also on the weak field

---

[29] For further general discussion of the relation of 'low energy quantum gravity' to fundamental theory see Burgess (2004) and Wallace (2022b).

THE EMERGENCE OF SPACETIME IN STRING THEORY   331

approximation to GR, in the sense that it is in this limit that the EFE becomes a linear equation, and classical plane wave solutions can be found, to which gravitons correspond on quantization. The situation is just as in typical QFTs in which particles are defined for the free, linear equations of motion.[30] In perturbative, interacting field theory—including quantum GR—one assumes that the 'Fock space' of arbitrary numbers of such states is a valid approximation, even when interactions, sources, and background fields are present.[31] This is a significant assumption in the case of gravity, since the weak field approximation is not even empirically adequate to the solar system, as it predicts 4/3 of the correct perihelion of Mercury (Misner et al. 1973, 183f). Thus understanding realistic astronomical or cosmological models as graviton coherent states requires the approximate validity of the graviton picture *beyond* the weak field.

We emphasize here the crucial point that the derivation of the EFE in (D) does *not* depend on the weak field limit, but on the small value of the expansion parameter α' in (10.5). We introduced α' as the reciprocal of string tension, but (with $\hbar = c = 1$) it has the units of $length^2$, so is also equivalent to a fundamental length scale in the theory: $\ell_s \equiv \sqrt{\alpha'}$, the 'string length' (familiar from T-duality). Then the rightmost sum in (10.5) can be rewritten (as it strictly should be) as an expansion in a *dimensionless* parameter ℓ_s/r, where r is the only other length scale, the 'radius of curvature' of the background spacetime (see Green et al. 1987, §3.4.2). Physically then, the approximation that leads to the EFE is valid as long as spacetime is approximately flat over string-length sized regions. That is potentially a vastly greater regime of applicability than the weak field regime: the string length is usually taken to be within a few orders of magnitude of the Planck length, so one only expects significant (perturbative) stringy deviations from GR close to the scales at which spacetime has to be treated quantum mechanically.[32] Thus the procedure of taking a background GR model, and computing stringy quantum perturbations around it (to calculate scattering amplitudes, say), is on a solid theoretical footing (notwithstanding its critics' objections). Rather the issue is how the classical metric field emerges in string theory: does the account given in (C) make sense for arbitrary spacetimes?

---

[30] The general picture in QFT was sketched in appendix A to chapter 9: the linearity of the free equations mean that general solutions can be treated as a sum of independent plane waves; on quantization they correspond to particles of definite momentum, and hence general free quantum fields are built from superpositions of many quanta. The specific application to GR is nicely reviewed in Kiefer (2004, §2.1).

[31] Even though Haag's theorem tells us that it cannot be exact: see Duncan (2012, §10.5) and Fraser (2006).

[32] This expectation may not always be realized. For instance, the 'fuzzballs' proposed for the interior of black holes are not thought to be restricted to a small region around the singularity, but to extend to the horizon. Additionally, the quantum (as opposed to stringy) nature of gravity will be manifest at far lower energies: Huggett et al. (2023).

332   OUT OF NOWHERE

Before we address that question, let us briefly discuss what can be said outside the perturbative regime, in situations in which spacetime is curved over regions that are small on the scale of ℓ_s. First, as we saw in chapter 8, T-duality shows that the classical conception of length only has operational significance above the string length ℓ_s; in the familiar sense of the word, 'length' is an effective concept, valid only to the string scale. Second, even the notion of target space is predicated on the perturbative approach to string theory; one expands around a supposed classical solution of an unknown exact theory. When $\ell_s/r > 1$ the approach breaks down, and one cannot simply assume target space. This is not necessarily to say that perturbative string theory has a non-spacetime phase[33] but rather to say that the theory does not apply in such situations. The nature of this regime has of course been subject to considerable scrutiny, because it may offer clues to 'M-theory' and because of its cosmological importance. Horowitz (1990) raises the question, Greene (1997) and Horowitz (2005) explore it using string dualities to look for non-perturbative signs of sub-ℓ_s physics, and Gasperini (2007) specifically investigates the possible 'string phase' physics of the big bang; in this work there are strong signs that the fundamental degrees of freedom of exact string theory will not be spatiotemporal, and that there will be states in which spacetime concepts are not well defined. However, our investigation is into emergence in perturbative string theory, so we will postpone these questions for another occasion, and return to the issue of what GR solutions can be understood in terms of dynamical string states.

The situation we have seen thus far is this: the derivation (D) of the EFE is valid without assuming the weak field limit; however, the description (C) of the metric field in terms of coherent graviton states depends on the validity of the graviton picture, which in turn does depend on the weak field approximation. One should not, however, conclude that only in the weak field limit is the metric derived, and otherwise is a given in the fundamental theory. Instead, one should take the success of the graviton picture to indicate that the classical metric should be taken to describe a string state *whatever* the GR solution (as long as ℓ_s/r is small), but that the description of that state in terms of gravitons is a better or worse approximation depending on how far the spacetime is from the weak field. We know of nothing significant written on the topic; however, in our conversations with string theorists while we have found a range of attitudes about how seriously to take the account given in (C), we have found unanimity that it does not hold in general, but that when it fails, the metric represents some string state not well described by gravitons. The nature of this state is not understood, and likely requires moving beyond perturbative string theory to the sought-for M-theory.

---

[33] See Oriti (2021) for discussion of the general idea.

10.5.2 Graviton coherent states

The central question then becomes whether it is possible to construct classical spacetimes as quantum field states at all. In appendix A to chapter 9 we saw: (i) a coherent state corresponds to a Gaussian wavefunction(al) over canonical position (and momentum), with minimal simultaneous uncertainty for both; (ii) a free, linear system will remain coherent, but an interacting, non-linear system will lose coherence.[34] One can question whether (i) is sufficient for an approximately classical state. But this is a general question of the interpretation of QM, so we will pass over this issue, and make the standard assumption that it is, because it minimizes quantum uncertainty. (Indeed, we assume that for this reason even approximately coherent states correspond to classical states.) Our focus instead is on (ii). How long do coherent states of non-linear fields remain coherent? In particular, will a coherent state of the gravitational field persist for the duration of a spacetime model, say the life of a universe like ours, so that it can be understood quantum mechanically in terms of a graviton field?

In the first place, as far as we have been able to determine, from the literature, and from discussions with physicists, although coherent states are generally understood to represent a classical limit of QFT, little or nothing has been proven in full QFT regarding fields. This is a significant lacuna in the literature (though see Rosaler 2013).

However, in the context of quantum gravity, one can approach the problem using the 'superspace' approximation, using a restricted set of degrees of freedom, instead of field quanta. In particular, one can study models of gravity in a 'minisuperspace', a finite dimensional configuration space: for instance, for a Friedmann universe described by a scale factor $a(t)$, with a massive scalar field $\phi(t)$. One quantizes by postulating a wavefunction $\psi(a, \phi)$ over this space (see Kiefer 2004, §8.1.2 for more details). From the QFT point of view, one has thus aggregated the microscopic state of field quanta into two degrees of freedom, so a coherent state cannot be given explicitly as an excited state, but rather will be described by a Gaussian $\psi(a, \phi)$. Just as for a particle, such a state corresponds to a maximally classical quantum state, with equally localized canonical position and momentum, and saturating the uncertainty relations. (From a quantum gravity point of view, one has restricted attention to states which can be described by—superpositions of—suitable classical states.)

In that formulation, one can pose the question of the persistence of coherence of the state: how long would it remain (approximately) an uncertainty minimizing

---

[34] Note the confusing terminology: the 'cohere' in 'coherent state' and 'decoherence' does not have the same meaning. Losing coherence in our sense is not the same as decoherence in the sense of approaching a mixed state—it generally means evolving into a pure state which is not coherent as we have defined it.

334 OUT OF NOWHERE

Gaussian? The duration of a universe? This question has indeed been studied, with mixed results. On the one hand, Kiefer and Louko (1999) show, for instance, that a Schwarzschild black hole coherent state will take around the Hawking evaporation time to disperse. It follows that over that time period one can rely on the classicality of the stringy graviton background, and hence (ceteris paribus) on results obtained from string theoretic models of black holes. Since such models have been an active area of research, this result is significant.

On the other hand, Kiefer (1988) investigates 'big crunch' Friedmann universes, which start and end with a singularity. The corresponding quantum state has, as it were, components describing the epochs including each singularity as coherent states, which interfere around the 'turnaround' in which the universe starts to recontract, leading to a loss of coherence for that epoch. That is, this quantum model does not allow one to understand the appearance of a full classical spacetime in terms of coherent states.

Moreover, Wallace (2012, §3.3) (drawing on Zurek and Paz 1994, 1995a, 1995b) argues that coherence will typically not be preserved in realistically complex systems; he considers particle systems, but since these couple to fields, the same should apply to fields as well. Non-linear effects in the classical dynamics lead to chaotic behavior: initially close states will become observably distinct. But since coherent states track classical states, such behavior will, on quantization, entail the spreading of wavepackets, and hence the loss of coherence. A relevant example for our purposes, since it involves the gravitational field, is the chaotic motion of Hyperion around Saturn: Zurek and Paz (1995b) argue that coherence will be observably lost with a few decades.

These studies and arguments are carried out within the study of 'decoherence' and the many worlds interpretation. For instance, Kiefer (2004, §8.1.2; 2013) argues that in mini-superspace models interactions between the aggregate and the microscopic degrees of freedom of the quanta lead to decoherence into an approximate mixture of coherent states. The point of Wallace's discussion is that the coherent state concept *alone* is not sufficient to recover a classical world; decoherence is also needed. If correct, although coherent states indeed represent classical states for the reasons explained above, one cannot simply identify a realistic, diachronic classical world as a unitarily evolving coherent state. If this general line of interpretation is pursued, in some way different possible classical worlds will have to be identified with decoherent branches—each consisting of a diachronic series of coherent states. One could follow Wallace, and view each branch as an equal world; one could adopt the view that a particular branch is picked out by 'hidden' variables; or perhaps something else. It is not the business of this discussion to adjudicate this question: we have bracketed questions of the interpretation of QM. However, we do point out the scope and ultimate limit of the coherent state concept in representing the world of classical fields in quantum mechanics. We commend this important issue to readers for investigation.

10.5.3 GR from QFT

The final issue concerns the relation of the string theoretic derivation of the metric to the treatment of gravity in QFT, and in particular to derivations of universal coupling (hence weak equivalence) and the (quantized) EFE by Weinberg, in his (1964) and (1965), respectively. Working within the S-matrix approach (see Cushing 1990), Weinberg first shows that Lorentz invariance entails that all *low-energy* massless spin-2 quanta must couple to matter fields with the same coupling strength, so that Newton's constant is indeed constant! (The restriction to low energy of course allows for high energy quantum corrections to the S-matrix.) This derivation is examined in a nice paper, Salimkhani (2018), arguing that the need to unify GR and QM follows from relativistic quantum mechanics: (i) from Wigner we have the classification of possible particles, including gravitons; while (ii) from Weinberg we have the necessity of quantum fields (this, along with the other results discussed here, is explained in Weinberg 1995), hence of second quantizing gravitons; and then (iii) we have Weinberg's derivation of weak equivalence, describing the graviton field's interactions with other fields; overall, entailing a unified QFT treatment of gravity and matter. (Of course, this unification is limited by the non-renormalizability of the resulting field theory, as discussed in footnote 35.)

In the later paper Weinberg goes further, and working in Dyson-Feynman perturbation theory shows that Lorentz invariance means that any massless spin-2 quantum field will satisfy the quantized EFE as its Heisenberg equation of motion (i.e., with fields replaced with operators); plus possible 'Fermi terms.'[35] As explained in appendix A to chapter 9, it follows that the expectation values of the coherent states of these fields will obey the classical EFE, essentially the same result that we found for strings. Does this mean that after all there is nothing so remarkable about finding gravity in string theory?

That conclusion would require accepting that quantum fields are a good description of states of many excited strings. Given the correspondence in symmetry properties, they must be if string theory is to make good physical sense, but that conditional is not trivial. Thus the stringy derivation of the EFE is an important consistency check on the interpretation of string theory in terms of quantum fields—they do satisfy the same equations of motion. Moreover, there is the very

---

[35] These include couplings to derivatives of curvature, which would violate minimal coupling. Since gravity is non-renormalizable, such terms will have to appear as counterterms at increasing orders of perturbation theory. See Kiefer (2004, §2.2) for a full discussion of this point. (Note that QFT gravity has predictive physical content, despite its non-renormalizability; like any such theory, only a finite number of constants need be empirically fixed at any given perturbative level in order to make renormalized predictions. See Wallace 2022b.) If these terms have small couplings at low energy, then the EFE will be the approximate equation of motion, and they will be quantum corrections. (We should also mention that the non-renormalizability of the theory might not be problematic at all, if it is 'asymptotically safe': e.g., Kiefer 2004, §2.2.5.)

336   OUT OF NOWHERE

interesting question of the different sources of the result in QFT and string theory: Lorentz invariance, and worldsheet Weyl symmetry, respectively. Why do a space-time symmetry and a symmetry of the string lead to the same place? (Note that we did see in §7.3 that these symmetries are traded off in lightcone quantization.)

10.6 Conclusions

This chapter framed the issue of spacetime emergence in terms of the origin of local and global topologies, and of the metric of spacetime; insofar as these are not baked into the formalism of string theory but derived, spacetime can be said to 'emerge' in a broad sense. (After all, their appearance only at the level of classical spacetime theory has considerable novelty.) We emphasize that these questions can be meaningfully engaged without having to address the question of spacetime substantialism, which we view as a methodological virtue.

Duality has been seen to entail that both local and global topology is derived, hence emergent. In the first case, there is no fact of the matter about how observable processes—specifically scattering events—are localized in target space. In the latter case, various dualities exist linking global topologies. And we take the stance, developed in chapter 8, that duals are physically equivalent, so that only the 'common core', in some sense, has physical significance. Thus the topology is derived.

The metric has been seen to arise from an excited—graviton—state of strings. Moreover, we have seen that it is indeed the metric of GR, since it must satisfy the EFE (to lowest order). We have sketched the scope of this derivation, with respect to different target space geometries, and the range of spacetime models that can be described. But the most important point is that the spacetime part of a string solution is arbitrary, and can be traded for gravitons: hence even the partial metric is not fundamental, but conventional—a choice of perturbative framework. In this sense only the full metric, including graviton field, is physical, but it is derived because of the graviton contribution. (These considerations are supported by models, mentioned in passing, of the big bang and black holes in which there is a non-spatiotemporal string phase.)

In other words, these three structures—local and global topologies, and metric—are all derived, hence emergent in the broad sense: not fundamentally present, and of considerable novelty in effective physical theories. Moreover, the derivations give a formal account of how the basic posits of perturbative string theory can give rise to these structures, and hence perform the spacetime functions that they play: making sense of 'where', 'how big', and what shape. We have discharged this part of the functionalist's charge.

Now, as we explained in chapter 2, such a formal derivation is only one part of the story; one also needs to accept that the derivation is 'physically salient', describing a physically real correspondence. As we have now seen, more specifically, the derivation requires accepting as 'principles' of physical salience: the identification of string excitations with quanta (through the identification of representations of the spacetime symmetries); the use of the tools of QFT (such as coherent states, anomalies, and renormalization); the operationalization of spacetime magnitudes (through scattering amplitudes); and the use of perturbation theory. As we argued, ultimately it must be the empirical success of a theory that convinces us to accept such principles. Indeed, most of them are already accepted because of their successful application in QFT.

However, we should acknowledge a limitation on what has been achieved. In the case of string theory, the perturbative nature of the derivation to an *unknown* underlying theory places a limitation on truly understanding from what non-spatiotemporal physics, and how, classical spacetime emerges; making our investigation incomplete until string theory is better understood. What really are we accepting when we accept perturbative string theory? What is M-theory, and how does the theory in our hands approximate it? Our current state of physical knowledge thus limits our understanding of spacetime emergence from string theory. We will return to this point in chapter 11.

Chapter 11
Conclusion: whence spacetime?

In the introduction we explained our notion of 'emergence' as the appearance of robust, novel entities or structure, and argued by example that a number of programs in quantum gravity strongly suggest the emergence of spacetime from the non-spatiotemporal (or not fully spatiotemporal). We then proposed that this emergence be understood through functional reduction, in a two-step program of 'spacetime functionalism':

(SF1) Spacetime entities/properties/states are functionalized by specifying their identifying roles, such as spacetime localization, dimensionality, interval, etc.

(SF2) An explanation is given of how the fundamental entities/properties/states postulated by the theory of quantum gravity fill these roles.

In the following chapters we have demonstrated our claim of functional emergence in more detail in the contexts of causal set theory (CST), loop quantum gravity (LQG), and string theory. In each case we explained the various ways in which the fundamental theory is less than fully spatiotemporal, and described how each proposes that central spacetime roles are realized—how spacetime emerges functionally. In this conclusion we will briefly review the main points of that analysis.

But the introduction also raised the issue of 'physical salience' (Maudlin 2007a). (SF2) will involve a formal derivation of structure 'isomorphic' (broadly speaking) to the spacetime that we observe and describe within general relativity (GR) and quantum field theory (QFT). Though hard in itself (to the point that it has only been partially achieved), a 'mere' derivation leaves open the possibility that the correspondence between the derived structure and spacetime that it establishes is nothing more than that—a coincidence of patterns in two very different things. (SF2) requires that we accept not only that the fundamental entities/properties/states play roles isomorphic to the spacetime roles, but that they are indeed *the very things* playing those roles.

We explained that to establish such a claim scientific practice demands that a derivation satisfy principles of physical salience. That poses a problem in the present case for, arguably, currently accepted principles are spatiotemporal: for instance, but especially importantly, we typically accept the reduction of one process to another more fundamental one only if they are spatiotemporally

Out of Nowhere. Nick Huggett and Christian Wüthrich, Oxford University Press. © Nick Huggett and Christian Wüthrich (2025). DOI: 10.1093/oso/9780198758501.003.0011

coincident. But such a criterion cannot be satisfied if the fundamental is non-spatiotemporal. The solution, we argued, was that principles of physical salience are historically contingent, changing and developing as science progresses, a process typically characterized by philosophical debate: consider, for instance, the changing fate of local action as a criterion for physical acceptability. Such change is not, of course, merely a matter of theoretical fashion, or of purely a priori reasoning, but rather part of a package confirmed by the empirical success of a theory.

In short, we should expect theories of quantum gravity, to the extent that they posit a non-spatiotemporal fundamental ontology, to come with new conceptions of physical salience, to be satisfied by formally successful derivations, if they are to show that spacetime roles are in fact played by their entities/properties/states. For example, in §2.3 we discussed the emergence of space from non-commutative geometry, arguing that an ansatz identifying an algebraic structure with spacetime was required, which appears quite ad hoc by any current accepted principles. However, we have no doubt that it would be accepted, were the theory to prove the empirical victor. Of course, today we are far from having the experimental evidence sufficient to firmly establish any such theory or principles. But it still makes sense to consider the known derivations (or partial derivations) of spacetime structure from theories of quantum gravity, and ask what principles would have to be accepted *if* the derivations are to fulfill (SF2). If and when empirical evidence is convincing, we will come to accept the functional reduction of spacetime, and also of the principles themselves.

So the third goal of this conclusion—aside from reviewing the non-spatiotemporality of our examples, and the derivation of spacetime structure in each case—is to tease out from our case studies of spacetime emergence what novel principles of physical salience are invoked. We will see, as expected from the historical record, that in many cases their application is a matter of philosophical debate.

11.1 Causal set theory

Let us start with causal set theory (CST). CST is an attempt to formulate a theory of quantum gravity inspired by the fact that, for causally sufficiently well-behaved spacetimes, the causal structure of spacetime fully determines its geometry, up to a conformal factor. From this insight, it starts by formulating a theory of a discrete version of classical spacetime. From this, it is hoped that a quantum theory of these structures can be found—a hope which currently remains unfulfilled.

Perhaps surprisingly, a mere 'discretization' of spacetime along the lines of CST suffices to eliminate some of its core spatiotemporal features. As for time, the basic

340 OUT OF NOWHERE

physical relation in CST is *causal*, rather than *temporal*. It is clear, however, that this relation is at least analogously temporal, and that consequently there is a sense in which time is directly and fundamentally present. If one cares about the distinction between temporal and causal relations, however, there are good reasons to not think of the fundamental relation as temporal—least not because the presumably quasi-temporal order in which the basic dynamical process of sequential growth stands is distinct from the explicitly causal partial order of a causal set, or 'causet' for short. While time may or not be fundamentally present in CST, space is absent. A natural analog of space resulting from a (highly non-unique) partition of any causet into maximal antichains is completely structureless beyond the cardinality of these sets. That will be a problem however we identify space through its functions at step (SF1).

We should of course expect the collection of functions identified in (SF1) to be the same for any research program in quantum gravity (or indeed anywhere), since the functions of spacetime either for physics or for the structuring of our empirical world do not depend on the approach to quantum gravity taken. Having said that, we should tolerate some variation because different sub-collections of functions may be sufficient for all these tasks. As far as CST is concerned, relevant efforts have concentrated on spatial dimensions, (spatial) topology, and some metric properties, such as spatial distance. In the latter part of chapter 3, we discussed a number of ways in which these functions can be executed by causets even though they are not spatiotemporal, realizing (SF2).

Given the incomplete state of the theory, it remains difficult to conclusively identify the principles of physical salience at work in these derivations. To the extent to which they can, for instance for dimensionality (§3.3), we have identified some properties of causets which are likely to play a key role in recovering dimension, such as those relevant in principles of embeddability of causets into relativistic spacetimes. Specifically, causal structure is central to these principles and, by virtue of the role it plays in emergence, it is not a mere formal partial order, but endowed with physical salience. Somewhat similar is the case of topology (§3.4), where causal structure also turns out to be physically salient as a result of the principles applied in extending the maximal antichains naturally associated with 'space' to ever more encompassing parts of causets necessary to construct a physically meaningful spatial topology from the fundamental structure. In the final example (§3.5), spatial distances are extracted again from the causal structure of a causet, using different techniques.

These three cases highlight efforts to identify spatiotemporal functions rather directly in a causet. Significant efforts in CST are also undertaken to understand the general conditions for the full emergence of spacetime. In chapter 4, we identified necessary conditions for the physical salience of the derivations involved. Central for this project are again principles of embeddability, which govern the relation between the fundamental causets and emergent spacetimes. The demand that the embedding be 'faithful' (i.e., that it preserves causal structure, that it entails a uniform distribution, and that there is no structure at sub-Planckian scales) turned out to be particularly critical. Furthermore, a crucial ingredient is also the addition of a dynamics. As a result, the dynamical principles involved (whatever they turn out to be) must also be physically salient. For example, in the classical sequential growth dynamics, principles of internal temporality, discrete general covariance, Bell causality, and the Markov sum rule are the dynamical principles at play. Finally, we may also choose to impose additional conditions governing the physical salience of spacetime emergence, such as principles of the 'physical reasonableness' of causets, or of the approximate uniqueness of the emerging spacetime.

Part of the project—here but particularly also as undertaken by physicists in the field—is to understand the relationship between these demands, such as to what extent satisfying one demand automatically (or contingently) satisfies another. In other words, much of the project is to identify the base assumptions needed to secure physical salience. Importantly, the regulative demand that causets are well approximated by relativistic spacetime is an example of how physical salience is endowed from 'top' to 'bottom', i.e., starts from the macroscopic, emergent level and percolates downward from there to the fundamental level.

The physical salience of the connection between causets and spacetimes is at the same time illustrated and deepened by the interpretive questions we broached in the latter part of chapter 4, where we discussed the metaphysics of time and of becoming in CST, as well as whether a form of non-locality that arises in discrete, Lorentz invariant structures threatens the physical viability of causets. For example, if the dynamical principles at play in classical sequential dynamics can in fact be connected to a phenomenologically relevant sense of becoming, such a connection would be compelling grounds to accept them as physically salient. The derivation of other phenomenological signatures such as Sorkin's derivation (or rather: estimate) of the value of the cosmological constant would have similar implications.

11.2 Loop quantum gravity

Loop quantum gravity (LQG) proceeds by attempting a canonical quantization of GR. Its proclaimed strategy is to take seriously GR's central lesson, which it identifies with background independence. Although this condition is often considered equivalent to general covariance, the latter is a strictly stronger principle, as we argued in chapter 5. LQG's vantage point is a reformulation (of the globally hyperbolic sector) of GR into a Hamiltonian theory, followed by a technically and conceptually demanding case of canonical quantization in the presence of constraint equations. One of these constraints turns out to be the Hamiltonian, which directly leads to the problems of time (i.e., the absence of a time parameter) and of change (all physical quantities are constants of motion, and so cannot change over time). In addition to these puzzling interpretive consequences, the Hamiltonian constraint cannot be solved, leaving the canonical program incomplete. The kinematic Hilbert space of the quantum theory obtained at this point admits a basis of spin network states. As the dynamics in the canonical program remains unresolved, the covariant perspective adds to kinematic LQG a dynamics of transition amplitudes between spin network states.

These spin network states afford a geometric interpretation, central to the emergence of spacetime in LQG, according to which they represent discrete, combinatorial, background-independent structures linking indivisible, in some sense spatial entities by relations of adjacency. However, even on this geometric interpretation, the fundamental structures of LQG fail to be spatiotemporal, in at least three ways, as argued in §6.1.

The first failure of full spatiotemporality arises from the quantum indeterminacy of the spin networks: generically, the quantum states will be superposition states of geometrically interpretable states and so do not instantiate a determinate geometry. Of course, it is expected that this failure of spatiotemporality affects any quantum theory of gravity. The second failure results from disordered locality, i.e., from the fact that even on a spatial interpretation of the fundamental adjacency relations these do not neatly map onto the locality structure of the corresponding emerging relativistic spacetime. Thus, even if the fundamental structures are spatial in some sense, physical space(time) is emergent. The third failure follows from the problem of time (and of change), i.e., the apparent complete absence of physical time from the fundamental ontology.

In order to ascertain the emergence of spacetime, we thus need to cut through quantum indeterminacy, disordered locality, and the problem of time. As spacetime functionalism maintains, this can be accomplished if it can be shown that generic quantum states, i.e., superpositions of spin network states, or spinfoams, execute all spatiotemporally relevant roles specified in (SF1). The problem of time can be successfully addressed by a relational interpretation, which construes physical quantities, including time, as relative to one another rather than as in some absolute standing. In this interpretation, time is just another physical quantity ontologically on a par with others. Time's relationality, like the relationality of each and every physical quantity, also reintroduces change, conceived, of course, as relational change.

Apart from relationalism as a key to solving the problem of time, two further key interpretative elements paved the way for our functionalist account of spacetime emergence: the geometric interpretation of the spin network states and the covariant perspective. The covariant approach was introduced in response to technical (and arguably conceptual) difficulties with completing the canonical quantization program and suggested a shift in perspective from a global to a regional interpretation of LQG, i.e., to an understanding on which LQG describes up to medium-sized regions of our world. However, even on a geometric reading of spin network states and a suitably regional understanding of covariant LQG, the generic fundamental structures still suffered from quantum indeterminacy and disordered locality and so the emergence of spacetime was not yet fully accounted for.

Accordingly, we need to understand the classical and continuum limits of the theory. In chapter 6, we have suggested that what we called the 'Butterfield-Isham scheme of emergence' fits the functionalist bill. The Butterfield-Isham scheme casts emergence as a combination of approximating and limiting procedures. Roughly, an approximating procedure selects some physical quantities or states as the physically salient ones and uses approximations in order to establish a connection between approximating theory and theory to be approximated, while a limiting procedure is taking mathematical limits of physical variables of the fundamental theory in order to establish a connection to the emergent theory.

We have argued that in the present case, the scheme works in two steps (not to be conflated with (SF1) and (SF2)). First, a physical mechanism—possibly a form of decoherence—drives generic quantum states to semi-classical states, which resemble classical states. These 'weave states' are spin network states which are eigenstates of the geometric operators of area and volume with eigenvalues approximating the corresponding classical values. This first step is thus described by an approximating procedure. Second, a limiting procedure relates these semi-classical states to the classical theory. If successfully completed, these two steps together demonstrate that GR is at least in some cases a very close realization to LQG.

Returning to spacetime functionalism, for (SF1) we focus on area and volume as central geometric quantities, in line with existing work in LQG. Of course, geometric operators constitute convenient but important examples of functions to be recovered from the fundamental physics. Other aspects such as durations and distances will also have to be shown to be functionally reducible to states in LQG. The Butterfield-Isham scheme is deployed in (SF2), providing the necessary tools of a suitable combination of approximating and limiting procedures to show that the fundamental structures fill at least some relevant geometric roles. Thus, the Butterfield-Isham scheme becomes an integral part of the spacetime functionalist program. Once again, spacetime functionalism casts the work to be done in a coherent strategic framework.

11.3 String theory

Finally, we turn to string theory. Here (SF1) is accomplished much as for the other cases. Chapter 10 discussed the local open set structure, the global topology, and the metric and their functional roles: respectively grounding localization (giving a minimal answer to the question 'where?'), the overall 'shape' of space,

344   OUT OF NOWHERE

and determining (in part) scattering amplitudes. A major question at this point is whether these things are emergent, or simply postulated as part of the basic furniture of the theory, rendering a functional identification trivial. After all, perturbative string theory as presented is constructed in Minkowski target space.

We rejected a worldsheet interpretation as a serious alternative, and the arguments of chapter 8 lead us to conclude that target space is not identical with classical, relativistic space, because target space has no determinate metric while space manifestly does. Thus (SF2) is a non-trivial project, and the story of how it has been accomplished by string theorists is told across chapters 7–10. A substantial part of that story involves the development of quantum string theory in target space: we have emphasized on the one hand the derivation of the string particle spectrum and gravitational field, but on the other the question of how something with no determinate metric could play the role of something with a determinate metric (and indeed determinate local open-set structure).

The first part of the project relies on the application of largely standard methods of quantum field theory (QFT) in a non-standard framework, of a conformal field with a sigma action on a 2-dimensional worldsheet. Thus this part of the explanation of how the spacetime roles are played is vindicated by the validity of those methods as parts of physical explanations: by their physical salience. To make clear that substantive assumptions are involved here, consider some of the key ones explicitly. In the derivation of the string particle spectrum: the standard application of quantum mechanics, the standard methods for quantizing gauge fields, the legitimacy of renormalization (especially zeta renormalization), and Wigner's identification of particle species with representations of spacetime symmetries. (There is also the novel, but reasonable assumption that at long range strings can be described as particles, which we will address below.) In the construction of the classical gravitational field: the applicability of Feynman-style perturbation theory (both to the conformal field on the string worldsheet and to the sum over worldsheets), standard methods of conformal field theory (including state-operator correspondence), standard lore concerning quantum anomalies, the theory of coherent states, lore concerning their aptness as a classical limit (or, if that fails, the ability of decoherence to provide a classical limit), that equivalence of actions is equivalence of physics, the validity of weak gravitational field QFT, and the idea of changes of 'frame' in dilaton gravity (strictly a part of spacetime physics rather than QFT). To this list we should add the novel assumption that there are many-string states which can be described by quantum fields at large distance scales.

We do not at all question (most) of these assumptions, or the physical salience of derivations that rely on them: they are (mostly) well-supported by their successful applications in QFT. Rather, the points are, first, that while they are to a greater-or-lesser degree understood mathematically, they involve assumptions that go beyond what follows logically from any mathematical formulation of the theory; and second, that they must be assumed if one accepts that the theory explains the phenomena. But again, those assumptions are justified, given their success in QFT. (Dawid 2007 similarly stresses that the success of QFT supports the use of its techniques in string theory: though he goes further than we do here in arguing that QFT is evidence for string theory.)

Before discussing the second, harder part of the justification of the emergence of spacetime in string theory, it is instructive to reflect on the question of background independence, discussed in §5.2 and §10.4, as an issue of whether string theory *fails* to satisfy a principle of physical salience. The many proponents of background independence certainly see it as a desideratum of an explanatory spacetime theory, and one learned from the discovery (and success) of GR: just as we characterized principles of physical salience in chapter 1. The problem with string theory is supposed to be that it assumes a spacetime background, thereby undermining its claim to offer a fundamental account of spacetime, since such a background should itself be explained in more fundamental terms. There are two kinds of response: first, one could dispute the charge of background independence (since the 'background' fields include stringy coherent states, for instance), essentially questioning the interpretation of the principle. Of course, another option, if string theory were to be sufficiently successful empirically (and absent comparable alternatives) is that physics would conclude that background independence was not a desideratum after all—just as it revised its view on action at a distance. Whatever view one takes, and however things turn out, the issue is 'philosophical', conceptual rather than purely formal or experimental. The example is instructive because it further illustrates the role of physical salience, not in the harder-to-see application of standard assumptions listed previously, but in the more visible debate over the possible violation of a generally agreed principle.

Let us turn now to the second, ineliminable part of the project, explaining how target space can play the role of determinate spacetime, since it is a novel structure, not identical with spacetime. So doing requires the acceptance of novel principles of physical salience if the derivation is to be accepted as physically salient—if, that is, (SF2) is to be successfully accomplished. Let us review what the explanation involves, so that we can identify what new principles are accepted when one accepts it.

At the root of the issue is the indeterminacy established in chapter 8, between a large target space and small winding space on the one hand, and a small target space and large winding space on the other (where sizes are relative to the minuscule string length). The result is that neither target space nor winding space can be identical with space, even though the former is standardly used to represent space in calculations. Granted metric indeterminacy, the resources of string theory are restricted to a 'common core' of the two descriptions, and similarly any stringy account of spacetime functions. However, there is no known formulation of string

346 OUT OF NOWHERE

theory explicitly in terms of such a core: presumably it would require rather more knowledge of the postulated M-theory. Instead, we need to be careful to appeal only to those features on which the two descriptions in fact agree.

Brandenberger and Vafa (1989) propose how this should be done when they explain why the universe is observed to have a large radius. Recall that they noted that in the large target space description low energy modes correspond to momentum excitations—represented by plane waves in target space; while in the small target space description they correspond to winding excitations—represented by plane waves in winding space. In other words, by plane waves in whichever space is larger in the representation: indeed, just as large as we observe space to be. Then when we understand that the probes which we use to measure the properties of space are always low energy by the scale of the theory, we see that they will always be sensitive to whichever of target or winding space is larger in the description. And of course, crucially, this derivation only appeals to the common core, because (modulo a free choice of units), the descriptions agree on the size of the largest space—just not on which it is. Moreover, in this case one is always free to choose the description in which target space is the larger, vindicating practice, as long as one recalls that one could have chosen winding space instead.

The same proposal works in the derivation of topology. For local open set topology, recall from §10.3 the observation that according to one dual scattering is a localized process of string annihilation and creation in a large radius target space, while according to the other non-local—in fact everywhere—in a small radius target space, since it is a process of winding creation and annihilation. But once again, in the latter dual, the process *is* localized in a large radius winding space. (Also once again, accessible processes are low energy, so that they do not involve winding around a large radius target space, or localization in a small radius target space.) Thus we can describe the observed localization of scattering using whichever of target or winding space is larger, something common to the duals.

We emphasize that using Brandenberger and Vafa's ansatz to describe the metrical and local topological properties of space is not a matter of *identifying* them with the metric and topology of either target or winding space, since space is not identical with either. Instead, it is a recipe for extracting them from the common core of the duals (even though that core has no independent formulation). It is of course very natural to use the correspondence between target (or winding) space and space to explain these spacetime functions, but the fact remains that the roles are played by something that does not include our classical spacetime as fundamental—rather the combination of target and winding spaces, *up to target duality*. Hence, accepting this ansatz in an explanation of how string theory plays spacetime roles involves accepting that the correspondence between the common core and classical spacetime is not only a formal correspondence, but also physically salient. No existing principle will tell us that, and so indeed explanations like Brandenberger and Vafa's involve new principles of physical salience.

Once we accept the physical salience of the correspondence, (SF2) can be completed: we can *explain* how the role of scattering localization, and how the role of the metric in scattering are played in string theory. (Moreover, these explanations help vindicate the effective description of strings as particles at long ranges, by explaining the sense in which 'longer' scales are actually meaningful!) In the (partial) project of completing (SF2) in string theory, as we have discussed, it is precisely the roles of localization and metricity *in scattering* that one aims to explain.

11.4 Out of nowhere

Throughout the course of the book we have therefore made good on our promise to explicate the functional reduction of spacetime to the less-than-fully-spatiotemporal in a number of promising candidates for quantum gravity—in short, showing how spacetime might come from 'out of nowhere'.

Bibliography

Stephen L Adler. Axial-vector vertex in spinor electrodynamics. *Physical Review*, 177(5): 2426–2438, 1969.

Maqbool Ahmed, and David Rideout. Indications of de Sitter spacetime from classical sequential growth dynamics of causal sets. *Physical Review D*, 81:083528, 2010.

Maqbool Ahmed, Scott Dodelson, Patrick B. Greene, and Rafael Sorkin. Everpresent Λ. *Physical Review D*, 69:103523, 2004.

Ian J R Aitchison. *An Informal Introduction to Gauge Field Theories*. Cambridge University Press, Cambridge, 2007.

David Z Albert. Elementary quantum metaphysics. In James T Cushing, Arthur Fine, and Sheldon Goldstein, editors, *Bohmian Mechanics and Quantum Theory: An Appraisal*, volume 184 of *Boston Studies in the Philosophy and History of Science*. Springer, Dordrecht, 1996.

David Z Albert. *After Physics*. Harvard University Press, Cambridge, MA, 2015.

Gerardo Aldazabal, Diego Marqués, and Carmen Núñez. Double field theory: a pedagogical review. *Classical and Quantum Gravity*, 30(16):163001, 2013.

H G Alexander, editor. *The Leibniz-Clarke Correspondence: Together With Extracts from Newton's Principia and Opticks*. Manchester University Press, Manchester, 1956.

Victor Ambarzumian, and Dmitri Iwanenko. Zur Frage nach Vermeidung der unendlichen Selbstrückwirkung des Elektrons. *Zeitschrift für Physik*, 64(7–8):563–567, 1930.

Aristidis Arageorgis. Spacetime as a causal set: universe as a growing block? *Belgrade Philosophical Annual*, 29:33–55, 2016.

Richard Arnowitt, Stanley Deser, and Charles W Misner. The dynamics of general relativity. In Louis Witten, editor, *Gravitation: An Introduction to Current Research*. John Wiley and Sons, New York, London, 1962.

Abhay Ashtekar. New variables for classical and quantum gravity. *Physical Review Letters*, 57:2244–2247, 1986.

Abhay Ashtekar and Jerzy Lewandowski. Quantum theory of geometry I: Area operators. *Classical and Quantum Gravity*, 14:A55–A81, 1997.

Abhay Ashtekar and Jerzy Lewandowski. Quantum theory of geometry II: Volume operators. *Advances in Theoretical and Mathematical Physics*, 1:388–429, 1998.

Abhay Ashtekar and Jerzy Lewandowski. Quantum field theory of geometry. In Tian Yu Cao, editor, *Conceptual Foundations of Quantum Field Theory*, Cambridge University Press, Cambridge, 1999.

Abhay Ashtekar and Jerzy Lewandowski. Relation between polymer and Fock excitations. *Classical and Quantum Gravity*, 18:L117–L128, 2001.

Abhay Ashtekar and Parampreet Singh. Loop quantum cosmology: a status report. *Classical and Quantum Gravity*, 28:213001, 2011.

Abhay Ashtekar, Carlo Rovelli, and Lee Smolin. Weaving a classical metric with quantum threads. *Physical Review Letters*, 69:237–240, 1992.

Abhay Ashtekar, Stephen Fairhurst, and Joshua L Willis. Quantum gravity, shadow states, and quantum mechanics. *Classical and Quantum Gravity*, 20:1031–1062, 2003.

Guido Bacciagaluppi. The role of decoherence in quantum theory. In Edward N. Zalta, editor, *Stanford Encyclopedia of Philosophy*, 2020. URL http://plato.stanford.edu/entries/qm-decoherence/.

John C Baez. Spin foam models. *Classical and Quantum Gravity*, 15:1827–1858, 1998.

David J Baker. Interpreting supersymmetry. *Erkenntnis*, 87:2375–2396, 2020.

David J Baker. Knox's inertial spacetime functionalism (and a better alternative). *Synthese*, 199:277–298, 2021.

Jeffrey A Barrett. Empirical adequacy and the availability of reliable records in quantum mechanics. *Philosophy of Science*, 63(1):49–64, 1996.

John D Barrow and Douglas J Shaw. The value of the cosmological constant. *General Relativity and Gravitation*, 43(10):2555–2560, 2011.

Katrin Becker, Melanie Becker, and John H Schwarz. *String Theory and M-Theory*. Cambridge University Press, Cambridge, 2006.

John S Bell. *Speakable and Unspeakable in Quantum Mechanics*. Cambridge University Press, Cambridge, 1987.

John S Bell. The theory of local beables. In *Speakable and Unspeakable in Quantum Mechanics*. Cambridge University Press, Cambridge, 2004.

John S Bell and R Jackiw. A PCAC puzzle: $\pi^0 \to \gamma\gamma$ in the σ-model. *Nuovo Cimento A*, 60(2):47–61, 1969.

Gordon Belot. Background-independence. *General Relativity and Gravitation*, 43(10):2865–2884, 2011.

Dionigi M. T. Benincasa and Fay Dowker. The scalar curvature of a causal set. *Physical Review Letters*, 104:181301, 2010.

Eugenio Bianchi, Marios Christodoulou, Fabio D'Ambrosio, Hal M Haggard, and Carlo Rovelli. White holes as remnants: a surprising scenario for the end of a black hole. *Classical and Quantum Gravity*, 35(225003), 2018.

John Bickle. Multiple realizability. In Edward N. Zalta, editor, *Stanford Encyclopedia of Philosophy*, 2020. URL https://plato.stanford.edu/archives/sum2020/entries/multiple-realizability/.

Ralph Blumenhagen, Dieter Lüst, and Stefan Theisen. *Basic Concepts of String Theory*. Springer, Berlin, 2012.

James Bogen and James Woodward. Saving the phenomena. *Philosophical Review*, 97(3):303–352, 1988.

Martin Bojowald. Follow the bouncing universe. *Scientific American*, 299(4):44–51, October 2008.

Martin Bojowald. *Once Before Time: A Whole Story of the Universe*. Alfred A Knopf, New York, 2010.

Martin Bojowald. *Quantum Cosmology: A Fundamental Description of the Universe*. Springer, New York, 2011a.

Martin Bojowald. *Canonical Gravity and Applications: Cosmology, Black Holes, Quantum Gravity*. Cambridge University Press, Cambridge, 2011b.

Béla Bollobás and Graham Brightwell. The structure of random graph orders. *SIAM Journal on Discrete Mathematics*, 10(2):318–335, 1997.

Luca Bombelli. *Space-Time as Causal Set*. PhD thesis, Syracuse University, 1987.

Luca Bombelli, Joohan Lee, David Meyer, and Rafael Sorkin. Spacetime as a causal set. *Physical Review Letters*, 59:521–524, 1987.

Luca Bombelli, Joohan Lee, David Meyer, and Rafael D Sorkin. Reply. *Physics Review Letters*, 60:656, 1988.

Luca Bombelli, Joe Henson, and Rafael D Sorkin. Discreteness without symmetry breaking: a theorem. *Modern Physics Letters A*, 24:2579–2587, 2009.

Luca Bombelli, Johan Noldus, and Julio Tafoya. Lorentzian manifolds and causal sets as partially ordered measure spaces. Withdrawn manuscript, 2012. URL https://arxiv.org/abs/1212.0601.

Robert Brandenberger and Cumrun Vafa. Superstrings in the early universe. *Nuclear Physics B*, 316:391–410, 1989.

Graham Brightwell. Partial orders. In Lowell W Beineke and Robin J Wilson, editors, *Graph Connections: Relationships between Graph Theory and other Areas of Mathematics*. Clarendon Press, Oxford, 1997.

Graham Brightwell and Ruth Gregory. Structure of random discrete spacetime. *Physics Review Letters*, 66:260–263, 1991.

350 BIBLIOGRAPHY

Graham Brightwell, and Douglas B West. Partially ordered sets. In Kenneth H Rosen, editor, *Handbook of Discrete and Combinatorial Mathematics*. CRC Press, Boca Raton, 2000.

Graham Brightwell, Fay Dowker, Raquel S Garcia, Joe Henson, and Rafael D Sorkin. 'Observables' in causal set cosmology. *Physical Review D*, 67:084031, 2003.

Graham Brightwell, Joe Henson, and Sumati Surya. A result in 2d causal set theory: the emergence of spacetime. *Journal of Physics: Conference Series*, 174(012049), 2009.

Cliff P Burgess. Quantum gravity in everyday life: general relativity as an effective field theory. *Living Reviews in Relativity*, 7(1):1–56, 2004.

Wit Busza, Robert L Jaffe, Jack Sandweiss, and Frank Wilczek. Review of speculative 'disaster scenarios' at RHIC. *Reviews of Modern Physics*, 72(4):1125–1140, 2000.

Jeremy Butterfield. Stochastic Einstein locality revisited. *British Journal for the Philosophy of Science*, 58:805–867, 2007.

Jeremy Butterfield. Emergence, reduction and supervenience: A varied landscape. *Foundations of Physics*, 41:920–959, 2011a.

Jeremy Butterfield. Less is different: emergence and reduction reconciled. *Foundations of Physics*, 41:1065–1135, 2011b.

Jeremy Butterfield. On dualities and equivalences between physical theories. In Christian Wüthrich, Baptiste Le Bihan, and Nick Huggett, editors, *Philosophy Beyond Spacetime*. Oxford University Press, Oxford, 2021.

Jeremy Butterfield and Henrique Gomes. Functionalism as a species of reduction. In Cristián Soto, editor, *Current Debates in the Philosophy of Science: In Honor of Roberto Torretti*. Springer Nature, Cham, 2023.

Jeremy Butterfield and Christopher Isham. On the emergence of time in quantum gravit. In Jeremy Butterfield, editor, *The Arguments of Time*. Oxford University Press, Oxford, 1999.

Jeremy Butterfield and Christopher Isham. Spacetime and the philosophical challenge of quantum gravity. In Craig Callender and Nick Huggett, editors, *Physics Meets Philosophy at the Planck Scale*. Cambridge University Press, Cambridge, 2001.

Curtis G Callan, Daniel Friedan, Emil J Martinec, and Malcolm J Perry. Strings in background fields. *Nuclear Physics B*, 262(4):593–609, 1985.

Craig Callender. Shedding light on time. *Philosophy of Science*, 67:S587–S599, 2000.

Craig Callender. *What Makes Time Special?* Oxford University Press, Oxford, 2017.

Craig Callender and Nick Huggett. Why quantize gravity (or any other field for that matter)? *Philosophy of Science*, 68:S382–94, 2001a.

Craig Callender and Nick Huggett. *Physics Meets Philosophy at the Planck Scale*. Cambridge University Press, Cambridge, 2001b.

Tian Yu Cao. *Conceptual Developments of 20th Century Field Theories*. Cambridge University Press, Cambridge, 1998.

Peter Carlip, Steven Carlip, and Sumati Surya. Path integral suppression of badly behaved causal sets. *Classical and Quantum Gravity*, 40(095004), 2023.

Sean M Carroll, Jeffrey A Harvey, V Alan Kostelecky, Charles D. Lane, and Takemi Okamoto. Noncommutative field theory and Lorentz violation. *Physical Review Letters*, 87:141601, 2001.

Nancy Cartwright. *How the Laws of Physics Lie*. Cambridge University Press, Cambridge, 1983.

Elena Castellani. Duality and 'particle' democracy. *Studies in History and Philosophy of Modern Physics*, 59:100–108, 2017.

Masud Chaichian, Andrey Demichev, and Peter Presnajder. Quantum field theory on noncommutative space-times and the persistence of ultraviolet divergences. *Nuclear Physics B*, 567(1–2):360–390, 2000.

David Chalmers. Finding space in a nonspatial world. In Christian Wüthrich, Baptiste Le Bihan, and Nick Huggett, editors, *Philosophy Beyond Spacetime: Implications from Quantum Gravity*. Oxford University Press, Oxford, 2021.

Rob Clifton and Mark Hogarth. The definability of objective becoming in Minkowski spacetime. *Synthese*, 103:355–287, 1995.

Monique Combescure and Didier Robert. *Coherent States and Applications in Mathematical Physics*. Springer, Dordrecht, 2012.

BIBLIOGRAPHY 351

Karen Crowther. *Effective Spacetime: Understanding Emergence in Effective Field Theory and Quantum Gravity*. Springer, Cham, 2016.

Karen Crowther. Inter-theory relations in quantum gravity: Correspondence, reduction, and emergence. *Studies in History and Philosophy of Modern Physics*, 63:74–85, 2018.

Erik Curiel. Against the excesses of quantum gravity: A plea for modesty. *Philosophy of Science*, 68(3):S424–S441, 2001.

Erik Curiel. The many definitions of a black hole. *Nature Astronomy*, 3:27–34, 2019.

James T Cushing. *Theory Construction and Selection in Modern Physics: The S-matrix*. Cambridge University Press, Cambridge, 1990.

Richard Dawid. Scientific realism in the age of string theory. *Physics and Philosophy*, 2007.

Richard Dawid. String dualities and empirical equivalence. *Studies in History and Philosophy of Modern Physics*, 59:21–29, 2017.

Sebastian de Haro. Dualities and emergent gravity: Gauge/gravity duality. *Studies in History and Philosophy of Modern Physics*, 59:109–125, 2017.

Sebastian de Haro. The heuristic function of duality. *Synthese*, 196(12):5169–5203, 2019.

Sebastian de Haro. Spacetime and physical equivalence. In Nick Huggett, Keizo Matsubara, and Christian Wüthrich, editors, *Beyond Spacetime: The Foundations of Quantum Gravity*. Cambridge University Press, Cambridge, 2020.

Sebastian de Haro and Jeremy Butterfield. On symmetry and duality. *Synthese*, 198:2973–3013, 2019.

Sebastian de Haro, Daniel R. Mayerson, and Jeremy N. Butterfield. Conceptual aspects of gauge/gravity duality. *Foundations of Physics*, 46(11):1381–1425, 2016a.

Sebastian de Haro, Nicholas Teh, and Jeremy Butterfield. On the relation between dualities and gauge symmetries. *Philosophy of Science*, 83(5):1059–1069, 2016b.

René Descartes. *Principia Philosophiae*. Ludovicum Elzevirium, Amsterdam, 1644.

Bryce S DeWitt. Quantum theory of gravity. II. The manifestly covariant theory. *Physical Review*, 162(5):1195–1239, 1967.

Dennis Dieks, Jeroen van Dongen, and Sebastian de Haro. Emergence in holographic scenarios for gravity. *Studies in History and Philosophy of Modern Physics*, 52B:203–16, 2015.

Paul A M Dirac. The fundamental equations of quantum mechanics. *Proceedings of the Royal Society A: Mathematical, Physical and Engineering Sciences*, 109:642–653, 1925.

Robert DiSalle. *Understanding Spacetime: The Philosophical Development of Physics from Newton to Einstein*. Cambridge University Press, Cambridge, 2006.

Juliusz Doboszewski. Non-uniquely extendible maximal globally hyperbolic spacetimes in classical general relativity: a philosophical survey. In Gábor Hofer-Szabó and Leszek Wroński, editors, *Making it Formally Explicit: Probability, Causality and Indeterminism*, European Studies in Philosophy of Science. Springer, Cham, 2017.

Juliusz Doboszewski. Epistemic holes and determinism in classical general relativity. *British Journal for the Philosophy of Science*, 71:1093–1111, 2020.

Cian Dorr. Of numbers and electrons. *Proceedings of the Aristotelian Society*, 110:133–181, 2010.

Fay Dowker. Real time. *New Scientist*, (2415):36–39, 2003.

Fay Dowker. Spacetime discreteness, Lorentz invariance and locality. *Journal of Physics: Conference Series*, 306(012016), 2011.

Fay Dowker. Introduction to causal sets and their phenomenology. *General Relativity and Gravitation*, 45:1651–1667, 2013.

Fay Dowker. The birth of spacetime atoms as the passage of time. *Annals of the New York Academy of Sciences*, 1326:18–25, 2014.

Fay Dowker. Being and becoming on the road to quantum gravity: or, the birth of a baby is not a baby. In Nick Huggett, Keizo Matsubara, and Christian Wüthrich, editors, *Beyond Spacetime: The Foundations of Quantum Gravity*. Cambridge University Press, Cambridge, 2020.

Fay Dowker, Joe Henson, and Rafael D Sorkin. Quantum gravity phenomenology, Lorentz invariance and discreteness. *Modern Physics Letters A*, 19:1829–1840, 2004.

Fay Dowker, Steven Johnston, and Rafael D Sorkin. Hilbert spaces from path integrals. *Journal of Physics A: Mathematical and Theoretical*, 43(27):275302, 2010.

352 BIBLIOGRAPHY

Benjamin F Dribus. On the axioms of causal set theory. *Manuscript*, 2013. URL http://arxiv.org/abs/1311.2148.

Benjamin F Dribus. *Discrete Causal Set Theory: Emergent Spacetime and the Causal Metric Hypothesis*. Springer, Cham, 2017.

Anthony Duncan. *The Conceptual Framework of Quantum Field Theory*. Oxford University Press, Oxford, 2012.

Ben Dushnik and E W Miller. Partially ordered sets. *American Journal of Mathematics*, 63:600–610, 1941.

John Earman. Notes on the causal theory of time. *Synthese*, 24:74–86, 1972.

John Earman. *World Enough and Space-Time: Absolute versus Relational Theories of Space and Time*. MIT Press, Cambridge, MA, 1989.

John Earman. *Bangs, Crunches, Whimpers, and Shrieks: Singularities and Acausalities in Relativistic Spacetimes*. Oxford University Press, New York, 1995.

John Earman. The Penrose-Hawking singularity theorems: History and implications. In Hubert Goenner, Jürgen Renn, Jim Ritter, and Tilman Sauer, editors, *The Expanding Worlds of General Relativity*, volume 7 of *Einstein Studies*. Birkhäuser, Boston, 1999.

John Earman. Tracking down gauge: An ode to the constrained Hamiltonian formalism. In Katherine Brading and Elena Castellani, editors, *Symmetries in Physics: Philosophical Reflections*. Cambridge University Press, Cambridge, 2003.

John Earman. Determinism: what we have learned and what we still don't know. In Joseph Keim Campbell, Michael O'Rourke, and David Shier, editors, *Freedom and Determinism*. MIT Press, Cambridge, MA, 2004.

John Earman. Two challenges to the requirement of substantive general covariance. *Synthese*, 148:443–468, 2006a.

John Earman. The implications of general covariance for the ontology and ideology of spacetime. In Dennis Dieks, editor, *The Ontology of Spacetime*. Elsevier, Amsterdam, 2006b.

John Earman. Aspects of determinism in modern physics. In Jeremy Butterfield and John Earman, editors, *Handbook of the Philosophy of Science. Vol. 2: Philosophy of Physics*. Elsevier, Amsterdam, 2006c.

John Earman. Reassessing the prospects for a growing block model of the universe. *International Studies in the Philosophy of Science*, 22:135–164, 2008.

John Earman and John Norton. What price spacetime substantivalism? The hole story. *The British Journal for the Philosophy of Science*, 38(4):515–25, 1987.

Philipp Ehrlich. Are points (necessarily) unextended? *Philosophy of Science*, 89:784–801, 2022.

Astrid Eichhorn, Sumati Surya, and Fleur Versteegen. Induced spatial geometry from causal structure. *Classical and Quantum Gravity*, 36(105005), 2019.

Harold Erbin. *String Field Theory*. Lecture Notes in Physics. Springer Nature, Cham, 2021.

Stefan Felsner, Peter C Fishburn, and William T Trotter. Finite three dimensional partial orders which are not sphere orders. *Discrete Mathematics*, 201:101–132, 1999.

Stefan Felsner, Irina Mustață, and Martin Pergel. The complexity of the partial order dimension problem—closing the gap. *SIAM Journal on Discrete Mathematics*, 31:172–189, 2017.

Richard P Feynman. *The Character of Physical Law*. MIT Press, Cambridge, MA, 1967.

Dana Fine and Arthur Fine. Gauge theory, anomalies and global geometry: The interplay of physics and mathematics. *Studies in History and Philosophy of Modern Physics*, 28(3):307–323, 1997.

Kit Fine. Tense and reality. In Kit Fine, editor, *Modality and Tense: Philosophical Papers*. Oxford University Press, Oxford, 2005.

David Finkelstein. The space-time code. *Physical Review*, 184:1261–1271, 1969.

Doreen L Fraser. *Haag's Theorem and the Interpretation of Quantum Field Theories with Interactions*. PhD thesis, University of Pittsburgh, 2006.

Daniel Friedan. Nonlinear models in $2+\epsilon$ dimensions. *Physical Review Letters*, 45(13):1057–1060, 1980.

Michael Friedman. *Dynamics of Reason*. CSLI Publications, Stanford, CA, 2001.

BIBLIOGRAPHY   353

Rodolfo Gambini and Jorge Pullin. *Loops, Knots, Gauge Theories and Quantum Gravity*. Cambridge Monographs on Mathematical Physics. Cambridge University Press, Cambridge, 1996.

Rodolfo Gambini and Jorge Pullin. *First Course in Loop Quantum Gravity*. Oxford University Press, Oxford, 2011.

T Garidi. What is mass in desitterian physics? *arXiv preprint hep-th/0309104*, 2003. Accessed 10 February 2024.

Maurizio Gasperini. *Elements of String Cosmology*. Cambridge University Press, Cambridge 2007.

Israel M Gelfand and Mark A Naimark. On the embedding of normed rings into the ring of operators in hilbert space. *Matematicheskii Sbornik*, 12:197–217, 1943.

Robert Geroch. Einstein algebras. *Communications in Mathematical Physics*, 26:271–275, 1972.

Roscoe Giles. Reconstruction of gauge potentials from wilson loops. *Physical Review D*, 24:2160, 1981.

Lisa Glaser and Sumati Surya. Towards a definition of locality in a manifoldlike causal set. *Physical Review D*, 88:124026, 2013.

Roy J Glauber. The quantum theory of optical coherence. *Physical Review*, 130(6):2529, 1963.

Henrique Gomes and Jeremy Butterfield. Geometrodynamics as functionalism about time. In Claus Kiefer, editor, *From Quantum to Classical: Essays in Honour of H.-Dieter Zeh*, volume 204 of *Fundamental Theories of Physics*. Springer, Cham, 2022.

Ernesto Graziani, Francesco Orilia, Elena Capitani, and Roberto Burro. Common-sense temporal ontology: An experimental study. *Synthese*, 202(6):1–39, 2023.

Michael B Green, John H Schwarz, and Edward Witten. *Superstring Theory*, volume I. Cambridge University Press, Cambridge, 1987.

Brian Greene. String theory on Calabi-Yau manifolds. *arXiv preprint hep-th/9702155*, 1997. Accessed 10 February 2024.

Brian Greene. *The Elegant Universe: Superstrings, Hidden Dimensions, and the Quest for the Ultimate Theory*. W. W. Norton, New York, 1999.

Adolf Grünbaum. *Philosophical Problems of Space and Time*. Alfred A. Knopf, New York, 1963.

Adolf Grünbaum. *Modern Science and Zeno's Paradoxes*. George Allen and Unwin, London, 1967.

Stan Gudder. A dynamics for discrete quantum gravity. *International Journal of Theoretical Physics*, 53:3575–3586, 2014.

Amit Hagar and Meir Hemmo. The primacy of geometry. *Studies in History and Philosophy of Modern Physics*, 44(3):357–364, 2013.

Jonathan J Halliwell and Stephen W Hawking. Origin of structure in the universe. *Physical Review D*, 31:1777–1791, 1985.

Daniel Harlow. What black holes have taught us about quantum gravity. In Nick Huggett, Keizo Matsubara, and Christian Wüthrich, editors, *Beyond Spacetime: The Foundations of Quantum Gravity*. Cambridge University Press, Cambridge, 2020.

Stephen W Hawking. Black hole explosions? *Nature*, 248(5443):30–31, 1974.

Stephen W Hawking. Zeta function regularization of path integrals in curved spacetime. *Communications in Mathematical Physics*, 55(2):133–148, 1977.

Stephen W Hawking and George F R Ellis. *The Large Scale Structure of Space-time*. Cambridge University Press, Cambridge, 1979.

Stephen W Hawking, A R King, and P J McCarthy. A new topology for curved space-time which incorporates the causal, differential, and conformal structures. *Journal of Mathematical Physics*, 17:174–181, 1976.

Richard Healey. On the reality of gauge potentials. *Philosophy of Science*, 68(4):432–455, 2001.

Richard Healey. *Gauging What's Real*. Oxford University Press, Oxford, 2007.

Marc Henneaux, and Claudio Teitelboim. *Quantization of Gauge Systems*. Princeton University Press, Princeton, 1992.

Joe Henson. The causal set approach to quantum gravity. In Daniele Oriti, editor, *Approaches to Quantum Gravity: Toward a New Understanding of Space, Time and Matter*. Cambridge University Press, Cambridge, 2009.

354   BIBLIOGRAPHY

Joe Henson. Causal sets: discreteness without symmetry breaking. In Jeff Murugan, Amanda Weltman, and George F R Ellis, editors, *Foundations of Space and Time: Reflections on Quantum Gravity*. Cambridge University Press, Cambridge, 2012.

Mary B Hesse. *Forces and Fields*. T. Nelson, Edinburgh, 1961.

David Hilbert and Nick Huggett. Groups in mind. *Philosophy of Science*, 73:765–777, 2006.

Toshio Hiraguchi. On the dimension of partially ordered sets. *Science Reports of Kanazawa University*, 1:77–94, 1951.

Gary T Horowitz. String theory as a quantum theory of gravity. In N. Ashby, D.F. Bartlett, and W. Wyss, editors, *General Relativity and Gravitation*. Cambridge University Press, Cambridge, 1990.

Gary T Horowitz. Spacetime in string theory. *New Journal of Physics*, 7(1):201, 2005.

Nick Huggett. Renormalization and the disunity of science. In Meinard Kuhlmann, Holger Lyre, and Andrew Wayne, editors, *Ontological Aspects of Quantum Field Theory*. World Scientific, Singapore, 2002.

Nick Huggett. What did Newton mean by 'absolute motion'? In Andrew Janiak and Eric Schliesser, editors, *Interpreting Newton: Critical Essays*. Cambridge University Press, Cambridge, 2012.

Nick Huggett. Skeptical notes on a physics of passage. *Annals of the New York Academy of Sciences*, 1326:9–17, 2014.

Nick Huggett. Target space ≠ space. *Studies in History and Philosophy of Modern Physics*, 59:81–88, 2017.

Nick Huggett. Spacetime 'emergence'. In Eleanor Knox and Alastair Wilson, editors, *The Routledge Companion to Philosophy of Physics*. Routledge, New York and Oxford, 2022.

Nick Huggett. Commentary:'physical time within human time' and'bridging the neuroscience and physics of time'. *Frontiers in Psychology*, 14:1087695, 2023.

Nick Huggett and Keizo Matsubara. Lost horizons. *In preparation*, 2020.

Nick Huggett and Tiziana Vistarini. Deriving general relativity from string theory. *Philosophy of Science*, 82(5):1163–1174, 2015.

Nick Huggett and Robert Weingard. On the field aspect of quantum fields. *Erkenntnis*, 40(3):293–301, 1994.

Nick Huggett and Christian Wüthrich. Emergent spacetime and empirical (in)coherence. *Studies in History and Philosophy of Modern Physics*, 44:276–285, 2013a.

Nick Huggett and Christian Wüthrich. Special issue on 'the emergence of spacetime in quantum theories of gravity'. *Studies in History and Philosophy of Modern Physics*, 44(3), 2013b.

Nick Huggett and Christian Wüthrich. The (a)temporal emergence of spacetime. *Philosophy of Science*, 85:1190–1203, 2018.

Nick Huggett, Tiziana Vistarini, and Christian Wüthrich. Time in quantum gravity. In Heather Dyke and Adrian Bardon, editors, *A Companion to the Philosophy of Time*. Wiley-Blackwell, Chichester, 2013.

Nick Huggett, Keizo Matsubara, and Christian Wüthrich, editors. *Beyond Spacetime: The Foundations of Quantum Gravity*. Cambridge University Press, Cambridge, 2020.

Nick Huggett, Fedele Lizzi, and Tushar Menon. Missing the point in noncommutative geometry. *Synthese*, 199:4695–4728, 2021.

Nick Huggett, Niels Linnemann, and Mike D Schneider. *Quantum Gravity in a Laboratory?* Cambridge University Press, Cambridge, 2023.

Raluca Ilie, Gregory B Thompson, and David D Reid. A numerical study of the correspondence between paths in a causal set and geodesics in the continuum. *Classical and Quantum Gravity*, 23:3275–3285, 2006.

Ted Jacobson and Lee Smolin. Non-perturbative quantum geometries. *Nuclear Physics B*, 299:295–345, 1988.

Claus Kiefer. Wave packets in minisuperspace. *Physical Review D*, 38(6):1761–1772, 1988.

Claus Kiefer. *Quantum Gravity*. Oxford University Press, Oxford, 2004.

Claus Kiefer. Decoherence in quantum field theory. In Erich Joos, H Dieter Zeh, Claus Kiefer, Domenico JW Giulini, Joachim Kupsch, and Ion-Olimpiu Stamatescu, editors, *Decoherence and the appearance of a classical world in quantum theory*. Springer, Heidelberg, 2013.

BIBLIOGRAPHY 355

Claus Kiefer and Jorma Louko. Hamiltonian evolution and quantization for extremal black holes. *Annalen der Physik*, 8(1):67–81, 1999.

Jaegwon Kim. *Physicalism, or Something Near Enough*. Princeton University Press, Princeton, 2005.

Elias Kiritsis. *String Theory in a Nutshell*. Princeton University Press, Princeton, 2011.

JR Klauder and BS Skagerstam, editors. *Coherent States: Applications in Physics and Mathematical Physics*. World Scientific, Singapore, 1985.

Daniel J Kleitman and Bruce L Rothschild. Asymptotic enumeration of partial orders on a finite set. *Transactions of the American Mathematical Society*, 205:205–220, 1975.

Eleanor Knox. Effective spacetime geometry. *Studies in History and Philosophy of Modern Physics*, 44:346–356, 2013.

Eleanor Knox. Spacetime structuralism or spacetime functionalism? Manuscript, 2014. URL https://www.eleanorknox.com/research.html.

Eleanor Knox. Physical relativity from a functionalist perspective. *Studies in History and Philosophy of Modern Physics*, 67:118–124, 2019.

E H Kronheimer and Roger Penrose. On the structure of causal spaces. *Mathematical Proceedings of the Cambridge Philosophical Society*, 63:481–501, 1967.

Thomas S Kuhn. *The Structure of Scientific Revolutions*. University of Chicago Press, Chicago, 1962.

Vincent Lam and Michael Esfeld. A dilemma for the emergence of spacetime in canonical quantum gravity. *Studies in History and Philosophy of Modern Physics*, 44:286–293, 2013.

Vincent Lam and Christian Wüthrich. Spacetime is as spacetime does. *Studies in History and Philosophy of Modern Physics*, 64:39–51, 2018.

Vincent Lam and Christian Wüthrich. Spacetime functionalism from a realist perspective. *Synthese*, 199:335–353, 2021.

Tom Lancaster and Stephen J Blundell. *Quantum Field Theory for the Gifted Amateur*. Oxford University Press, Oxford, 2014.

Nicolaas P Landsman. Between classical and quantum. In Jeremy Butterfield and John Earman, editors, *Handbook of the Philosophy of Science. Vol. 2: Philosophy of Physics*. Elsevier B.V., Amsterdam, 2006.

A L Larsen and N Sánchez. Mass spectrum of strings in anti-de sitter spacetime. *Physical Review D*, 52(2):1051–64, 1995.

Baptiste Le Bihan. String theory, loop quantum gravity and eternalism. *European Journal for Philosophy of Science*, 10(17):1–22, 2020.

Baptiste Le Bihan. Spacetime emergence in quantum gravity: Functionalism and the hard problem. *Synthese*, 199(2):371–393, 2021.

Baptiste Le Bihan and Niels Linnemann. Have we lost spacetime on the way? Narrowing the gap between general relativity and quantum gravity. *Studies in History and Philosophy of Modern Physics*, 65:112–121, 2019.

Baptiste Le Bihan and James Read. Duality and ontology. *Philosophy Compass*, 13(12):e12555, 2018.

Jonathan Leake. Big bang machine could destroy earth. *The Sunday Times*, July 18 1999.

Koen Lefever. A century of axiomatic systems for ordinal approaches to special relativity theory. *Manuscript*, 2013.

David Lewis. Psychophysical and theoretical identifications. *Australasian Journal of Philosophy*, 50(3):249–258, 1972.

Fedele Lizzi. Noncommutative spaces. *Lecture Notes in Physics*, 774:89–109, 2009.

Claud Lovelace. Pomeron form factors and dual Regge cuts. *Physics Letters B*, 34(6):500–506, 1971.

Claud Lovelace. Dual amplitudes in higher dimensions: A personal view. In A Cappelli, E Castellani, F Colomo, and P Di Vecchia, editors, *The Birth of String Theory*. Cambridge University Press, Cambridge, 2012.

Seth Major, David Rideout, and Sumati Surya. Spatial hypersurfaces in causal set cosmology. *Classical and Quantum Gravity*, 23:4743–4752, 2006.

356 BIBLIOGRAPHY

Seth Major, David Rideout, and Sumati Surya. On recovering continuum topology from a causal set. *Journal of Mathematical Physics*, 48:032501, 2007.

David B Malament. The class of continuous timelike curves determines the topology of space-time. *Journal of Mathematical Physics*, 18:1399–1404, 1977.

David B Malament. In defense of dogma: Why there cannot be a relativistic quantum mechanics of (localizable) particles. In Rob Clifton, editor, *Perspectives on Quantum Reality*. Springer, Dordrecht, 1996.

David B Malament. *Topics in the Foundations of General Relativity and Newtonian Gravitational Theory*. Chicago University Press, Chicago, 2012.

Juan Maldacena. The large N limit of of superconformal field theories and cosmotopology. *Advances in Theoretical and Mathematical Physics*, 2:231–252, 1998.

JB Manchak. What is a physically reasonable space-time? *Philosophy of Science*, 78:410–420, 2011.

JB Manchak. General relativity as a collection of collections of models. In Judit Madarász and Gergely Székely, editors, *Hajnal Andréka and István Németi on the Unity of Science: From Computing to Relativity Theory Through Algebraic Logic*, Outstanding Contributions to Logic. Springer, Cham, 2021a.

JB Manchak. On Feyerabend, general relativity, and 'unreasonable' universes. In Karim Bschir and Jamie Shaw, editors, *Interpreting Feyerabend: Critical Essays*. Cambridge University Press, Cambridge, 2021b.

JB Manchak, Chris Smeenk, and Christian Wüthrich. *Time and Again: On the Logical, Meta-physical, and Physical Possibility of Time Travel*. Oxford University Press, Oxford, under contract.

Fotini Markopoulou and Lee Smolin. Disordered locality in loop quantum gravity states. *Classical and Quantum Gravity*, 24:3813–3823, 2007.

Samir D Mathur. Black holes and beyond. *Annals of Physics*, 327(11):2760–2793, 2012.

Keizo Matsubara. Realism, underdetermination and string theory dualities. *Synthese*, 190(3):471–489, 2013.

Keizo Matsubara and Lars-Göran Johansson. Spacetime in string theory: A conceptual clarification. *Journal for General Philosophy of Science*, 49(3):333–353, 2018.

Tim Maudlin. *Quantum Non-locality and Relativity: Metaphysical Intimations of Modern Physics*. Blackwell, Oxford, 1994.

Tim Maudlin. Three measurement problems. *Topoi*, 14:7–15, 1995.

Tim Maudlin. Thoroughly muddled McTaggart: Or how to abuse gauge freedom to generate metaphysical monstrosities. *Philosophers' Imprint*, 2(4), 2002.

Tim Maudlin. Completeness, supervenience, and ontology. *Journal of Physics A: Mathematical and Theoretical*, 40:3151–3171, 2007a.

Tim Maudlin. *The Metaphysics Within Physics*. Oxford University Press, Oxford, 2007b.

Kerry McKenzie. Relativities of fundamentality. *Studies in History and Philosophy of Science Part B: Studies in History and Philosophy of Modern Physics*, 59:89–99, 2017.

David McMahon. *String Theory Demystified*. McGraw Hill, New York, 2008.

Henryk Mehlberg. Essai sur la théorie causale du temps I. *Studia Philosophica*, 1:119–258, 1935.

Henryk Mehlberg. Essai sur la théorie causale du temps II. *Studia Philosophica*, 2:111–231, 1937.

Tushar Menon. Taking up superspace: the spacetime structure of supersymmetric field theory. In Christian Wüthrich, Baptiste Le Bihan, and Nick Huggett, editors, *Philosophy Beyond Spacetime*. Oxford University Press, Oxford, 2021.

David A Meyer. *The Dimension of Causal Sets*. PhD thesis, Massachusetts Institute of Technology, 1988.

David A Meyer. Spherical containment and the Minkowski dimension of partial orders. *Order*, 10:227–237, 1993.

Charles W Misner, Kip S Thorne, and John Archibald Wheeler. *Gravitation*. Freeman, San Francisco, 1973.

Cristopher Moore. Comment on 'space-time as a causal set'. *Physical Review Letters*, 60:655, 1988.

BIBLIOGRAPHY 357

Luboš Motl. What is background independence and how important is it?, 2012. URL http://motls.blogspot.com/2012/12/what-is-background-independence-and-how.html.

Luboš Motl. Could we be on the inside of a concave hollow universe? 2015. URL https://physics.stackexchange.com/questions/201722/could-we-be-on-the-inside-of-a-concave-hollow-universe. Accessed 10 February 2024

Jan Myrheim. Statistical geometry. Technical Report TH-2538, CERN preprint, 1978.

Mikio Nakahara. *Geometry, Topology and Physics*. CRC Press, Bristol, 2003.

Horatiu Nastase. The RHIC fireball as a dual black hole. 2005. URL https://https://arXiv preprint hep-th/0309104. Accessed 10 February 2024.

Isaac Newton. *Philosophiae naturalis principia mathematica*. Guil. & Joh. Innys, Regiæ Societatis typographos, 3 edition, 1726.

Isaac Newton. *Opticks: or, A Treatise of the Reflexions, Refractions, Inflexions and Colours of Light.* William Innys, London, 4 edition, 1730.

Theodore D Newton. A note on the representations of the de Sitter group. *Annals of Mathematics*, 1950.

Alyssa Ney. Fundamental physical ontologies and the constraint of empirical coherence. *Synthese*, 192:3105–3124, 2015.

Hrvoje Nikolić. Bohmian mechanics in relativistic quantum mechanics, quantum field theory and string theory. *Journal of Physics: Conference Series*, 67(012035):1–6, 2007.

Emmy Noether. Invariante Variationsprobleme. *Nachrichten von der Königlichen Gesellschaft der Wissenschaften zu Göttingen. Mathematisch-physikalische Klasse*, 1918. Reprinted in Nathan Jacobson, editor, *Emmy Noether: Gesammelte Abhandlungen—Collected Papers*. Springer-Verlag, Berlin, 1983.

Joshua Norton. Loop quantum ontology: spacetime and spin-networks. *Studies in History and Philosophy of Modern Physics*, 71:14–25, 2020.

Peter Øhrstrøm and Per Hasle. Future contingents. In Edward N. Zalta, editor, *The Stanford Encyclopedia of Philosophy*. Metaphysics Research Lab, Stanford University, 2020. URL https://plato.stanford.edu/archives/sum2020/entries/future-contingents/.

Daniele Oriti. Disappearance and emergence of space and time in quantum gravity. *Studies in History and Philosophy of Modern Physics*, 46:186–199, 2014.

Daniele Oriti. Levels of spacetime emergence in quantum gravity. In Christian Wüthrich, Baptiste Le Bihan, and Nick Huggett, editors, *Philosophy Beyond Spacetime: Implications from Quantum Gravity*. Oxford University Press, Oxford, 2021.

L A Paul and Ned Hall. *Causation: A User's Guide*. Oxford University Press, Oxford, 2013.

Amanda W Peet. Black holes in string theory. In Eva Silverstein, Jeffrey Harvey, and Shamit Kachru, editors, *Strings, Branes, and Gravity: TASI 99*. World Scientific, Singapore, 2001.

Roger Penrose. *The Road to Reality: A Complete Guide to the Laws of the Universe*. Vintage Books, New York, 2004.

Lydia Philpott, Fay Dowker, and Rafael D Sorkin. Energy-momentum diffusion from spacetime discreteness. *Physical Review D*, 79:124047, 2009.

J Brian Pitts. A first class constraint generates not a gauge transformation, but a bad physical change: the case of electromagnetism. *Annals of Physics*, 351:383–406, 2014.

Henri Poincaré. *Science and Hypothesis*. Walter Scott Publishing Company, London and Newcastle-on-Tyne, 1905.

Joseph Polchinski. Dualities of fields and strings. *Studies in History and Philosophy of Modern Physics*, 59:6–20, 2017.

Joseph Gerard Polchinski. *String Theory*. Cambridge University Press, Cambridge, 1998.

Oliver Pooley. Substantive general covariance: another deacde of dispute. In Mauricio Suárez, Mauro Dorato, and Miklós Rédei, editors, *EPSA Philosophical Issues in the Sciences: Launch of the European Philosophy of Science Association*, volume 2. Springer, Dordrecht, 2010.

Oliver Pooley. Relativity, the open future, and the passage of time. *Proceedings of the Aristotelian Society*, 113:321–363, 2013.

358 BIBLIOGRAPHY

Oliver Pooley. Background indepedence, diffeomorphism invariance and the meaning of coordinates. In Dennis Lehmkuhl, Erhard Scholz, and Gregor Schiemann, editors, *Towards a Theory of Spacetime Theories*, volume 13 of *Einstein Studies*. Birkhäuser, New York, 2017.

Oliver Pooley and David Wallace. First-class constraints generate gauge transformations in electromagnetism (reply to Pitts). Manuscript, 2022. URL https://arxiv.org/abs/2210.09063.

James Read. The interpretation of string-theoretic dualities, December 2014. URL http://philsci-archive.pitt.edu/11205/.

James Read. Background independence in classical and quantum gravity. Master's thesis, University of Oxford, 2016.

James Read. On miracles and spacetime. *Studies in History and Philosophy of Modern Physics*, 65:103–111, 2019.

James Read and Tushar Menon. The limitations of inertial frame spacetime functionalism. *Synthese*, 199:229–251, 2021.

James Read and Thomas Møller-Nielsen. Motivating dualities. *Synthese*, 197:263–291, 2020.

Michael Redhead. Symmetry in intertheory relations. *Synthese*, 32:77–112, 1975.

Michael Redhead. A philosopher looks at quantum field theory. In Harvey R Brown and Rom Harré, editors, *Philosophical Foundations of Quantum Field Theory*. Clarendon Press, Oxford, 1990.

Hans Reichenbach. *Axiomatik der relativistischen Raum-Zeit-Lehre*. Friedrich Vieweg und Sohn, Braunschweig, 1924.

Hans Reichenbach. *Philosophie der Raum-Zeit-Lehre*. Walter de Gruyter, Berlin and Leipzig, 1928.

Hans Reichenbach. *Experience and Prediction*. University of Chicago Press, Chicago, 1938.

Hans Reichenbach. *The Direction of Time*. University of California Press, Berkeley, 1956.

David D. Reid. Discrete quantum gravity and causal sets. *Canadian Journal of Physics*, 79:1–16, 2001.

Michael P Reisenberger and Carlo Rovelli. 'Sum over surfaces' form of loop quantum gravity. *Physical Review D*, 56:3490–3508, 1997.

Dean Rickles. A philosopher looks at string dualities. *Studies in History and Philosophy of Modern Physics*, 42(1):54–67, 2011.

Dean Rickles. AdS/CFT duality and the emergence of spacetime. *Studies in History and Philosophy of Modern Physics*, 44(3):312–320, 2012.

Dean Rickles. Mirror symmetry and other miracles in superstring theory. *Foundations of Physics*, 43(1):54–80, 2013.

Dean Rickles. *A Brief History of String Theory*. Springer, Berlin and Heidelberg, 2014.

David Rideout and Rafael D Sorkin. Classical sequential growth dynamics for causal sets. *Physical Review D*, 61:024002, 1999.

David Rideout and Petros Wallden. Emergence of spatial structure from causal sets. *Journal of Physics: Conference Series*, 174:012017, 2009.

Bernhard Riemann. Über die Hypothesen, welche der Geometrie zugrunde liegen. *Abhandlungen der königlichen Gesellschaft der Wissenschaften zu Göttingen*, 13:133–152, 1868.

Alfred A Robb. *A Theory of Time and Space*. Cambridge University Press, Cambridge, 1914.

Alfred A Robb. *The Geometry of Time and Space*. Cambridge University Press, Cambridge, 1936.

Joshua S Rosaler. *Inter-theory relations in physics: case studies from quantum mechanics and quantum field theory*. PhD thesis, University of Oxford, 2013.

Carlo Rovelli. Halfway through the woods: Contemporary research on space and time. In John Earman and John Norton, editors, *The Cosmos of Science*. University of Pittsburgh Press, Pittsburgh, 1997.

Carlo Rovelli. Partial observables. *Physical Review D*, 65:124013, 2002.

Carlo Rovelli. *Quantum Gravity*. Cambridge University Press, Cambridge, 2004.

Carlo Rovelli. A new look at loop quantum gravity. *Classical and Quantum Gravity*, 28:114005, 2011.

Carlo Rovelli. Why gauge? *Foundations of Physics*, 44:91–104, 2014.

Carlo Rovelli. Neither presentism nor eternalism. *Foundations of Physics*, 49:1325–1335, 2019.

Carlo Rovelli. Space and time in loop quantum gravity. In Nick Huggett, Keizo Matsubara, and Christian Wüthrich, editors, *Beyond Spacetime: The Foundations of Quantum Gravity*. Cambridge University Press, Cambridge, 2020.

Carlo Rovelli and Lee Smolin. Knot theory and quantum gravity. *Physical Review Letters*, 61:1155–1158, 1988.

Carlo Rovelli and Lee Smolin. Loop representation of quantum general relativity. *Nuclear Physics B*, 331:80–152, 1990.

Carlo Rovelli and Lee Smolin. Discreteness of area and volume in quantum gravity. *Nuclear Physics B*, 442:593–622, 1995a. Erratum: *Nuclear Physics B*, **456**:734.

Carlo Rovelli and Lee Smolin. Spin networks and quantum gravity. *Physical Review D*, 52:5743–5759, 1995b.

Carlo Rovelli and Francesca Vidotto. *Covariant Loop Quantum Gravity: An Elementary Introduction to Quantum Gravity and Spinfoam Theory*. Cambridge University Press, Cambridge, 2015.

Carlo Rovelli and Francesca Vidotto. Philosophical foundations of loop quantum gravity. In Cosimo Bambi, Leonardo Modesto, and Ilya Shapiro, editors, *Handbook of Quantum Gravity*. Springer Nature, Singapore, 2024.

Laura Ruetsche. *Interpreting Quantum Theories: The Art of the Possible*. Oxford University Press, Oxford, 2011.

Hanno Sahlmann, Thomas Thiemann, and Oliver Winkler. Coherent states for canonical quantum general relativity and the infinite tensor product extension. *Nuclear Physics B*, 606:401–440, 2001.

Kian Salimkhani. Quantum gravity: A dogma of unification? In Alexander Christian, David Hommen, Nina Retzlaff, and Gerhard Schurz, editors, *Philosophy of Science Between the Natural Sciences, the Social Sciences, and the Humanities*, volume 9 of *European Studies in Philosophy of Science*. Springer, Cham, 2018.

Amitabha Sen. Gravity as a spin system. *Physics Letters B*, 119:89–91, 1982.

Theodore Sider. *Four-Dimensionalism: An Ontology of Persistence and Time*. Clarendon Press, Oxford, 2001.

Theodore Sider. Against vague existence. *Philosophical Studies*, 114:135–146, 2003.

Theodore Sider. *The Tools of Metaphysics and the Metaphysics of Science*. Oxford University Press, Oxford, 2020.

Lawrence Sklar. Prospects for a causal theory of space-time. In Richard Swinburne, editor, *Space, Time and Causality*. D. Reidel Publishing Company, Dordrecht, 1983.

Chris Smeenk and Christian Wüthrich. Time travel and time machines. In Craig Callender, editor, *The Oxford Handbook of Philosophy of Time*. Oxford University Press, Oxford, 2011.

Chris Smeenk and Christian Wüthrich. Determinism and general relativity. *Philosophy of Science*, 88:638–664, 2021.

Lee Smolin. The case for background independence. In Dean Rickles, Steven French, and Juha Saatsi, editors, *The Structural Foundations of Quantum Gravity*. Oxford University Press, Oxford, 2006.

Lee Smolin. *The Trouble With Physics: The Rise of String Theory, The Fall of a Science, and What Comes Next*. Mariner Books, Boston, 2007.

Hartland S Snyder. Quantized space-time. *Physical Review*, 71:38–41, 1947.

Rafael Sorkin. A specimen of theory construction from quantum gravity. In Jarrett Leplin, editor, *The Creation of Ideas in Physics: Studies for a Methodology of Theory Construction*, University of Western Ontario Series in Philosophy of Science. Kluwer Academic Publishers, Dordrecht, 1995.

Rafael D Sorkin. Spacetime and causal sets. In Michael P Ryan, Luis F Urrutia, Juan Carlos D'Olivo, Federico Zertuche Mones, M Rosenbaum, and Eduardo Nahmad-Achar, editors, *Relativity and Gravitation: Classical and Quantum (Proceedings of the SILARG VII Conference)*. World Scientific, Singapore, 1991.

Rafael D Sorkin. Quantum measure theory and its interpretation. In B L Hu and Feng Da Hsuan, editors, *Quantum Classical Correspondence: Proceedings of the 4th Drexel Symposium*

360   BIBLIOGRAPHY

on Quantum Nonintegrability, Drexel University, Philadelphia, USA, September 8-11, 1994. International Press, Cambridge, MA, 1997a.

Rafael D Sorkin. Forks in the road, on the way to quantum gravity. *International Journal of Theoretical Physics*, 36(12):2759–2781, 1997b.

Rafael D Sorkin. An example relevant to the Kretschmann-Einstein debate. *Modern Physics Letters* A, **17**:695–700, 2002.

Rafael D Sorkin. Causal sets: Discrete gravity. In Andrés Gomberoff and Donald Marolf, editors, *Lectures on Quantum Gravity*. Springer, New York, 2005.

Rafael D Sorkin. Geometry from order: causal sets. *Einstein Online*, 2:1007, 2006.

Rafael D Sorkin. Relativity theory does not imply that the future already exists: A counterexample. In Vesselin Petkov, editor, *Relativity and the Dimensionality of the World*. Springer, Dordrecht, 2007.

Rafael D Sorkin. Does locality fail at intermediate length scales? In Daniele Oriti, editor, *Approaches to Quantum Gravity: Toward a New Understanding of Space, Time and Matter*. Cambridge University Press, Cambridge, 2009.

Marshall Spector. *Concepts of Reduction in Physical Science*. Temple University Press, Philadelphia, 1978.

Howard Stein. On relativity theory and openness of the future. *Philosophy of Science*, 58:147–167, 1991.

C R Stephens, G't Hooft, and B F Whiting. Black hole evaporation without information loss. *Classical and Quantum Gravity*, 11(3):621–647, 1994.

Raymond F Streater and Arthur S Wightman. *PCT, Spin, Statistics and All That*. Benjamin, New York, 1964.

Andrew Strominger, Shing-Tung Yau, and Eric Zaslow. Mirror symmetry is T-duality. *Nuclear Physics B*, 479(1-2):243–259, 1996.

Sumati Surya. The causal set approach to quantum gravity. *Living Reviews in Relativity*, 22(5), 2019.

Sumati Surya and Stav Zalel. A criterion for covariance in complex sequential growth models. *Classical and Quantum Gravity*, 37(195030), 2020.

Leonard Susskind. Lecture 1 | string theory and m-theory, March 2011. URL https://www.youtube.com/watch?v=25haxRuZQUk. Accessed 10 February 2024.

Leonard Susskind and James Lindesay. *An Introduction to Black Holes, Information and the String Theory Revolution*. World Scientific, Singapore, 2005.

Gerard't Hooft. Quantum gravity: A fundamental problem and some radical ideas. In Maurice Lévy and Stanley Deser, editors, *Recent Developments in Gravitation: Cargèse 1978*. Plenum Press, New York and London, 1979.

Washington Taylor. String field theory. In Daniele Oriti, editor, *Approaches to Quantum Gravity: Toward a New Understanding of Space, Time and Matter*. Cambridge University Press, 2009.

Nicholas J Teh. Holography and emergence. *Studies in History and Philosophy of Modern Physics*, 44(3):300–311, 2013.

Karim P Y Thébault. The problem of time. In Eleanor Knox and Alastair Wilson, editors, *The Routledge Companion to Philosophy of Physics*. Routledge, New York and London, 2022.

Thomas Thiemann. Anomaly-free formulation of non-perturbative, four-dimensional Lorentzian quantum gravity. *Physics Letters B*, 380:257–264, 1996.

Thomas Thiemann. *Modern Canonical Quantum General Relativity*. Cambridge University Press, Cambridge, 2007.

Llewellyn H Thomas. On unitary representations of the group of de Sitter space. *Annals of Mathematics*, 42(1):113–126, 1941.

David Tong. Lectures on quantum field theory, 2006. URL http://www.damtp.cam.ac.uk/user/tong/qft.html. Accessed 10 February 2024.

Bas C van Fraassen. *An Introduction to the Philosophy of Time and Space*. Random House, New York, 1970.

Bas C van Fraassen. *The Scientific Image*. Oxford University Press, Oxford, 1980.

BIBLIOGRAPHY 361

Madvahan Varadarajan. Fock representations from $U(1)$ holonomy algebras. *Physical Review D*, 61:104001, 2000.

Tiziana Vistarini. Holographic space and time: Emergent in what sense? *Studies in History and Philosophy of Modern Physics*, 59:126–135, 2017.

Tiziana Vistarini. *The Emergence of Spacetime in String Theory*. Routledge, Abingdon and New York, 2019.

Robert M Wald. *General Relativity*. University of Chicago Press, Chicago, 1984.

David Wallace. *The Emergent Multiverse: Quantum Theory According to the Everett Interpretation*. Oxford University Press, Oxford, 2012.

David Wallace. Stating structural realism: mathematics-first approaches to physics and metaphysics. *Philosophical Perspectives*, 36(1):345–378, 2022a.

David Wallace. Quantum gravity at low energies. *Studies in History and Philosophy of Science*, 94:31–46, 2022b.

Sean Walsh and Tim Button. *Philosophy and Model Theory*. Oxford University Press, Oxford, 2018.

James Owen Weatherall. Equivalence and duality in electromagnetism. *Philosophy of Science*, 87:1172–1183, 2020.

Steven Weinberg. Photons and gravitons in s-matrix theory: Derivation of charge conservation and equality of gravitational and inertial mass. *Physical Review*, 135(4B):B1049–B1056, 1964.

Steven Weinberg. Photons and gravitons in perturbation theory: Derivation of maxwell's and einstein's equations. *Physical Review*, 138(4B):B988–B1002, 1965.

Steven Weinberg. *The Quantum Theory of Fields*, volume 1. Cambridge University Press, Cambridge, 1995.

Steven Weinberg. Effective field theory, past and future. *International Journal of Modern Physics A*, 31(06):1630007, 2016.

Robert Weingard. A philosopher looks at string theory. *PSA: Proceedings of the Biennial Meeting of the Philosophy of Science Association*, 1988:95–106, 1988.

Steven Weinstein. Gravity and gauge theory. *Philosophy of Science*, 66:S146–S155, 1999.

Eugene Wigner. On unitary representations of the inhomogeneous Lorentz group. *Annals of Mathematics*, 40(1):149–204, 1939.

Clifford M Will. The confrontation between general relativity and experiment. *Living Reviews in Relativity*, 17(4), 2014.

Mark Wilson. *Wandering Significance: An Essay on Conceptual Behavior*. Oxford University Press, Oxford, 2008.

John A Winnie. The causal theory of space-time. In John Earman, Clark Glymour, and John Stachel, editors, *Foundations of Space-Time Theories*, volume 8 of *Minnesota Studies in the Philosophy of Science*. University of Minnesota Press, Minneapolis, 1977.

Edward Witten. Non-commutative geometry and string field theory. *Nuclear Physics B*, 268(2):253–294, 1986.

Edward Witten. Reflections on the fate of spacetime. *Physics Today*, 49(4):24–30, April 1996.

Christian Wüthrich. To quantize or not to quantize: fact and folklore in quantum gravity. *Philosophy of Science*, 72:777–788, 2005.

Christian Wüthrich. To quantize or not to quantize: fact and folklore in quantum gravity. *Philosophy of Science*, 72(5):777–788, 2005.

Christian Wüthrich. *Approaching the Planck Scale from a Generally Relativistic Points of View: A Philosophical Appraisal of Loop Quantum Gravity*. PhD thesis, University of Pittsburgh, 2006.

Christian Wüthrich. Challenging the spacetime structuralist. *Philosophy of Science*, 65:1039–1051, 2009.

Christian Wüthrich. The structure of causal sets. *Journal for General Philosophy of Science*, 43:223–241, 2012.

Christian Wüthrich. The fate of presentism in modern physics. In Roberto Ciuni, Kristie Miller, and Giuliano Torrengo, editors, *New Papers on the Present: Focus on Presentism*. Philosophia Verlag, Munich, 2013.

362 BIBLIOGRAPHY

Christian Wüthrich. Raiders of the lost spacetime. In Dennis Lehmkuhl, Gregor Schiemann, and Erhard Scholz, editors, *Towards a Theory of Spacetime Theories*, volume 13 of *Einstein Studies*. Birkhäuser, New York, 2017.

Christian Wüthrich. Time travelling in emergent spacetime. In Judit Madarász and Gergely Székely, editors, *Hajnal Andréka and István Németi on the Unity of Science: From Computing to Relativity Theory Through Algebraic Logic*, number 19 in Outstanding Contributions to Logic. Springer, Cham, 2021.

Christian Wüthrich. One time, two times, or no time? In Alessandra Campo and Simone Gozzano, editors, *Einstein vs Bergson: An Enduring Quarrel on Time*. De Gruyter, Berlin and Boston, 2022.

Christian Wüthrich. The philosophy of causal set theory. In Cosimo Bambi, Leonardo Modesto, and Ilya Shapiro, editors, *Handbook of Quantum Gravity*. Springer Nature, Singapore, 2024.

Christian Wüthrich and Craig Callender. What becomes of a causal set. *British Journal for the Philosophy of Science*, 68:907–925, 2017.

Christian Wüthrich, Baptiste Le Bihan, and Nick Huggett, editors. *Philosophy Beyond Spacetime: Implications from Quantum Gravity*. Oxford University Press, Oxford, 2021.

David Yates. Thinking about spacetime. In Christian Wüthrich, Baptiste Le Bihan, and Nick Huggett, editors, *Philosophy Beyond Spacetime: Implications from Quantum Gravity*. Oxford University Press, Oxford, 2020.

Eric Zaslow. Duality. In Timothy Gowers, June Barrow-Green, and Imre Leader, editors, *The Princeton companion to mathematics*. Princeton University Press, Princeton, 2008.

E Christopher Zeeman. Causality implies the Lorentz group. *Journal of Mathematical Physics*, 5:490–493, 1964.

Wojciech Hubert Zurek and Juan Pablo Paz. Decoherence, chaos, and the second law. *Physical Review Letters*, 72(16):2508–2511, 1994.

Wojciech Hubert Zurek and Juan Pablo Paz. Quantum chaos: a decoherent definition. *Physica D: Nonlinear Phenomena*, 83(1):300–308, 1995a.

Wojciech Hubert Zurek and Juan Pablo Paz. Zurek and Paz reply. *Physical Review Letters*, 75(2):351–351, 1995b.

Barton Zwiebach. *A First Course in String Theory*. Cambridge University Press, Cambridge, 2004.

Barton Zwiebach. Quantum dynamics, 2013. URL https://ocw.mit.edu/courses/physics/8-05-quantum-physics-ii-fall-2013/lecture-notes/MIT8_05F13_Chap_06.pdf. Accessed 10 February 2024.

Index

≪-isomorphism 39–40, 43–4
26 dimensions 200, 216–7, 220, 261, 274

adjacency relation 146, 148, 151, 153–7, 161–3, 342
ADM variables 134–6, 165
AdS-CFT duality *see* duality, AdS-CFT
Aggregates 8–9, 26, 33
Albert, David 16, 30
anomaly (quantum) 277–9, 299
  Weyl 212, 278, 280–1, 322
antichain *see* causal set theory, antichain
approximating procedure 173–5, 177, 187, 343
approximation
  double field 224n3, 249, 252, 255
  of causal set by spacetime 62, 81, 86
  of strings by quantum field theory 263, 320, 328, 330–2
area operator *see* loop quantum gravity, area operator
Ashtekar, Abhay 134, 179
Ashtekar variables 134–135
astrophysical scope *see* loop quantum gravity, astrophysical scope
asynchronous becoming *see* becoming, asynchronous
auxiliary string metric *see* string, auxiliary metric

background independence 12, 113–5, 118–9, 121, 145, 322–5, 341, 345
background spacetime
  in causal set theory 42, 107
  in loop quantum gravity 111, 114–115, 118–21, 160–1, 341–2
  in string theory 269–83, 302, 308–10, 322–9, 331, 345
Baker, David 29n15, 219
becoming 91–100, 147, 162, 184, 186–7, 341
  asynchronous 94–5
  hypersurface 94–5
  relativistic 70, 92, 95
  worldline 94–5, 99
Bell causality 76, 341
black hole 1–3, 7, 50, 166–7, 228, 255–7, 276, 334
Bohm's theory 229

Brandenberger, Robert 228, 232–4, 237, 241–2, 248, 257–8, 318–9, 346
Butterfield, Jeremy 31, 96, 169, 171–4, 177–8, 243, 245–6

Callender, Craig 93, 187
Canberra plan 31–3
canonical quantization 5, 111–3, 122, 130–1, 135–7, 159, 205, 211, 259–61, 285
canonical quantum gravity 111–3, 119, 121, 125, 130–7, 159, 167, 171–2
Cartesian gravity 20–1
Cauchy surface 123, 127, 158
causal
  connectibility 37–8, 40–1, 53
  isomorphism 39–40, 43–4
  precedence 5, 36, 47, 49, 55
  relation 5, 35–42, 46–7, 52, 69, 77–8, 340
causal set theory (CST)
  antichain 58–61, 63–9, 74, 78–9, 93, 340
  background spacetime *see* background spacetime, in causal set theory
  chain 58–9, 67, 69, 74, 77–9, 96–7, 107
  classical sequential growth dynamics 75–7, 79, 92, 95–7, 108, 341
  distance 64–5, 67–9, 81, 90, 100, 340
  Einstein field equation *see* Einstein field equation, and causal set theory
  geodesic 67–8, 107, 194
  Hauptvermutung 67, 83, 85–6
  height 59, 74
  kinematic axiom 12, 35, 47–9, 73, 75, 84
  matter on 107–8
  Minkowski dimension *see* dimension, Minkowski
  non-Hegelian pair 64
  non-locality *see* non-locality, in causal set
  physical salience *see* physical salience, of causal set theory
  Poisson sprinkling 82, 101–2, 104
  post 98
  topology *see* topology, of a causal set
  transitive percolation 77–8, 80, 83, 98–9
  width 59, 74
causal theory of time 36–8, 38–43

364 INDEX

causation (philosophical concept) 38, 52–4, 55n33
chain *see* causal set theory, chain
chronological future or past 40
classical sequential growth dynamics *see* causal set theory, classical sequential growth dynamics
closed timelike curves 41–42, 72, 83–4
coherent state 178, 183, 267–76, 282–96
    of gravitons 306–7, 322–35
conformal 4, 42n13, 43–5, 62, 196–8, 200, 202–3
    field theory (CFT) 255–7, 344
    invariance or symmetry 199–203, 208–9, 264–8, 275, 280
    isometry 43–4
    transformation 202–3, 265
constrained Hamiltonian system 127, 129
constraint 123–31, 134–7, 142, 149, 203–4, 206, 209–11, 213
    first-class 126–7, 129
    Hamiltonian 134, 136–7, 139–40, 142, 152, 341–2
    primary and secondary 125–6
continuum limit 35, 45–6, 183, 343
coordinate order 60–1
cosmological constant Λ 39, 51, 106, 108, 341
cosmological scope 165, 176
covariant dynamics (of loop quantum gravity) *see* loop quantum gravity, covariant dynamics

Dawid, Richard 230, 345
decoherence 175–6, 181–2, 334, 343–4
definitional extension 172–4
de Haro, Sebastian 236, 243–4, 246
Descartes, René 11, 20–1
diffeomorphism 43, 129, 136–9, 149–50, 195–6, 200, 261, 324
    definition of 116–7
    invariance or symmetry 118, 195, 199, 214, 280
dilaton 214, 264, 273–4, 344
Dilworth's theorem 59
dimension
    Minkowski 62–3, 68–9, 82, 211
    of partial order 60–3, 74–5
Dirac observable see observable, Dirac
directed acyclic graph 48
Dirichlet condition 198, 253–4
DiSalle, Robert 30
discrete general covariance *see* general covariance, discrete
discreteness 50–2, 69, 72, 93, 100–1, 106, 151–2

disordered locality 146, 148, 155–8, 161, 163, 181, 185
distance (in causal set theory) *see* causal set theory, distance
distinguishing spacetime 41–2, 44–5, 81, 122
Dowker, Fay 55n33, 94–5, 104–7
Dp-branes 194, 204, 254–5
duality 220, 222–58 309, 313–21, 325–6, 329, 332, 336
    AdS-CFT 255–7
    mirror symmetry 251
    S-duality 251

Earman, John 38–43, 96, 120, 123, 128, 162, 249
Ehrenfest theorem 293
Einstein field equation (EFE)
    and causal set theory 38, 72
    linear 214–6, 296
    and loop quantum gravity 117, 122
    and string theory 271–6, 287, 302, 308–9, 311, 321–2, 331–2, 335–6
emergence
    Butterfield-Isham scheme of 146, 172, 177–8, 343
    general concept of 8–9, 169–177, 338
empirical incoherence 16, 32, 89, 141, 145–6, 159, 169, 181
equivalence, physical 227–57 344
equivalence principle 2n2, 114–5, 335
eternalism 147, 184–7
Euler characteristic 196
Euler-Lagrange equation 23, 118–9, 124–5, 274
explanatory gap 28–9, 159

faithful embedding 81–2, 86, 341
Feynman's principle 260
Finkelstein, David 35, 45
foliation of spacetime 57–9, 93–5, 156
Friedan, Daniel 272
Friedman, Michael 10, 22
functional definition 26, 31–2
functionalism 27–31, 171, 180
    functional reduction 27–8, 32, 110, 169, 186, 338–9, 347
    in the philosophy of mind 9, 28–9
    spacetime *see* spacetime functionalism
future distinguishing 41, 44, 51–2, 122
fuzzball 257n34, 331n32

gauge 120, 123–7, 129, 166, 206, 235, 278–81
    anomaly *see* anomaly
    constraints 203, 209
    fixing 127, 197, 199–201, 218, 261
    -gravity duality *see* duality, AdS-CFT

orbit 126, 129
symmetry or invariance 247–9, 277–80
of loop quantum gravity 96, 118, 129–30, 131–2, 137–8, 149–50
of string theory 198, 214
transformation 118, 126, 138, 279
Gelfand-Naimark theorem 23
general covariance 92, 105, 113, 115–21, 127, 130, 145, 324
definition of 115–6
discrete 76, 94, 341
principle of 116, 118, 129
Geneva plan 32, 34
genidentity 37–8, 54n31
geodesic (in causal set theory) *see* causal set theory, geodesic
ghosts (negative norm states) 206, 208–9, 212, 217n25
Glauber, Roy 294n39
globally hyperbolic 57–8, 67, 123, 127–8, 132–3, 167, 175
Gomes, Henrique 31, 32n21
graviton 13, 214–5, 259, 267, 305–7, 330–6
coherent state of *see* coherent state
Green, Michael 188n1, 262n4, 271, 303, 307, 331
Greene, Brian 231, 332
Gromov-Hausdorff function 81, 86
group field theory 6, 149
growing block view 79, 92, 147, 184–5
Grünbaum, Adolf 37, 40

haecceity 64, 78, 241n19
Hamiltonian constraint *see* constraint, Hamiltonian
Hamiltonian general relativity 12, 113, 122–4, 127–30, 145–6
Hamiltonian system with constraints *see* constrained Hamiltonian system
Harlow, Daniel 256–7
Hasse diagram 48, 49–50
Hauptvermutung *see* causal set theory, Hauptvermutung
Hawking-King-McCarthy theorem 43
Healey, Richard 247, 277n19
heterotic string see string, heterotic
holomorphic 199n12, 199n12, 217, 265
holonomy 134, 138
Hooke's law 18, 194, 208
Huggett, Nick 16, 24, 234, 236, 243, 248, 324
hypersurface becoming *see* becoming, hypersurface

ideal gas law 18–20
indeterminate existence 98–9
indeterminate space
in loop quantum gravity 152–4, 161, 163, 168
in string theory 233–7, 247–50, 319–321
induced string metric *see* string, induced metric
internal temporality 76–7, 341

Kalb-Raymond 214, 273
Kiefer, Claus 176n26, 215n23, 276n17, 331n30, 334, 335n35
Kim, Jaegwon 27
kinematical Hilbert space (of LQG) *see* loop quantum gravity, kinematical Hilbert space
kinematic axiom (of causal set theory) *see* causal set theory, kinematic axiom
Kleitman-Rothschild order or theorem 73–4
Knox, Eleanor 28, 31
Kuhn, Thomas 10, 22, 323n21

Lam, Vincent 27, 29, 89
Landsman, Klaas 175–6
Laplace equation 199–200
Le Bihan, Baptiste 29–30, 159, 185–6
Leibniz, Gottfried Wilhelm 11, 20–1, 36
Lewis, David 25–8, 30n18, 31–3, 241n19
lightcone 53, 66, 93–5, 99, 102–4
coordinates 202, 207–9, 211, 213, 267
gauge 208–9
quantization 206, 209n20, 212, 259, 261, 336
structure 214, 308
limiting procedure 174–5, 177–8, 181–3, 187, 343
linear gravity *see* Einstein field equation, linear
Linnemann, Niels 159
loop quantum cosmology 136, 139–40
loop quantum gravity (LQG)
area operator 150–3, 157, 162, 167, 174, 179–81, 343
astrophysical scope 165, 176
covariant dynamics 119, 136, 139, 142, 146, 152, 342
Einstein field equation *see* Einstein field equation, and loop quantum gravity
gauge symmetry *see* gauge, symmetry, of loop quantum gravity
indeterminate space 152–4, 161, 163, 168
kinematical Hilbert space 136–7, 149, 153, 178
physical Hilbert space 131, 135–7, 142, 178
physical salience *see* physical salience, of loop quantum gravity

366 INDEX

loop quantum gravity (LQG) (*Continued*)
  presentism *see* presentism, in loop quantum
    gravity
  regional perspective 165–8, 176, 181
  spinfoam 137, 142–5, 164, 166–8, 173, 183–4
  spin network state 137–9, 142–3, 146–4, 168,
    171, 178–8
  volume operator 150–2, 157, 162, 167, 174,
    179–81, 343
  weave state 162, 174, 179–83, 343
Lorentz invariance or symmetry 44, 51, 100–8,
  208–9, 212, 244–5, 263, 335–6

Malament, David 40, 43–4, 46–7
Malament's theorem 43–7, 51–2, 62, 86n18
Maldacena, Juan 255–6
manifoldlike 60, 80, 82–3
Markov sum rule 76, 341
Matsubara, Keizo 231, 235n15, 251
Maudlin, Tim 16–9, 25, 132, 167
Mehlberg, Henryk 37
Menon, Tushar 24, 329
Meyer, David 35, 47, 62
Minkowski dimension (of a causal set) *see*
  dimension, Minkowski
mirror symmetry *see* duality, mirror symmetry
model of GR 72, 117–8, 122, 124, 133, 327
  definition of 38
Motl, Luboš 326–8
M-theory 263, 270, 276, 304, 332
  and duality 220–1, 245–6, 252, 325, 345–6
Myrheim, Jan 35, 45, 47, 51, 68
Møller-Nielsen, Thomas 231, 244–6

Nambu-Goto action 193, 195
Neumann condition 192, 194n7, 197, 220
Newton, Isaac 10, 20–1, 30, 111, 231, 245
Newtonian gravity 10, 20–1, 246
Noether's theorem 23, 118, 119n13, 277, 281,
  296–300
non-commutative geometry 6, 22–5, 339
  physical salience *see* physical salience, of
    non-commutative geometry
non-Hegelian pair *see* causal set theory,
  non-Hegelian pair
non-linear sigma model *see* sigma model
non-locality
  in causal set theory 51, 55, 70, 72, 100–6, 110
  quantum 21, 88, 100–1
normal ordering 211, 213, 216, 293–4
Norton, John 218n26
Norton, Joshua 160–1
n-thickening 65, 66–67

observable 123, 133, 135, 138, 151, 225–30,
  238–41, 300
  Dirac 129, 131–2, 152, 159, 167
  partial 152, 159, 167

partial order 35–6, 46–9, 52–3, 54n29, 58–63, 77
past distinguishing 41, 44, 51, 52–3, 122
path integral 260–6, 277–8, 298–301, 303–8,
  326–7
Penrose, Roger 43, 322–4
philosophy of physics, role of 3–4, 9–11
physical Hilbert space (of loop quantum gravity)
  see loop quantum gravity, physical
    Hilbert space
physically reasonable spacetime 72–3
physical salience 17–22, 29, 33, 34
  of causal set theory 62, 71, 73, 80, 87, 90, 100
  of loop quantum gravity 151, 171, 177, 181,
    187
  of non-commutative geometry 24–5
  of string theory 302, 338–41, 344–7
Planck length 7, 46, 104, 179–80, 204, 227, 331
Poincaré, Henri 56–7
Poincaré
  invariance or symmetry 195, 198, 254, 272,
    280, 328
  group 119–20, 214
Poisson sprinkling *see* causal set theory, Poisson
  sprinkling
Polchinski, Joseph 188n1, 200, 205n17, 252n25,
  260n2, 280–1, 307
pre-order 49, 63
presentism 92, 147
  emergent 186
  in loop quantum gravity 184–6
primitive ontology 87, 89
problem of change 127, 129, 131, 136, 146, 148,
  158–63, 168
problem of space 35, 54, 56–60, 63, 83
problem of time 127–8, 131–2, 136, 146, 148,
  158–63, 168
Pythagoreanism 15n1, 89, 109

quantum
  anomaly *see* anomaly
  non-locality *see* non-locality

Read, James 244–6, 325, 329
Regge slope 204, 271, 308
regional perspective (on loop quantum gravity)
  see loop quantum gravity, regional
    perspective
regularization or renormalization 13, 50, 137,
  272–3, 277–80, 337

zeta 217, 344
Reichenbach, Hans 37, 170
Rickles, Dean 204, 230
Rideout, David 65, 68–9, 75–6, 79, 98
Riemann, Bernhard 45, 51
Robb, Alfred A 36, 43
Rosaler, Joshua 282, 293
Rovelli, Carlo 142, 151–2, 161–2, 165–7, 174–5, 179, 186–7

Salimkhani, Kian 215n23, 335
Schwarz, John 188n1, 262n4, 271, 303, 307, 331
S-duality *see* duality, S-duality
Sider, Ted 97
sigma action 195–8, 200, 203, 260, 265, 273, 281, 305–7
(non-linear) sigma model 271–2, 275, 277, 279, 308
Sklar, Larry 15, 29
s-knot state 150, 160, 161n7, 179n33
Smolin, Lee 137, 179, 309–10
Sorkin, Rafael 35, 47, 51, 75–9, 85, 92, 94–5, 98, 105–6, 108, 119–21
space
 (no) essence of 56
 phenomenal 14, 29, 231–2
 relativistic 45, 258, 344
spacetime
 causal structure of 36–45, 49, 53–4, 63, 89
 definition of relativistic spacetime 38
spacetime functionalism 8–9, 25–33, 63–7, 87–91, 177–84, 310–3, 338–9
 SF1 27–8, 31, 73, 90, 302, 338–40, 342–3
 SF2 27, 90, 302, 338–40, 343–7
spinfoam *see* loop quantum gravity, spinfoam
spin network state *see* loop quantum gravity, spin network state
stress-energy 38, 120, 201–3, 275, 280
string
 auxiliary metric 195–6, 198, 200, 260–1
 background spacetime *see* background spacetime, in string theory
 characteristic length ℓ_s 191, 203–4, 223, 227, 263, 320, 331–2, 345
 covariant quantization 206
 Einstein field equation *see* Einstein field equation, and string theory
 field theory 210, 262, 270, 276, 304, 327
 gauge symmetry *see* gauge, symmetry, of string theory
 heterotic 275
 induced metric 193, 195, 198, 261, 307n6
 indeterminate space *see* indeterminate space, in string theory

lightcone quantization, *see* lightcone, quantization
mass spectrum 191, 204–5, 210–5, 217–8, 226, 305
physical salience *see* physical salience, of string theory
scattering 262*f*, 265–6, 268–9, 304, 312–3, 320, 324
target space 324–6, 332, 336
 and curved background spacetime 259–62, 268–70, 304–16
 and T-duality 227, 232–7, 241–3, 247–53, 316–21, 329, 344–6
tension 191–5, 208, 224, 228, 256, 271, 305, 308
vertex operator 266–71, 273, 276, 290, 282, 287, 290, 306–7
Virasoro algebra or constraints 203, 206, 210, 211, 213
winding 223–8, 230–3, 241–2, 248–50, 252–4, 317–21, 34–6
(mini-)superspace 136, 139, 333
supersymmetry 218–21
supervenience 172–4, 186
Susskind, Leonard 188n1, 207

tachyon 212–3, 218, 220, 266, 268–9, 305–6, 309
target space *see* string, target space
Teh, Nic 231
temporal orientability 39
temporal precedence 5, 55
Thiemann, Thomas 136, 178
timelike relation 39–40, 53, 67
topology 123, 128, 167, 250–2, 281, 311–3, 316–9, 336
 definition 63
 discrete 64–5
 indiscrete 64–5
 of a causal set 63–7, 340
total order 47, 49, 58, 61, 74, 94
transitive percolation *see* causal set theory, transitive percolation
triad field 134–5, 179

unification 275, 335

Vafa, Cumrun 228, 232–4, 237, 241–2, 248, 257–8, 318–9, 346
van Fraassen, Bas 37–8, 40–1, 236
vertex operator *see* string theory, vertex operator
Virasoro algebra or constraints *see* string theory, Virasoro algebra or constraints
Vistarini, Tiziana 231, 324

368 INDEX

volume operator *see* loop quantum gravity, volume operator

Wallace, David 2, 30, 334
wavefunction realism or monism 16, 30
Weatherall, James 231, 240
weave state *see* loop quantum gravity, weave state
Weinberg, Steven 215–6, 335
Weingard, Robert 219
Weyl
   anomaly *see* anomaly, Weyl

   invariance or symmetry 196–200, 212, 217–8, 261, 272–82, 307–9, 322, 336
   representation 14, 23–4
   transformation 196, 200, 261, 265, 279
Wheeler-DeWitt equation 131, 134n41, 142, 158, 168
Witten, Edward 188n1, 225, 262n4, 271, 303, 307, 314–5, 326, 331
worldline becoming *see* becoming, worldline
worldsheet interpretation 199, 314–5, 326, 344
Wüthrich, Christian 16, 27, 29, 93, 128, 170
't Hooft, Gerard 35, 46–7, 255